The Open World, Hackbacks and Global Justice

A. Jean Thomas

The Open World, Hackbacks and Global Justice

palgrave
macmillan

A. Jean Thomas
Washington, DC, USA

ISBN 978-981-19-8131-9 ISBN 978-981-19-8132-6 (eBook)
https://doi.org/10.1007/978-981-19-8132-6

This Palgrave Macmillan imprint is published by the registered company Springer Nature Singapore Pte Ltd.
The registered company address is: 152 Beach Road, #21-01/04 Gateway East, Singapore 189721, Singapore

For my incredible family,
especially my mother and father

Inez Hooper Thomas, M.S.W., L.I.C.S.W.
Claudius Thomas, M.D.

Acknowledgments

Open world discourse is a complex system of rules, presumptions, expectations, stakeholders, beneficiaries and victims. Unfortunately, many bad faith "winners" in the global economy use open world discourse as a sword and a shield. When used as a sword, open world discourse enables all participants to transcend unnecessary or inefficient jurisdictional restrictions, encourage cross-cultural communication and appreciation for the rule of law and hopefully improve the lives of the most vulnerable members of society. In the same way that "a rising tide lifts all boats," bad faith actors take advantage of the streamlining affects of open world norms and rules. Unfortunately, bad faith actors also ally with extreme cyber-libertarians who believe that the benefits of open world discourse always outweigh the costs, and opportunists who use the specter of moral ubiquity (i.e., "everyone is doing it") to ignore persistent pockets of abuse or discrimination. Both strategies treat open world discourse as a shield that protects people from authoritarian restrictions, as well as reasonable institutional, political and legal reforms.

This book is part of growing body of literature on the long-term implications of persistent bad behavior by individuals, organization and nation-states in cyberspace. The continuing appeal of retaliatory hackbacks is a sign that blind faith in the open world shield is eroding, but loss of faith in the idea that the preferences of those who benefit from bad faith behavior in cyberspace is now the "new normal" should not lead to the passage of new laws that severely restrict access to the internet. Solving the problem of persistent, bad faith behavior in cyberspace requires a scalpel, not a hammer, even when the signs of that abuse are everywhere and the abusers are entrenched and largely unrepentant.

Dismantling the systems of power that persistent cyber abusers rely on requires a thorough re-examination of the theoretical underpinnings of the "permissible cyber irresponsibility" that erodes trust in the digital domain and amplifies political and economic disparities between the Global North and the Global South, racial groups and gender communities. Debates about fairness and welfare affect virtually every aspect of life in the digital age, from simple computer access rules to the complex, vexing issue of retaliatory hackbacks. The fairness/welfare debate also dominates cyber discourse, providing bad faith actors with a large arsenal of rationales for their conduct. Hopefully, this book, along with many others, will demonstrate that capabilities theory is well-positioned to challenge the theoretical status quo in cyberspace.

Many people helped me to complete this project, including Richard Banks, Edwin Lo and Rohan Mukherjee. I also benefitted immensely from a variety of lectures and talks I attended at the Brookings Institute, the Center for New American Security, the Bipartisan Policy Center, the Center for Strategic and International Studies and the American Constitution Society. I'd like to thank Jacob Meyer, Ashwini Elango and Melvin Lourdes, three incredible editors at Palgrave-Macmillan who took a chance on this project. Finally, I'd like to thank my incredible family (Inez and Claudius Thomas, Claudia Thomas-Navolio and John Navolio, Carol Thomas-Knipes, Joseph Knipes, Ally Navolio and Aedan Knipes, Sharon Samuels), a close family friend (Fay Benjamin), and my patient, best friend, Thomas Edward Burton, Jr.

Contents

List of Tables

CHAPTER 1

Introduction

One of the greatest gifts bestowed upon the global community by the Western world is the belief that persons, information, good and services, and access to commonly cherished natural resources should be available to everyone regardless of jurisdiction, language, diaspora or even religion. By almost every measure, the open world ethos has accelerated the pace of human development, but even the strongest proponents of open world thinking can't ignore obvious examples of abuse.[1] Proponents of the free flow of information in one country, for example, may hire someone to commit crimes abroad that they could not commit at home; proponents of open borders may take improper advantage of large migration events to sneak terrorists or dangerous individuals into a country; and proponents of unfettered speech on the internet may promote opinions on both sides of an important issue in order to confuse the public.

[1] See. e.g., Clausing (2019) (pro-open world policies must account for widespread economic inequality), Lissner and Hooper (2020) (encourages the United States to defend open world values through alliance-building and the protection of the rule of law), Soros (2019) (open societies require protection from authoritarianism and unchecked market operations), Van Der Vossen and Brennan (2018) (in spite of its flaws, free markets are the best way to eliminate global poverty), and Philippon (2019) (balances faith in free markets against recognition that economic inequality is pervasive).

A. J. Thomas, *The Open World, Hackbacks and Global Justice*,
https://doi.org/10.1007/978-981-19-8132-6_1

This book analyzes a particular self-help response to open world abuses in cyberspace: retaliatory hackbacks (also referred to as 'mitigative counter-strikes' or 'active defense' measures).[2] Hackbacks occur when victims, frustrated by the slow pace of law enforcement in cyberspace or the magnitude of damage caused by an attack, choose to 'take the law into their own hands' by retaliating against those who have violated access restrictions, stolen valuable data or damaged network operations. A classic hack scenario involves an aggressor, unknown to the victim, who purchases 1000 stolen passwords from a third party, uses those passwords to enter the victim's computer system, circumvents their security, retrieves files and eventually incorporates the contents of those files into their own computer system files. The victim discovers that the files have been appropriated but immediately runs into three familiar obstacles: (1) hardening one's system (i.e., increasing cyber security) may, or may not, deter future attacks, but it will not retrieve appropriated files or necessarily punish the aggressor at all; (2) after informing local police and discussing the issue with the local district attorney, the victim discovers that even though there probably was a technical violation of state/federal laws, the district attorney declines to prosecute due to the value of the harm, the length of time required to bring a successful prosecution and the real possibility that even if the government eventually prevails in court, it won't have the authority to retrieve stolen files now; and (3) after consulting with a private attorney, the victim discovers that even though the suspected aggressor probably violated numerous state and federal data privacy laws, litigation could take years, anticipated damages will be low even if the case is successful and state/federal laws do not authorize the retrieval of stolen files. Given these possibilities, should the cyber victim consider a retaliatory hackback?

In order to answer this question, cyber victims have to consider the conflict between means and ends. Means, ranging from skills or income to access or food, provide someone with the opportunity to achieve certain goals, whereas ends are, with few exceptions, the goals themselves. In the search for equality, efficiency, fairness and distributive justice, political

[2] See DODNIST (2011) (defines active defense as "synchronized, real-time capability to discover, detect, analyze and mitigate threats and vulnerabilities.") and Kesan and Hayes (2012, 435) (defines 'mitigative counterstriking' as "taking active efforts to mitigate harm to a targeted system, in a manner strictly limited to the amount of force necessary to protect the victim from further damage.").

philosophers have devised numerous justifications for either prioritizing means over ends or ends over means. The former may be reasonably easy for the state (and even private civil society) to secure, but means-satisfaction may never achieve the desired end. The latter, on the other hand, is likely to approximate true happiness, but it is much more difficult to achieve.

There are two ways to conceptualize retaliatory hackbacks as 'means-ends' problems. Under a narrow view, law enforcement (criminal/civil), private lawsuits (civil) and increased cyber security ('hardening') are 'means' because each strategy is designed to eventually deter the suspected aggressor and satisfy the victim's demand for retributive justice. However, a retaliatory hackback that retrieves the victim's files from the perpetrators computer, or immediately prevents the perpetrator from launching comparable attacks on other parties, may satisfy the victim's desire for retributive justice immediately.

Under a broad view, retaliatory hackbacks, standing alone, are a metaphorical contest between means and ends. Some people believe that there is no reason for the victim of a cyber attack to engage in a retaliatory hackback if they possess important rights or 'primary' goods. Others believe that retaliatory hackbacks aren't necessary if the victim has sufficient wealth. But some cyber victims who possess valuable rights, primary goods and wealth, and still resort to a retaliatory hackback because they lack the capability to take advantage of available opportunities in their immediate environment. For these people, the most compelling justification for a retaliatory hackback is not the absence of primary goods/rights (attainable means), or the absence of wealth (an unattainable end because wealth accumulation is not available to most people), but the absence of capabilities (an attainable end).

Regardless of whether one adopts the narrow or broad view, the choice architecture available to many cyber victims is very similar to the choice architecture faced by other victims of open world abuse. Both views raise the same kinds of questions about the best way to protect vulnerable entities in open world zones. Many proponents of open world discourse believe that is neither efficient nor fair to deny cyber victims access to an immediate, attainable-ends solution to damaging cyber attacks, or to limit retaliatory hackback authorizations to nation-states. But the same proponents are justifiably concerned about possible spillover effects, escalation scenarios and a possible erosion of rule of law norms in cyberspace.

This book defends both the narrow and broad views. Retaliatory hackbacks should be considered a legitimate 'available-ends' alternative to law enforcements, private suits and hardening for all entities (individuals, organizations, governments and nation-states). It is also time for political decision-makers and cyber experts to re-examine the core assumption that people, organization and nation-states with primary goods/formal rights and relative wealth don't opt for retaliatory hackbacks after a cyberattack. This book proposes a retaliatory hackback model based on the capabilities approach, developed principally by Amartya Sen and Martha Nussbaum. Capabilities analysis identifies the actual possibilities of lives, organizations and nation-states, possibilities constrained by circumstances that measurably undermine freedom and flourishing. The capabilities model developed in this volume, however, does not focus exclusively on capability deficits caused by cognitive limitations, physical disabilities or existential threats comparable to starvation. Instead, my model focuses on capability deficits caused by situational constraints that can affect everyone, regardless of their station in life, constraints that may require someone to develop 'adaptive preferences' in order to avoid constant disappointment, harassment or even violence.

The capabilities-based defense of retaliatory hackbacks developed in this book, is called the 'violation-retaliation nexus' (VRN), and it refines the capabilities approach in three additional ways. First, VRN imposes a duty upon capability-deficient victims of cyber attacks to only hackback out of reasonable necessity, using proportional force and with sufficient attribution. Second, VRN creates a presumption that aggressive, repeated violations of reasonable open world expectations are undertaken deliberately and with sufficient information. Failure to rebut this presumption renders the aggressor vulnerable to a retaliatory hackback administered by the victim, a proxy or the state.

Third, VRN apportions responsibility for any damages caused by a retaliatory hackback. Capability-deficient individuals who retaliate even though they don't believe that retaliation is absolutely necessary, fail to exercise reasonable care in choosing retaliatory measures that are roughly proportional to the attack, or fail to exercise reasonable care in their attempts to determine the identity and/or location of a suspected attacker, must shoulder a portion of the costs for any damages caused by retaliation. This requirement holds even if reasonable care pushes the limits of a victim's capabilities. Individuals who engineer attacks that could be defined as particularly dangerous, on the other hand, must absorb the costs of all damages caused by the victim's retaliation.

Chapter 2 analyzes the structure of open world discourse. Section 1 identifies five open world zones that focus on immigration, green trade, access to foreign research or medical expertise and internet access. Section 2 proposes a scale, ranging from zealotry, competitive skepticism and criminal skepticism to opportunism. Points along the scale range from an expansive to a progressively narrower view of open world expectations.

Chapter 3 initially defends a pragmatic skeptical view of the enforceability of open world norms, rather than the radical skeptical view that reduces almost all open world policies to strategic opportunism, or an equally radical anti-skeptical view that ignores all but the most egregious open world violations. Section 2 distinguishes between deontological and consequentialist (or teleological) theories. Discussion of the former focuses on Rawls' understanding of rights, fairness, the difference principle and primary goods. Discussion of the latter focuses on utilitarianism, welfarism, wealth maximization, efficiency and the tensions between 'means-oriented' and 'ends-oriented' theories. Section 3 analyzes the legal elements of self-defense and self-help, and critiques rights-based and utilitarian defenses of self-defense. Section 4 considers the advantages of grounding self-defense in forced choice theory.

The analysis of efficiency requires additional comment. The capabilities critique of welfarism and utility usually focus on wealth maximization as a measure of happiness and human flourishing. However, a sizable number of capability theorists have also challenged the welfarist tendency to define efficiency either in terms of cost-savings, or desirable objectives ultimately reducible to cost-savings. Instead of focusing on costs in 'market spaces', capability theorists have proposed alternative theories that measure efficiency in terms of maximizing both the quality and quantity of reasonable opportunities. Opportunity-based efficiency models modify some of the core assumptions of the Pareto, Kaldor-Hicks or cost-benefit standards including, most notably, the relative importance of market transactions, self-interested behavior and preference satisfaction measured primarily in terms of wealth, as well as the relative unimportance of distributive justice and equality.[3] VRN reflects the capabilities critique of wealth maximization and efficiency.

Chapter 4 describes the substance and structure of VRN. Section 1 compares the capabilities models of Amartya Sen and Martha Nussbaum. Section 2 describes the limits of Sen and Nussbaum, particularly with regard to any potential substantive or practical overlaps between capabilities, primary goods and welfarism. Section 3 transforms the famous Melamed-Calabresi entitlement matrix into a capabilities-based framework that resolves entitlement conflicts over attacks and retaliations. Instead of using a property or liability rule to protect the entitlements of plaintiffs and defendants, VRN uses a presumptive and desert rule to protect capability-deficient people who retaliate after an attack. Also, each zone of the matrix corresponds to a different combination of initial attack and retaliation: *zone 1* (impermissible attack, permissible retaliation), *zone 2* (~permissible attack, permissible retaliation), *zone 3* (permissible attack, impermissible retaliation) and *zone 4* (impermissible attack, impermissible retaliation). Section 4 summarizes the most important elements of VRN.

Chapter 5 examines the requirements for holding either the victim or attacker responsible for damages caused by retaliation under VRN. Section 1 analyses three requirements for permitted retaliation: reasonable necessity, reasonable attribution and reasonable proportionality. If any of the requirements are not met, victims who have already satisfied the requirements for a presumptively legitimate retaliation claim must pay a portion of the damages caused by retaliation. Section 2 analyzes exacerbating factors that justify depriving attackers compensation for any retaliatory measures employed by the victim of an attack. Section 3 examines the way that open world zealots and open world opportunists use the logic of deterrence and non-escalation to discourage retaliatory self-defense or self-help by victims.

Chapter 6 examines the political, technological and legal landscape of retaliatory hackbacks. Section 1 analyzes the limits of global justice in cyberspace when inquiry is limited to primary goods, rights, utility or wealth maximization, the promise of what have been described as 'digital

[3] Kaplow and Shavell (2002, 5) : ("Under a common understanding of normative economics analysis, legal rules are assessed by reference to wealth maximization or efficiency, criteria that many construe as omitting important aspects of individuals' well-being and as ignoring distributive concerns.").

capabilities, and the political orientation of VRN.' Section 2 provides an empirical and statutory overview of cyber attacks and hackbacks. Section 3 conceptualizes a hypothetical in which a provoked victim of IP theft attacks a rival who steals trade secrets, followed by retaliation claim made by the thief, as a *zone 3* scenario in which an initial attack is deemed permissible, and a subsequent retaliation is deemed impermissible.

The last portion of Section 3 examines the efficiency of retaliatory hackbacks after IP theft. VRN's protection of capability-deficient IP theft victims who choose to hackback appears to be 'inefficient', from the standpoint of traditional economics, because VRN does not require proof that the victim was either the 'cheapest cost avoider' or effectively lowered the 'avoidance costs' of third parties by establishing firm control over intellectual property prior to executing a hackback. However, VRN does increase the number of remedies available to capability-deficient IP theft victims (including hackbacks), and if those hackbacks can be executed with 'due diligence', society in general will benefit. Fewer victims are likely to experience regret because the law deprived them of reasonable opportunities to recover lost property or prevent further theft. It is at least arguable, therefore, that an IP system that produces less regret on the part of theft victims is also efficient.

Section 4 analyzes a ransomware attack as a *zone 1* scenario in which the initial attack is deemed impermissible and the subsequent retaliation is deemed permissible. The last portion of Section 4 examines the efficiency of responding to a ransomware attack with a retaliatory hackback. Ransomware negotiations are often considered a classic Prisoner's Dilemma in which collective action problems may drive the parties apart, even though it is in their best interest to cooperate. The most 'Pareto-efficient' resolution to a ransomware crisis typically requires the victim to disclose his vulnerabilities to law enforcement, settle upon an agreed ransom quickly and pay the ransom. But if we examine ransomware crises in terms of maximizing opportunities rather than minimizing costs, the most efficient strategy might be to slow-walk negotiations and not disclose vulnerabilities. Such a strategy may increase the likelihood of locating the perpetrators and shutting down their operation using a retaliatory hackback.

Chapter 7 examines the moral and legal permissibility of retaliatory hackbacks in light of the US Supreme Court's recent decision in Van Buren v. United States. For many years, Congress has tried to regulate retaliatory hackbacks, but it was never clear if any formulation violated the Computer Fraud and Abuse Act (CFAA), arguably the most important anti-hacking federal statute. The central question in Van Buren was whether an individual who legally accessed one section of a local governmental database, and retrieved information from that one section for an improper purpose, violated the CFAA's prohibition on 'access exceeding authorization.' The Supreme Court concluded that Van Buren's action did not violate the Sect. 1030(a)(2) of the CFAA, but the majority did not address retaliatory hackbacks at all. Sections 1 through 5 of Chapter 7 examine the definition of 'exceeding authorized access' in the CFAA, the debate about 'narrowing' or 'broadening' the scope of the CFAA prior to Van Buren, and the majority and dissenting opinions in the Van Buren.

Sections 6 through 9 of Chapter 7 contend that Van Buren lays the groundwork for two models of hackback legislation in the future. The first model is symmetric, reciprocal legislation that authorizes hackbacks consistent with the fact pattern in Van Buren. A second model takes full advantage of VRN and recent guidelines from the Department of Justice on 'exceeding authorized access' cases. The Van Buren Court did not address the pervasiveness of extremely egregious cyber attacks engineered by malicious bad faith actors and dubious security experts who engineer 'particularly dangerous' hacks. 'Exceeding access' violations by the aforementioned parties are usually deliberate, not casual, mistaken or negligent. I argue that a retaliatory hackback exception to Van Buren should be made for the aforementioned individuals; and that this exception should consider purpose consistent with *zone 2* of the VRN, and the DOJ's recent guidelines on 'exceeding authorized access'. The last section considers the efficiency of retaliatory hackbacks against insiders who use their authorization credentials to acquire information for an improper purpose. Two aspects of retaliatory hackbacks are examined : 1) the efficiency of not disclosing vulnerability information to law enforcement officials before hacking back in response to insider abuse; 2) the efficiency of conceptualizing hackbacks as penalty enhancements, and then using retaliatory hackbacks to minimize 'social waste', as a way to deter insider abuse. With regard to the former, I argue that for some capability-deficient victims subjected to particularly dangerous cyberattacks engineered by insiders, the cost of not having a range of strategic options and opportunities available to them if they disclose their vulnerabilities before launching a retaliatory hackback far exceeds the benefits of full disclosure, whether they choose to hackback or not. From the standpoint of opportunity-freedom, it may be efficient not to disclose. In terms of the latter, research has shown that Paretian efficiency is more likely to produce 'social waste', in the form of 'perverse preferences and false beliefs,' than efficiency models that maximize opportunities, especially when we examine the efficiency of penalty enhancement regimes. Since 'Zone 2' hackbacks under VRN operate as a form of penalty enhancement, I argue that 'zone 2' hackbacks reduce waste more efficiently when they increase opportunities rather that minimize costs.

Chapters 8 and 9 examine the difficult task of justifying retaliatory hackbacks by the American government when many nation-states, including allies and rivals, routinely attack each other in cyberspace. In Chapter 8, I argue that the United States should justify retaliatory hackbacks not just on competitive or technological grounds, but also on the grounds that authoritarian countries enjoy more autonomy to hack than non-authoritarian countries, and therefore should be particularly vulnerable, both morally and strategically, to retaliatory hackbacks engineered or authorized by democratic, non-authoritarian countries. The disadvantage endured by the United States in cyberspace relative to authoritarian countries is a capability deficit, or 'situational constraint' that prevents the United States from taking advantage of opportunities in spite of its wealth, collective endowment in primary goods, and commitment to formal rights and the rule of law. Grounding America's active-defense policy in capabilities theory also strengthens the existential backbone of open world discourse—liberal internationalism and democracy—enabling US policy-makers to distinguish 'impermissible' hackbacks by authoritarian, illiberal, rogue nations and 'permissible' hackbacks by democratic, liberal nations that respect international law and human rights norms.

The first section of Chapter 8 analyzes the changing landscape of American strategy, focusing on the gradual decline of the liberal internationalist ethos established after World War II, and the distinction between primacy and restraint in international relations theory. Section 2 examines the perception that authoritarian countries enjoy certain advantages in the current geo-political climate due to their tenuous commitment to the international liberal order, international law, political accountability and human rights. This strategic advantage enables authoritarian countries to act quickly during crises, weather international sanctions and largely ignore international restrictions on armed conflict and cyber operations. America's geo-strategic 'burden,' broadly stated, is a capability deficit that forced the United States to develop an offensive cyber program. The chapter closes by considering and rejecting, two possible objections to applying VRN to American foreign policy and retaliatory attacks/hackbacks engineered by the US government.

Chapter 9 argues that populist American presidents are more likely to conceptualize America's foreign policy agenda as 'capability-deficient' than neo-liberal presidents. Populist presidents are particularly sensitive to vulnerabilities experienced by 'ordinary' citizens (as opposed to elites). That sensitivity can turn inward (domestic vulnerability translated into isolationism and exaggerated, inconsistent assessments of national strength) or outward (domestic vulnerability translates into 'connected internationalism' and accurate, honest assessments of national strength). The latter is better positioned to not only reconcile non-elite domestic preferences with a compelling, moral justification for continued American global dominance, but also justify national security priorities, including retaliatory hackbacks, on moral grounds.

Chapter 9 begins with an examination of the relationship between morality, American foreign policy and the capabilities critique of neoliberalism. That critique focuses on two concepts: (1) democratic participation over the rule of technocratic elites, and (2) minimizing unnecessary and psychologically toxic 'responsibilization' strategies that blame vulnerable people for their own problems. The next section examines cyberspace challenges to presidential leadership. Section 3 analyzes the Obama administration's response to cyber attacks.

The last two sections examine the retaliatory hackback policies of Presidents Trump and Biden, two populist presidents with dramatically different understandings of American foreign policy, but a similar professed interest in 'non-elite' domestic preferences. I argue that Biden's left-variant 'neo-populism' is a better platform for a capability-based active-defense program than Trump's 'right-variant' populism was during Trump's term in the White House.

References

Clausing, K. 2019. Open : The Progressive Case for Free Trade, Immigration, and Global Capital (Harvard University Press).

DODNIST. 2011. Department of Defense Strategy for Operating in Cyberspace, NIST : Computer Security Resource Ctr. 7 (July 2011). https://csrc.nist.gov/CSRC/media/projects/ISPAB/documents/DOD-Strategy-for-Operating-In-cyberspace.pdf

Kaplow, L., and S. Shavell. 2002. *Fairness Versus Welfare.* Cambridge: Harvard University Press.
Kesan, J. and Carol Hayes. 2012. Mitigative Counterstriking: Self-Defense and Deterrence in Cyberspace. *Harvard Journal of Law and Technology.* 25(2) : 429–543.
Lissner, R. and Mira Rapp Hooper. 2020. An Open World : How America Can Win The Content For the Twenty First Century Order (Yale University Press).
Philippon, T. 2019. The Great Reversal : How America Gave Up on Free Markets (Harvard University Press).
Soros, G. 2019. In Defense of Open Society (2019, Hachette Book Group).
Van Der Vossen, B. and J. Brennan. 2018. In Defense of Openness : Why Global Freedom is the Human Solution to Global Poverty (Oxford University Press).

PART I

Open World Skepticism and Attainable Ends

CHAPTER 2

The Open World Ethos

1 Structure of the Open World

Open World Zones

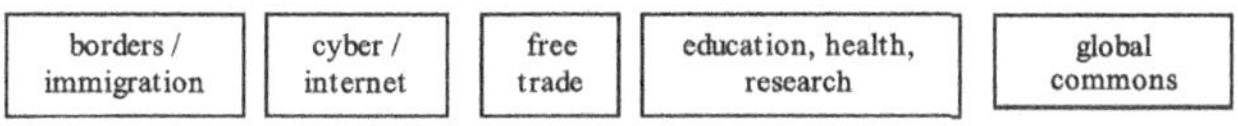

Open World Spectrum

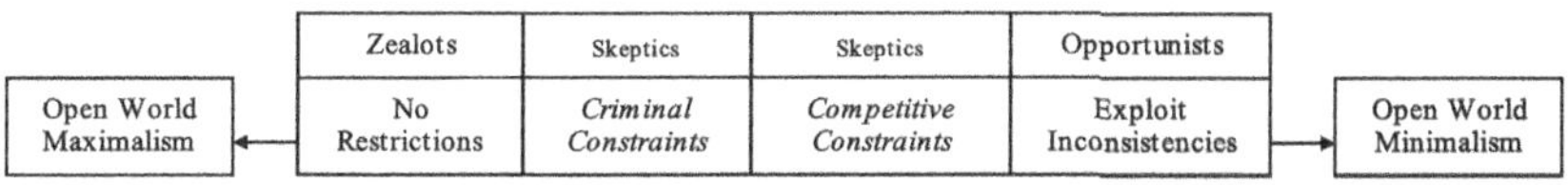

Since the liberal international order was established, ensuring an open world has been a central objective of international institutions and international humanitarian law. Virtually every military, strategic or diplomatic intervention undertaken by the United States and its allies since World War II has attempted to alleviate widespread humanitarian disasters, promote democracy[1] and the rule of law, protect vital routes necessary for transnational commerce and secure ‘open-world’ rights.

The ‘open world ethos’ contains two internal dynamics. The first, subject matter differentiation, tethers open world expectations to specific environments. There are at least five open world subject

[1] For more on the democracy promotion narrative, see Monten (2005).

A. J. Thomas, *The Open World, Hackbacks and Global Justice*,
https://doi.org/10.1007/978-981-19-8132-6_2

matter zones: borders/immigration, cyber/internet, free trade, education/health/research and global commons. Each zone involves the movement of persons, resources or information and expectations attached to some zones are more controversial than similar expectations attached to other zones. The second dynamic, scale, places open world sensibilities along a scale of beliefs about the desirability of open world policies. Sensibilities may persist across zones or only apply to a specific zone. For the sake of clarity, I refer to one end of the scale as 'open world minimalism' and the other end as 'open world maximalism.'

Within the 'open world minimalism'/'open world maximalism' spectrum, there are four sensibilities. The zealous sensibility, occupying the position on the scale closest to open world maximalism, calls for open borders, unfettered free trade, universal, uncensored, guaranteed access to the internet, broad global commons protection, and unencumbered travel for education and medical care. Criminal skepticism, located closer to the center of the scale, calls for limitations on open world expectations that discourage criminal activity. Competitive skepticism questions whether open world expectations are attainable due to competitive pressures exerted by allies and enemies in either global or market environments. Finally, the opportunistic sensibility capitalizes on the symbolic value of open world thinking, while taking advantage of inconsistencies, conflicts and uncertainty in the open world community.

2 Open World Governance

2.1 The Global Commons

Promulgating rules and norms for the use and/or exploitation of resources located in the global commons is one of the central achievements of the liberal international order.[2] Zealots define the global commons widely to include space, most of the Arctic/Antarctic, and areas designated as the 'common heritage of mankind.'[3] Zealots oppose giving states unilateral authority to significantly develop or undermine

[2] Shifrinson and Lalwani (2014).

[3] UNCLOS (1982). UNCLOS only designates a limited number of areas (sea bed, ocean floor and subsoil outside of territorial control) as 'common heritage of mankind' (CHM), which means that the areas have to be managed, controlled and potentially exploitable by everyone. Use of CHM-designated areas must also be limited to peaceful activities, and any scientific research must be available to all. Joyner (1986).

wild areas considered essential to the health of the planet, or destroy wilderness preserves containing endangered species.[4] Echoing traditional fears of the 'tragedy of the commons',[5] criminal skeptics believe that over-utilized commons areas might attract illicit criminal activity like over-fishing, piracy and deforestation.[6] Competitive skeptics believe that global competitive pressures often make it difficult to fully protect global commons areas. Expanded international water designations might be the best way to protect depleted fish populations, for example, but weak states and international monitoring groups may lack the resources to protect vast areas of ocean from the aggressive encroachments of strong

[4] Waarden (2010, 11–19), UNCLOS (1982, article 119) (defines factors that states must consider to avoid unreasonable species losses in the sea); Donoff (1992, 1408–22) (describes unilateral and multilateral efforts to protect endangered or heavily hunted species occupying commons areas from overly aggressive hunting tactics); Waarden (2010) (describes attempts to protect common pool resources like fish and forests from over-utilization and species loss in an area of economic interdependence and intense global competition using international instruments like the Convention on International Trade With Endangered Species of Wild Fauna and Flora (CITES) and eventually the Convention on Biological Diversity).

[5] Historically, the concept of a commons was associated with open access land, such as a forest or a common pool resource. See Hardin (1968, 1244) and Dagan and Heller (2001, 551). For a partial critique of Hardin's thesis, see Ostrom et al. (1984). The latter suggests a potentially large, but limited, number of users. Dagan and Heller (2001, 551, n.2) (distinguish between open access and common pool). The traditional problem with common areas has always been either over-utilization or under-utilization. Over-use occurs when an individual who has access to a common area, and may lack either the incentive or ability to prevent other users from exploiting the same area, takes maximum advantage of any resources in the common area. If other common area users adopt the same strategy, all resources will eventually be depleted, resulting in a wasted space. The second potential problem caused by commons regimes is under-utilization, as captured by the 'tragedy of the anti-commons.' In its purest form, the tragedy of the anti-commons occurs "when multiple owners each have a right to exclude others from a scarce resource and no one has an effective privilege of use." Heller and Eisenberg (2002, 160). Extreme fragmentation of interests can produce more efficiency and even a surplus as individuals try to exploit every conceivable usage of land, but when fragmentation leads to small parcels purchased either to exploit a use in the future or simply to prevent others from exploiting the land now, under-utilization of property can result. See Michelman (1982, 6–7) and Fennell (2009, 3).

[6] Clancy (1998, 604) ("Global commons are ...targets for over-exploitation."); Hsu (2005, 87–94) (critiques the violation of domestic statutes and international agreements due to the over-utilization of common pool resources) and Standing (2015, 175–97).

states.[7] In the same vein, scientific data demonstrating the ecological significance of vast rain forests might be compelling, but it is very difficult to enforce international environmental norms against powerful states without significant political or economic pressure.[8]

Opportunists profess symbolic support for the global commons, but exploit conflicting commons protection standards to enhance their reputations as global climate leaders. Between 2006–2018, for example, Brazilian Presidents Lula and Rouseff cultivated Brazil's image as a key player in the global climate change/deforestation movement, touting its influence over UNFCCC (United Nations Framework Convention on Climate Change) negotiations, challenging the United States, China and India for not adopting ambitious emissions targets, and bragging about the success of Brazil's anti-deforestation policies. While Brazil's carefully crafted image as a global climate change leader prior to President Bolsonaro did lead to modest reductions in deforestation, Brazil overstated its capacity to predict and control deforestation in the Amazon, mistakenly assumed that emissions reductions from the Lula era would continue into the future, surreptitiously pursued aggressive pro-development policies behind closed doors and quietly channeled public and private resources toward the development of an industry that actually contributes to global warming: oil production.[9]

[7] Due to competitive pressures, weak states may be unwilling to permit outsiders to exploit commons areas within their territory, but they may also require the assistance of international institutions to properly monitor the use of those common areas. Competitive pressures may also drive weak states to adopt neoliberal policies that discourage subsistence-based cooperative sharing programs in favor of corporate schemes that lead to over-exploitation. Maranganti (2012, 10–13) and Maynz (2002, 5) (international institutions charged with the protection of commons areas may fail if weak member states lack the power to protect their strategic interests and competitive priorities).

[8] Carvalhos (2019, 122–30) (Due to international cooperation, deforestation in the Amazon dropped dramatically from 2004–2012, as well from 2016–2017. Deforestation increased after President Bolsonaro announced his intention to withdraw from the UN Paris Agreement, which called for reduced deforestation and eventually reforestation, as well as a reduction in greenhouse emissions by 2030.); Hope (2019) (Bolsonaro's pro-development and pro-deforestation stance enhances Brazil competitive position in the global agribusiness industry); Franchin and Vida (2019) (Bolsonaro official opposes climate change agenda because it "(increases) the power of international organizations over national societies ... [and favors] China's economy.").

[9] Franchin and Vida (2019).

2.2 *Access to Educational Facilities and Research*

Ensuring that individuals have access to educational facilities, research and experts in other countries is another open world priority. Zealots believe that states should always guarantee foreign access to domestic educational, medical and research resources. The zealous sensibility on foreign access is actually the default view of the federal government and specific scientific organizations. National Security Decision Directive 189, for example, states that "It is the policy of this administration that, to the maximum extent possible, the products of fundamental research remain unrestricted NSDD (2006)." But federal regulations authorize greater restrictions on classified research, research covered by EAR (2006), ITAR (2006) and other controlled non-classified substances than fundamental research. JASON, a scientific community, is also supportive of broad openness regulations for academic research, even though JASON recently called for the adoption of robust disclosure rules and conflict of interest guidelines for foreign researchers due to allegations that foreign academics doing research in the United States are 'double dipping' (doing research at university while working for a private firm).

Criminal skeptics believe that ensuring constant access is difficult given the likelihood of intellectual property theft.[10] Between great power competition, and increasing demand for a limited supply of valuable technological and scientific goods, we shouldn't expect either powerful or emerging nations to provide all foreigners with unfettered access to local research institutions or medical experts.[11] Consequently, criminal skeptics might restrict travel by scholars from global competitor states when those

[10] Coats (2019, 14) ("We assess that China's intelligence services will exploit the openness of American society, especially academia and the scientific community, by a variety of means."); Seidel (2019) (some Chinese students/professors don't know that they are being used for espionage, while others might be coerced by the Chinese government). In response to accusations that they are using students to engage in espionage, the Chinese government has stated that "such statements are completely untrue and made with ulterior motives. People-to-people exchanges form a basis for the promotion of China-US cooperation in all areas, which is the common interest of both powers." Cohen and Marquard (2019).

[11] Widener (2019) (Numerous bills are under consideration in the House and Senate that impose new restrictions on foreign access to American researchers, databases and universities.).

states repeatedly violate international law.[12] Opportunists recognize that there is enormous political cachet to being perceived as a cosmopolitan, ultra-modern nation open to foreign ideas, filled with experts and institutions that foreigners desperately want to consult. But they are prepared to exploit loopholes, ambiguities and uncertainties in the open access community, particularly with regard to whether conflict of interest and disclosure rules should apply to all researchers.

2.3 *Free Trade*

Free trade is the metaphorical pivot point in open world discourse. For decades, America and allies linked free trade to capitalism, diplomacy, the rule of law, multilateralism,[13] the end of war and poverty.[14] Free trade was crucial to the establishment of the liberal international order,[15] and the embrace of free trade principles was a crucial requirement for emerging nation-states seeking financial and developmental assistance from international institutions. Many America presidents embraced the

[12] Adath (2019) (Iranian scholar cannot submit manuscript to US-based scientific journal because of IS sanctions against Iran. Trump's Muslim-majority travel ban has also made it difficult for Iranian scholars to visit the United States or interact with colleagues at American universities.).

[13] Low (2003) (compares 'first-best' multilateral trade agreements facilitated by the World Trade Organization to 'second-best' regional trade agreements in Asia); Tussie and Woods (2000, 5) (describes various critiques of the "liberal institutionalist view that openness and multilateralism are part and parcel of the same process with multilateralism, strengthening and deepening as more countries integrate into the world economy and more markets converge").

[14] Zeiler (1999, 2–3). Free trade alone cannot eradicate poverty in developing countries. The burgeoning movement of 'aid for trade' is based on the recognition that poor countries need free trade and developmental assistance, and that international institutions as well as wealthy nations should not condition aid solely on a poor country's willingness to blindly open up their markets. See Stiglitz and Charlton (2000) ("Aid for trade involves the flow of development finance from rich to poor countries for the purpose of enhancing the world trading system."); Hallaert and Munro (2009, 6) (Limited capacity and lack of infrastructure are domestic constraints that make it difficult for developing countries to take full advantage of global markets and trade opportunities.).

[15] Ikenberry (2011, 67) ("The liberal international order has succeeded over the decades because its rules and institutions have not just enshrined open trade and free markets but also provided tools for governments to manage economic and security interdependence."); Kundnani (2017, 3) (attributes widely held belief that that free trade and protection of the global commons are crucial components of the liberal international order to the Atlantic Charter, signed by Churchill and Roosevelt on 14 August 1941).

free trade mantra because it provided struggling workers and farmers with new markets, while other presidents championed protectionist policies like tariffs.[16] Zealots completely reject economic tariffs, quotas and economic subsidies because they believe that free trade always enhances social welfare.[17] Criminal skeptics believe that free trade requires careful monitoring to avoid criminal abuses like the import of unsafe products or illegal controlled substances.[18] Moderate competitive skeptics are prepared to support temporary measures to protect the fragile economies

[16] Zeiler (1999, 9). William McKinley was one of the most pro-tariff American presidents. As a strong proponent of reciprocity rather than 'free trade,' McKinley believed that strong tariffs were essential to protect American business interests, middle- and working-class Americans and trade unions. See generally Skrabek (2008, 8–11). Unlike McKinley, Andrew Jackson's pro-tariff sensibilities changed considerably over time due to the perception, widely held during the nineteenth century, that tariffs aren't the best way to reduce government debt, and that tariffs unnecessarily increase federal power over sovereignty and autonomy of states. Wilentz (2007, 89–91) and Hardin (2003, 288–89). For more on southern attempts to nullify tariff laws from 1828 and 1832. See also Hardin (2003, 288–89). The protection of 'infant' American industries through tariffs, originally defended by Hamilton, was a mainstay of nineteenth-century American trade policy. When protections were no longer needed due to the maturation and competitive success of 'infant industries,' 'reciprocity' in the form of reciprocal concessions and bilateral agreements, became the dominant American trade policy. After World War II, the United States sought to "achieve free trade through reciprocal concessions in the GATT and WTO." Lal (2004, 16).

[17] In effect, free trade is 'welfare enhancing.' Gupta (1997, 62) ("The mainstream defense of openness is rooted in the classical theory of trade. Free trade, the argument goes, results in a more efficient international division of labor as nations are compelled to specialize in the production of goods and services in which they have a comparative advantage. The primary beneficiary of free trade are consumers, who enjoy more choice, lower prices and, on average, higher real incomes. The gains from trade are still greater in models which assume increasing returns for scale and 'dynamic gains from trade.'"); (Panaganya (2019, 1–4) (defends free trade as an engine of reconstruction and recovery after World War II in Europe, and accelerated economic growth in the United States).

[18] Farzarnegan (2012, 93–94) (smuggling tends to increase as 'trade openness' (liberalization), or "the share of total trade (imports and exports) in GDP" increases because "it is easier to hide smuggling from enforcement agencies," thereby reducing the cost of smuggling operations); Uhm (2016, 17) (along with habitat loss and degradation, illegal wildlife trade undermine biodiversity); Staake and Fleish (2008, 16) ("Illegal trade denotes a wide variety of illegal or non-contractual activity. Trafficking in controlled substances, stolen and smuggled goods, trade of all kinds … infringing intellectual property rights, and even parallel imports may fall into this category.").

of weak states like short-term tariffs.[19] Radical competitive skeptics challenge the underlying logic of free trade in an increasingly global world. Instead of viewing free trade as a perfect vehicle for exporting western values and bolstering the liberal international order, radical competitive skeptics believe that free trade has provided multi-national companies with an excuse to ship jobs to countries with low wages and weak worker protections.[20] Growing economic inequality[21] has dramatically increased support for free trade minimalism. Opportunists recognize that openness to free trade, like openness to foreign ideas and scientific expertise from abroad, is tremendously valuable as a political and economic too, but they also know that states often move between protectionist and free trade policies for short-term political gain. In response to uncertainty in their domestic political environments, therefore, opportunists may proclaim their willingness to fix 'defective' free trade mechanisms that don't take advantage of their comparative strengths in specific areas like technology, agriculture or energy.[22]

2.4 *Internet Access and Cyber Sovereignty*

Another understanding of 'open world' is internet access. Dozens of organizations, states and technology firms have promulgated rules for navigating cyberspace, and encouraged companies and states to make massive quantities of internally generated data available to the public via the World Wide Web.[23] Zealots generally believe that the internet

[19] Hobson (1997, 19–20) (weak state dependence on temporary tariffs); Singh (2005, 90) (in developing countries, tariffs protect vulnerable local industries, but they also provide badly needed government revenue).

[20] Tonelson (2002).

[21] Meyer (2017) (new multinational agreements must include obligations to reduce economic inequality and widespread job losses).

[22] Schwabach and Cockfield (2009, 14) (developed countries welcome free trade in high technology products, but not agricultural products like sugar because developing countries have a comparative advantage); Dunkley (2000, 38) ("Free trade preachments" are common, but "the persistence of industry protection systems throughout the world suggests that such systems play a legitimate role in stabilization and minimizing social adjustment costs."); Ericksen (2007) (the "Americans are all for free trade, but against imports").

[23] Benedek (2008, 33) (self-governing regulatory bodies provide useful "guidelines for the fields of human rights and democracy, access, openness, diversity and security that

is more than a technology; it is a vehicle for personal and collective liberation.[24] Internet world should remain free and unfettered as much as possible,[25] corporations, governments and universities should be encouraged to 'publish' useful information on the web,[26] and governments should not censure, filter or restrict access to the global internet.[27] The internet is a common resource that belongs to everyone,[28] almost as vital to modern existence as food, shelter, education, health care and freedom from violence. Radical zealots believe that access to information on the internet is a fundamental human right, while moderate zealots believe that the right to internet access should be civil, rather

governments, in cooperation with stakeholders ... should promote"); Weber (2010, 179) (self-governing ICANN is primarily responsible for "the coordination and evolution of the domain name system root server frameworks"); Tambini et al. (2008, 3–17) (describes the architecture of self-governing regulation on the internet).

[24] Salter (2016, 19) (utopian actors characterize the internet as a "medium for personal freedom and collective liberation"); Shields (1996, 117) ("cyberspace time promises us liberation from constraints of space, time and materiality"); Turner (2010, 208) ("desktop computers and the internet became tools for personal and collective liberation").

[25] Shuman (2001, 91) (examines the affect of free internet access on general library use); Toulouse and Luke (2013, 12) ("US government and educational institutions, and some businesses, provide free internet access and in some cases free computers or at least workplace access."); Jupp (2001, 24) ("US government and educational institutions, and some businesses, provide free internet access and in some cases free computers or at least workplace access."); Scheidel (2014, 105) (free online access is exchanged for consumer or watching paid advertisements) (Jupp, 2001).

[26] Hernon (2003, 24) summarizes government regulations that either recommend of require the publication of information on governmental websites); Gaillet and Gigliemo (2014, 121) (some academic institutions require the 'open access' publication of theses and dissertations on the web); Swan (2012, 15–19) (describes content perfect suited for open access designations).

[27] Warf (2012, 47) (internet censorship may restrict the flow of digital information using a variety of mechanisms, or through context filtering using keyword matching algorithms); Deibert (2013, 72) (distinguishes between internet censorship indicating that targeted research does not exist and censorship indicating that requested research is merely blocked); Goodman (2015, 192) (governments justify internet censorship on many grounds, ranging from protecting national security, ensuring intellectual property, preserving religious values and saving children).

[28] Promoting (2002, 36) (the physical and content layers of the internet are not a common pool resource, but the code layer generally is); and Bertachini (2012, 197) (fluid digital communities are cultural commons resources). See also Lessig (2002) (end to end principle makes the internet a 'commons').

than fundamental.[29] Cyber-zealotry dominated the thinking of 70s and 80s era Silicon Valley pioneers steeped in the utopian sensibilities of the Woodstock generation.[30] Zealotry also served the business purposes of ambitious American communications companies who sought access to customers and resources across the globe.[31] Zealots tend to support net neutrality[32] as well as data sovereignty (i.e., individuals can control access to their personal data regardless of where they are or where the data is stored),[33] while opposing government censorship.

Criminal and competitive skeptics recognize that events have made it more difficult to defend unfettered internet access and cyber freedom. Edward Snowden's revelations about the extent to American intelligence gathering on both domestic and foreign targets after 9/11[34] severely

[29] Verma and Verma (2014, 2–3) (argues that internet access is a human right established by Article 19 of the ICCPR [International Convention on Civil and Political Rights]); Glen (2017, 82) (internet access should be a civil right, rather than a fundamental human right); Lambrects and Situba (2016, 13) ("internet connectivity is considered an official human right with reference to the United Nations (UN) Sustainable Development Goals").

[30] Hafner (1999).

[31] Diveloh and Dishpande (2004, 160) (customers assume that global brands are cutting edge, well-designed, innovative and cosmopolitan); Haggart (2014, 21) (US technology firms dominate foreign markets by pushing for the international adoption of copyright standards and expectations that reflect American copyright principles).

[32] Hazlett (2011) (arguments for open access to broadband network mirrors arguments for net neutrality); Gilroy (2011, 1) (net neutrality refers to the "general principle that owners of the networks that compose and provide access to the internet should not control how consumes lawfully use that network"); Stiegler and Sprumont (2013, 123) (corporate opponents of net neutrality are concerned about governmental interference with free enterprise).

[33] Armeur et al. (2012, 145–47) (suggest strategies for maintaining data, including restrictions on third-party access and control of personal data, liability rules for breaches and data personalization); Brkan (2017, 17) (distinguishes between data privacy and data protection laws).

[34] Greenwald (2013) These revelations stunned most Americans, including citizens who were generally supportive of ongoing US intelligence gathering and surveillance efforts. After these revelations, the Obama Administration acknowledged the existence of the NSA's massive data-mining operation, authorized ostensibly by Section 215 of the Patriot Act. The existence of NSA data-mining programs that targeted internet communications was revealed around the same time. In the wake of the Snowden revelations and Presidential responses, many lawsuits challenging the constitutionality of the NSA meta-data program were brought in the federal courts. See also Clapper (2013).

undermined America's credibility as the supposedly benevolent guarantor of internet freedom.[35] Criminal skeptics generally acknowledge that complete cyber freedom might be attainable when the architecture of the internet is fully decentralized and encrypted,[36] but they are deeply concerned about rampant criminal activity in the dark web,[37] fraudulent crypto-currencies, and attribution problems after hack attacks.[38] Competitive skeptics believe that the internet has always been constrained by state competition, self-interested forum shopping by computer firms, and the potential for both kinetic and non-kinetic military conflict.[39] They were not surprised when rival powers enacted a flood of data nationalist policies that not only restrict the internet world of their own citizens, but also provide them with new technological opportunities to export their authoritarian values into other countries. China's Belt and Road initiative,

[35] Resilience (2014, 114) (America's reputation as the primary steward and protector of internet freedom was severely undermined by the NSA); Stein (2006, 81–94) (discussing the role of First Amendment law on the development of global speech rights and internet access expectations, particular with regard to whether government intervention is necessary in order to protect speech); Schutte (2013, 76) (Reagan entrusted American corporations with the security of the internet to ensure America's continued dominance of cyberspace).

[36] At present, total anonymity in cyberspace is impossible. What appears to be anonymity is merely 'pseudonymity.' However, advances in complex encryption and obfuscation strategies like zero knowledge proofs, secure multiparty computation (SMPC), ring signatures and secret sharing may eventually defeat all law enforcement decryption techniques, even with the assistance of quantum computers.

[37] Kassab and Rosen (2018, 184–89) (the 'dark web' refers to networks used by criminals or web-savvy individuals to communicate and/or transfer illegal goods using difficult to detect or decrypt technologies); Ozkaya and Islam (2019) (distinguishes between surface web (legal), deep web (legal/illegal), dark web (illegal) and dark net (illegal). See also Shackleford (2014, 160) (discusses an earlier period in internet history when organized crime was not interested in cyber hacking).

[38] Tropina and Callahan (2015, 8) (attribution problems make it difficult to determine the identity and motivations of suspected cyber-criminals); Geers (2011, 157) (attribution problems can lead to "false-flagging operations – where the attacker tries to pin the blame on a third party"); Yannakngeous and Lowther (2013, 72) (distinguishes between technical and 'all source' attribution).

[39] Valeriano et al. (2018, 135–48) (cyberspace, rather than nuclear or terrorism, is the new focus of great power competition between the United States and China); Gray (2013, 40) (unlike nuclear competition, cyber competition is unlikely to result in mass destruction). See also Manzanec and Thayer (2014, 32) (American deterrence of Chinese cyber attacks emboldens regional actors in Asia looking for any perceived America weaknesses in cyberspace).

for example, has extended China's infrastructure and technological expertise deeply into the Third World (especially Africa and the Middle East), but participants are also expected to accommodate China's restrictive human rights and censorship policies.[40] Initially, according to competitive skeptics, America exploited the open world mantra to gain unfettered access to global markets, export western values and gather valuable intelligence information; now our rivals exploit open world norms to secure valuable resources, maintain non-kinetic strategic parity with the United States and export authoritarian values to fragile states.

Opportunists exploit chaos, uncertainty and the absence of trust in cyberspace, particularly in the midst intractable conflicts, resource shortages, intense rivalries or dramatic technological advances. As a result, cyber opportunists may profess support for internet freedom, but their policy proposals tend to careen unpredictably between more authoritarian control over cyber resources, and toleration for chaotic, risky experiments specifically designed to punish bad actors in cyberspace.

2.5 *International Migration*

Besides free trade, the free movement of persons over national or contested borders is the clearest example of the open world ethos.[41] Nation-states devote an enormous amount of resources to border issues, ranging from controlling points of legal and illegal entry, providing temporary housing for immigrants and refugees, screening for potential criminals and terrorists, targeting applicants likely to possess skills that the nation-states desperately needs, to monitoring immigrant flows for contraband and disease. The United States and its allies developed elaborate international agreements to protect displaced persons, refugees and immigrants from human rights abuses and exploitation by host countries.[42] Zealots generally support these agreements. Criminal skeptics express concerns about the possibility of increased civil and criminal illegality arising from open borders.[43] Fear of crime also partially explains

[40] Vaseli (2018, 73) (China's Belt and Road Initiative "does not explicitly promote authoritarianism," but it does "incline towards authoritarianism").

[41] Passi (2009, 28–30).

[42] Bruch (2007, 199) (human rights justifications for more open border policies).

[43] Hendrich (2001, 51–55) (In aging, mostly white countries like Germany, fears of increased crime caused by new immigrants are greatly exaggerated.); Wallace (2001, 58)

the belief that open borders don't work well when neighbors have dissimilar cultural traditions,[44] or that open borders might overwhelm the legal screening capabilities of receiving countries.[45]

Competitive skeptics believe that economic insecurity and backlashes against the perceived excesses of globalization make it difficult to fully protect the moral expectations of immigrants. Politicians, for example, frequently use open border policies to encourage nationalist and racist violence against targeted immigrants, comparing immigration waves to the spread of disease.[46] Like protectionism (free trade) and data nationalism (internet access), restrictive border policies reflect a desire for 'cultural cleansing,' but they also exacerbate competition between states.[47] Opportunists take advantage of immigration flows caused by unpredictable events like natural disasters, drug waves or communicable diseases to selectively challenge open border policies, using tactics like conservative amnesty rules for illegal immigrants,[48] treating illegal immigration as

(describes fear that immigrants have increased the amount of transnational, as opposed to purely domestic, crime). See also Shantz (2010, 3) (neoliberals encouraged the development of a 'national security state' that criminalizes entire segments of the population, including left-wing activists, racial minorities and immigrants).

[44] Borders (2019, 10) ("[S]ome borders could be open, but only if the people on both sides are generally similar in terms of wealth, culture and governance patterns. To them, this is why a relatively open border between Canada and the United States works but such an arrangement between Mexico and the United States would not.").

[45] Fairchild (2013, 4–8) (screening immigrants for diseases and defects served to additional purpose of preparing immigrants to be productive participants in the American economy); Fretsedenas (2012, 102) (aggressive immigration screening strategies are really designed to police national identity); Wong (2017, 84) (increasing detention capacity for immigrants after 9/11).

[46] Weber (2015, 2) (open borders and moral panic).

[47] Mitsilegas (2012).

[48] Young (2003, 209) (general discussion of amnesty); O'Brien (2005, 261) (basic definition of amnesty); Fretsedenas (2012, 145) (describes republican critiques of Obama-era immigration reforms that protected widespread deportation of non-citizens); Levey (2010, 7–8) (defines amnesty as "forgiving illegal immigrants for their illegal acts and creating a way for them to become legal immigrants"); (Laham 2008, 189–95) (criticizes amnesty provisions in the IRCA (Immigration and Control Act of 1986) because of the 'low socioeconomic status' of Central Americans and Mexicans, the primary beneficiaries of amnesty programs).

a criminal violation[49] and opposing the extension of social welfare benefits to undocumented aliens regardless of their status.[50]

3 Threshold Considerations

The last two sections describe the basic structure of open world discourse: five substantive zones, and a scale from open world maximalism to open world minimalism. The structure of the open world scale raises two important threshold questions,. The first issue concerns the meaning of what lies between the poles. If both criminal and competitive skepticism are located between the poles, why is competitive skepticism located closer to the maximalist pole than criminal skepticism? On this issue, the chart is somewhat misleading because there are actually compelling reasons to place competitive skepticism nearer to the maximalist pole, and criminal skepticism near the minimalist pole. One the one hand, competitive skepticism should be located near the maximalist pole because increased competition is the logistical and organizational byproduct of open borders, and in most democratic societies, competition of all kinds is clearly encouraged. Ensuring healthy competition in a border-less zone

[49] Illegal presence is a civil violation under the INA (Immigration Naturalization Act), but illegal entry is a criminal violation under 8 USC 1325. See also Peoria (1983) (federal immigration laws permits local and state enforcement of criminal violations in INA). For the alternative view that state/local officials can arrest aliens for potential violations of federal criminal and civil immigration laws, see Vasquez (1999) (discusses 8 USC 1242 violation—state/local officials can arrest/detain alien illegally present in the United States who has already been convicted of US felony and deported or left the country after conviction), and Preemption (2002, 8–12) (discussing 8 USC 1252c). See also Abbott and Hetznel (2010, 90–97); Santana-Garcia (2001) (states have implicit natural authority to ask about possible civil violations of federal immigration laws (i.e., illegal presence), but court didn't distinguish between criminal and civil violations); (illegal presence [i.e., without papers] in the United States is only a "minor civil violation"); and Bacon (2008, 5) (illegal presence [i.e., without papers] in the United States is only a "minor civil violation").

[50] Kivisto and Faist (2009, 206) (In California, both Proposition 187 and The Personal Responsibility Act of 1996 contained provisions that denied various welfare benefits to other undocumented immigrants or legal/illegal immigrants); Barbour (2010, 35) (illegal immigrants are not entitled to welfare benefits, and very few manage to obtain them using forged documents); Moloney (2012, 102) (due to confusion about federal immigration laws, including a 1999 revision of the IRCA that guarantees access to welfare benefits for recent immigrants and those who have not received permanent residency via a green card, "immigrants and their children, whether citizens or not, are less likely than others in the US to access welfare benefits for which they are eligible").

is also probably easier to achieve than eliminating most criminal acts from the same space. As a result, competitive skeptics are more likely to embrace at least the maximalist ideal because maximalism encourages competition. Criminal skeptics will perceive the competition fostered by maximalism as a fertile opportunity for more fraud, violence and misrepresentation, acts of illegality that will probably overwhelm already over-burdened cross-jurisdictional law enforcement efforts. On the other hand, however, competitive skepticism should be located closer to open world minimalism side than criminal skepticism because it is usually easier to prevent the abuse of open world activities through effective law enforcement (either unilaterally or multilaterally) than to steer global competitors away from self-interested policies that undermine open world expectations.[51]

The significance of what lies between the open world poles, however, is less important than characterizing specific perspectives on the open world scale as 'skeptical.' The most intuitive justification for characterizing certain open world views as skeptical is that skeptics fundamentally question 'truth claims' of all forms: whether there is always a gap between perception and reality, whether the human brain is capable of grasping complex realities, whether the customs and habits that we have developed over time actually perform the function that they were intended to perform when they were originally developed (obsolescence), whether there is a distinction between ideal and actual human knowledge, whether 'expert' communities are more concerned about the coherence and internal consistency of their discursive practices than the essence of their chosen subject matter, whether the stated objectives of policy proposals are nothing more than aspirational goals. All of these concerns inform open world skepticism, or the belief, broadly stated, that widely accepted 'truths' about open world policies are not reliable or even necessarily embraced in good faith by open world proponents.

Skeptics question the validity of truth-claims, but skeptical arguments in open world discourse must do more than simply challenge the accuracy of empirical data presented in support of open world policies, the purported neutrality or good faith of open world 'experts,' or even the rationality of open world policies that consistently either miss their stated targets or miscalculate costs and benefits. Skepticism in the open world

[51] Cortes (2015, 19) (describes dramatic increases in MLAT adoptions as responses to international cyber-crime).

has to be more than a negative critique; it must also propose a positive political and socio-legal agenda capable of satisfying credible demands for global justice.

References

Abbott, E., and O. Hetznel. 2010. *Homeland Security and Emergency Management: A Legal Guide for State and Local Governments*. Chicago: ABA.

Adath, D. 2019. Science Under Maximum Pressure in Iran. *Thescientist.Com*, September 3.

Armeur, E., S. Gamb, and A. Ho. 2012. Maintaining Sovereignty over Personal Data on Social Networking Sites. In *Managing Privacy Through Accountability*, ed. Calla Iffen, Inga Kroener, Daniel Heybard, and Hector Postigo. London: Palgrave Macmillan.

Bacon, D. 2008. *Illegal People: How Globalization Creates Migration and Criminalizes Immigration*. Boston: Beacon Press.

Barbour, S. 2010. *Does Illegal Immigration Harm Society?* San Diego: Reference Point Press.

Benedek, W., Veronica Bauer, and Matthias Ketterman. 2008. *Internet Governance and the Information Society: Global Perspectives and European Dimensions*. New Jersey: Eleven International Publishing.

Bertachini, E. 2012. *Cultural Commons: A New Perspective on the Production and Evolution of Cultures*. Edward Elgar Publishing.

Borders. 2019. *Open Borders: In Defense of Free Movement*, ed. Reece Jones. Athens: University of Georgia Press.

Brkan, M. 2017. The Courts of Justice on EU: Data Privacy and Data Protection—Judge-Made Law as a Leitmotif on Fundamental Rights Protection. In *Courts, Privacy and Data Protection in the Digital Environment*, ed. Maja Brkan and Evangelia Psychogropoulo. Cheltenham: Edqard Elgar Publishing.

Bruch, E. 2007. Open or Closed: Balancing Border Policy with Human Rights. *Kentucky Law Journal* 96: 197–224.

Carvalhos, W. 2019. Deforestation Control in the Brazilian Amazon: A Conservation Struggle Being Lost as Agreements and Regulations Are Subverted and Bypassed. *Perspectives in Ecology and Conservation* 17 (3): 122–130.

Clancy, E. 1998. The Tragedy of the Global Commons. *Indiana Journal of Global Legal Studies* 5 (2): 601–610.

Clapper. 2013. Clapper v Amnesty International USA, 133 S. Ct. 1138 (2013).

Coats, D. 2019. Director of National Intelligence, Senate Select Committee on Intelligence. *Statement for the Record: Worldwide Threat Assessment of the US Intelligence Community*, January 29.

Cohen, Z., and A. Marquard. 2019. Intelligence Warns China Is Using Student Spies to Steal Secrets. *CNN.com/politics*, February 2.

Cortes, S. 2015. MLAT Jiu-Jitsu and Jur: Mutual Legal Assistance Treaties in Surveillance. *Richmond Journal of Law and Technology* 22 (1): 2–19.

Dagan, H., and M. Heller. 2001. The Liberal Commons. *Yale Law Journal* 110 (4): 549–612.

Deibert, R. 2013. *Black Code: Surveillance, Privacy and the Dark Side of the Internet*. Toronto: McClellan and Stewart.

Diveloh, A., and R. Dishpande. 2004. *The Global Market: Developing a Strategy to Manage Across Borders*, ed. John A. Diveloh and Rahit Dishpande. New York: Wiley.

Donoff, J. 1992. Reconciling International Trade with Preservation of 'The Global Commons': Can We Prosper and Protect? *Washington and Lee Law Review* 49: 1408–1422.

Dunkley, G. 2000. *The Free Trade Adventure: The WTO, the Uruguay Round and Globalization: A Critique*. London: Zed Books.

EAR. 2006. Export Administration Regulations, 15 C.F.R. Section 734.8 (c).

Ericksen, T. 2007. *Globalization: The Key Concepts*. Oxford: Berg.

Fairchild, A. 2013. *Science at the Border, Immigrant Media Inspection and the Modern Industrial Work Force*. Baltimore: John Hopkins University Press.

Farzarnegan, M. 2012. Dark Side of Trade in Iran: Evidence from a Structural Equation Model. In *Illicit Trade and the Global Economy*, ed. Claudia Costa Sorti and Paul De Grauwe. Cambridge: MIT Press.

Fennell, L. 2009. Commons, Anticommons and Semicommons. Reprinted in *Research Handbook on the Economics of Property Law*, ed. Kenneth Ayotte and Henry E. Smith. Cheltenham: Edward Elgar.

Franchin, M., and E. Vida. 2019. Myths and Images in Global Climate Governance, Conceptualization and the Case of Brazil (1989–2019). *Revista Brasiliera De Politica International* 62 (2): 1–25.

Fretsedenas, P. 2012. *The Immigration Crucible: Transforming Race, Nation and the Limits of Law*. New York: Columbia University Press.

Gaillet, L., and L. Gigliemo. 2014. *Scholarly Publication in a Changing Academic Landscape: Models for Success*. New York: Springer.

Geers, K. 2011. *Strategic Cyber Security Blog*.

Gilroy, A. 2011. Access to Broadband Networks. *Net Neutrality Debate*. CRS 7-5700, R40616, March 15.

Glen, C. 2017. *Controlling Cyberspace: The Politics of Internet Governance and Regulation*. Santa Barbara: ABC-CLIO.

Goodman, M. 2015. *Future Crimes: Inside the Digital Underground and the Battle for Our Connected World*. New York: Random House.

Gray, C. 2013. *Making Strategic Sense of Cyber Power: Why the Sky Is Not Falling*. Lulu.com.

Greenwald, G. 2013. NSA Collects Phone Records of Millions of Verizon Customers Daily. *Guardian* (London), June 5.

Gupta, S. 1997. *The Political Economy of Globalization*. Berlin: Springer Science and Business Media.
Hafner, K. 1999. *Where Wizards Stay Up Late: The Origins of the Internet*. New York: Simon and Schuster.
Haggart, B. 2014. *Copyfight: The Global Politics of Digital Copyright Reform*. London: University of Toronto Press.
Hallaert, J., and L. Munro. 2009. Binding Constraints to Trade Expansion: Aid for Trade Objectives and Diagnostics Tools. OECD Trade Policy Working Paper, No. 94, December 2.
Hardin, G. 1968. The Tragedy of the Commons. *Science* 162: 1243–1252.
Hardin, R. 2003. *Liberalism, Constitutionalism and Democracy*. Oxford: Oxford University Press.
Hazlett, T. 2011. *The Fallacy of New Neutrality*. Encounter Books.
Heller, M., and R. Eisenberg. 2002. Can Patents Deter Innovation? The Anti-Commons in Biomedical Research. In *Perspectives on Property Law*, 3rd ed., ed. Robert C. Ellickson, Carol M. Rose, and Bruce A. Ackerman. New York: Aspen Law and Business.
Hendrich, K. 2001. The German States in Comparative Perspective in the Year 2000. In *Leviathan Transformed: Seven National States in the New Century*, ed. Theodore Kaplow, 51–55. Montreal: McGill-Queens University Press.
Hernon, P., R. Dugan, and J. Shuler. 2003. *US Government on the Web: Getting the Information You Need*. Englewood: Libraries Unlimited.
Hobson, J. 1997. *The Wealth of States: A Comparative Sociology of Institutional and Political Change*. Cambridge: Cambridge University Press.
Hope, M. 2019. The Brazilian Development Agenda Drives Amazon Devastation. *The Lancet/planetary Health* 3: 1–5.
Hsu, S. 2005. What Is a Tragedy and the Commons?: Over-Fishing and the Campaign Spending Problem. *Albany Law Review* 69: 75–94.
Ikenberry, J. 2011. The Future of the Liberal World Order. *Foreign Affairs* 90(3): 56–68.
ITAR. 2006. International Traffic in Arms Regulations 22 CFR Sect. 120.11(8).
Joyner, C. 1986. Legal Implications of the Concept of Common Heritage of Mankind. *International and Comparative Law Quarterly* 35: 190–201.
Jupp, B. 2001. *Divided by Information? The Digital Divide and the Use of the New Meritocracy*. London: Demos.
Kassab, H., and J. Rosen. 2018. *Illicit Markets, Organized Crime and Global Security*. Cham: Spring.
Kivisto, P., and T. Faist. 2009. *Beyond a Border: The Causes and Consequences of Contemporary Immigration*. London: Sage Press.
Kundnani, H. 2017. What Is the Liberal International Order? German Marshall Fund Policy Essay, No. 17.

Laham, N. 2008. *Ronald Reagan and the Politics of Immigration Reform*. Greenwich: Greenwich Publishing.

Lal, D. 2004. *In Defense of Empires*. Washington, DC: AEI Press.

Lambrects, W., and S. Situba. 2016. *Last-Mile Internet Access: Emerging Energy Economies*. Berlin: Springer.

Lessig, L. 2002. The Architecture of Innovation. *Duke Law Journal* 51 (6): 1783–1801.

Levey, J. 2010. *Illegal Immigration and Amnesty: Open Borders and National Security*. New York: The Rosen Publishing Group.

Low, L. 2003. Linda Low, Multilateralism, Regionalism, Bilateral and Cross-Regional Free Trade Agreements: All Paved with Good Intentions for Asean? *Asian Economic Journal* 17 (1): 65–86.

Manzanec, B., and B. Thayer. 2014. *Deterring Cyber Warfare: Bolstering Strategic Stability in Cyberspace*. Berlin: Springer.

Maranganti, A. 2012. Tragedy of the Commons Revisited (I): Granite Quarrying in Telangana. *Economic and Political Weekly* 47 (42): 10–13.

Maynz, R. 2002. National States and Global Governance, VII Inter-American Congress of Clad on State and Public Reform, October 8–11.

Meyer, T. 2017. Saving the Political Consensus in Favor of Free Trade. *Vanderbilt Law Review* 70 (3): 985–1026.

Michelman, F. 1982. Ethics, Economics and the Law of Property. In *Nomos Xxiv: Ethics, Economics and the Law*, ed. J.R. Pennock and J.W. Chapman, 3–40. New York: New York University Press.

Mitsilegas, V. 2012. Immigration Control in an Era of Globalization: Vetting Foreigners, Weakening Citizens, Strengthening the State. *Indiana Journal of Global Legal Studies* 19 (1): 3–60.

Moloney, D. 2012. *National Insecurities: Immigrants and US Deportation Policy Since 1882*. Chapel Hill: University of Carolina Press.

Monten, J. 2005. The Roots of the Bush Doctrine: Power, Nationalism, and Democracy Promotion in U.S. Strategy. *International Security* 29 (4): 112–156.

NSDD. 2006. National Security Decision Directive—189. *Export Controls and the Safe Harbor Exemption for Fundamental Research*, 1. Revised 12/21/2006.

O'Brien, R. 2005. Amnesty and International Law. *Nordic Journal of International Law* 74(2): 261–278.

Ostrom, E., Roy Gardner, and James Walker. 1984. *Rules, Games and Common Pool Resources*. Ann Arbor: The University of Michigan Press.

Ozkaya, E., and R. Islam. 2019. *Inside the Dark Web*. Boca Raton: CRC Press.

Panaganya, A. 2019. *Free Trade and Prosperity: How Openness Helps the Developing Countries Grow Richer and Combat Poverty*. New York: Oxford University Press.

Paasi, A. 2009. Bounded Spaces in a Borderless World? Border Studies, Power and the Anatomy of Territory. *Journal of Power* 2 (2): 213–224.
Peoria. 1983. Gonzales v City of Peoria, 722 F.2d 468 (9th Cir. 1983).
Preemption. 2002. US Dept. of Justice, Office of Legal Counsel. *Memo for the Attorney General Re: Non-Preemption of the Authority of States and Local Law Enforcement Officers to Arrest Aliens for Immigration Violations*, April 3.
Promoting. 2002. *Government's Role in Promoting the Future of the Telecommunications Industry and Broadband Development* (October 1), S. Hrg. 107-1148, Hearing Before the Committee on Commerce, Science and Transportation, US Senate, 107th Congress, Second Session.
Resilience. 2014. *Ensuring the Security, Stability, Resilience and Freedom of the Global Internet, Hearing Before the Subcommittee on Communication and Technology of the Committee on Energy and Commerce*. House of Representatives, 113th Congress, Second Session, April 2.
Salter, M. 2016. *Crime, Justice and Social Media*. Hertfordshire: Taylor and Francis.
Santana-Garcia. 2001. US v Santana-Garcia, 264 F.3d 1188 (10th Circ. 2001).
Scheidel, D. 2014. *Profits From the Data Economy: Understanding the Rules of Consumers, Innovators and Regulators in a Data Driven World*. London: Pearson Education.
Schutte, S. 2013. *Cached: Decoding The Internet in Global Popular Culture*. New York: New York University Press.
Schwabach, A., and A. Cockfield. 2009. *International Law and Institutions*. EOLSS Publications.
Seidel, J. 2019. Chinese Students Are Stealing Intellectual Property for Beijing, Warns US Intelligence Agency. *News.com.au*, February 3.
Shackleford, S. 2014. *Managing Cyber Attacks in International Law, Business and Relations: In Search of Cyber Peace*. Cambridge: Cambridge University Press.
Shantz, J. 2010. *Racism and Borders: Representation, Repression and Resistance*, ed. Jeff Shantz. New York: Algora Publishing.
Shields, R. 1996. *Cultures of the Internet: Virtual Spaces, Real Histories and Living Bodies*. London: Sage.
Shifrinson, Joshua, and Sameer Lalwani. 2014. It's a Common's Misunderstanding: The Limited Threat to American Command of the Commons. In *A Dangerous World?: Threat Perception and Us National Security*, ed. Christopher Preble and John Mueller, 229–244. Washington: Cato Institute.
Shuman, B. 2001. *Issues for Libraries and Information Science in the Internet Age*. Englewood: Libraries Unlimited.
Singh, K. 2005. *World Trade Organizations and the Third World*. New Delhi: Mittel Publications.

Skrabek, Q. 2008. *William Mckinley, Apostle of Professionalism*. New York: Algora Publishing.

Staake, T., and E. Fleish. 2008. *Counterfeit Trade, Best Practice Strategies and Management Toolbox*. Berlin: Springer.

Standing, A. 2015. Mirage of Pirates: State-Corporate Crime in West Africa's Fisheries. *State Crime Journal* 4 (2): 175–197.

Stein, L. 2006. *Speech Rights in America: The First Amendment, Democracy and the Media*. Urbana: University of Illinois Press.

Stiegler, Z., and D. Sprumont. 2013. Framing the Net Neutrality Debate. In *Regulating the Web: Network Neutrality and the Fate of the Open Internet*, ed. Zach Stiegler. Lanham: Rowman and Littlefield.

Stiglitz, J., and A. Charlton. 2000. Aid for Trade. *International Journal of Development Issues* 5 (2): 1–41.

Swan, A. 2012. *Policy Guidelines for the Development and Promotion of Open Access*. UNESCO.

Tambini, D., D. Leonardi, and C. Marsden. 2008. *Codifying Cyberspace: Community Self-Governance in the Age of Internet Convergence*. London: Routledge Press.

Tonelson, A. 2002. *Race to the Bottom: Why World-Wide Worker Surplus and Uncontrolled a Free Trade Are Sinking American Living Standards*. Boulder: Westview Press.

Toulouse, C., and T. Luke. 2013. *Politics of Cyberspace*. New York: Routledge Press.

Tropina, T., and C. Callahan. 2015. Self- and Co-Regulation. In *Cybercrime, Cyber Security and National Security*. Berlin: Springer.

Turner, F. 2010. *From Counterculture to Cyberculture: Steward Brand, the Whole Earth Network and the Rise of Digital Utopianism*. Chicago: University of Chicago Press.

Tussie, D., and N. Woods. 2000. Trade, Regionalism and the Threat to Multilateralism. *Latin American Trade Network*. Working Paper #2.

UNCLOS. 1982. *Third United Nations Conference On The Law Of The Sea, United Nations Convention On The Law Of The Sea*. opened for signature December 10, 1982.

Uhm, D. 2016. *The Illegal Wildlife Trade: Inside the World of Poachers, Smugglers and Traders*. Cham: Springer.

Valeriano, B., B. Jensen, and R. Maness. 2018. *Cyber Strategy: The Evolving Character of Power and Coercion*. Oxford: Oxford University Press.

Vaseli, A. 2018. The Normative Foundations of the Belt and Road Initiative. In *Normative Reading of the Belt and Road Initiative: Road to New Paradigm*, ed. Wenhua Shay, Kimao Nuotio, and Kangle Zhang. Cham: Springer.

Vasquez. 1999. US v Vasquez, 176 F.3d 1294 (10th Cir. 1999).

Verma, A., and A. Verma. 2014. Right to Internet. In *Human Rights and Ethics: Concepts, Methodologies, Tools and Applications*. Hershey: IGI Global.
Waarden, F. 2010. Governing Global Commons: Public-Private of Fish and Forests. *Jerusalem Papers in Regulation and Governance*. Working Paper No. 17.
Wallace, C. 2001. Xenophobia in Central and Eastern Europe. In *Holding the Course/East-West Institutes*, ed. Peter Rutland. New York: ME Sharpe.
Warf, N. 2012. *Global Geography*. Dordrecht: Springer Science and Business Media.
Weber, R. 2010. *Shaping Internet Governance: Regulatory Challenges*. Dordrecht: Spring Science and Business Media.
Weber, L. 2015. Leanne Weber, Peace at the Border—A Thought Experiment. In *Rethinking Border Control for a Globalizing World: A Preferred Future*, ed. Leanne Weber. London: Routledge Press.
Widener, A. 2019. New Restrictions on Foreign Scientists or Fundamental Research Not Needed, Defense Panel Says. *Chemicalandengineering News.Com* 9 (48) (February 22).
Wilentz, S. 2007. *Andrew Jackson: The American Presidents Series—7th President, 1829–1837*. New York: Macmillan Press.
Wong, T. 2017. *The Politics of Immigration: Partisanship, Demographic Change and American National Identity*. New York: Oxford University Press.
Yannakngeous, P., and A. Lowther. 2013. The Prospects for Cyber Deterrence: American Sponsorship of Global Norms. In *Conflict and Cooperation in Cyberspace: The Challenge for National Security*, ed. Pannayotis A. Yannakngeous and Adam B. Lowther. Boca Raton: CRC Press.
Young, G. 2003. All the Truth and as Much Justice as Possible. *University of California—Davis Journal of International Law and Policy* 9: 209–227.
Zeiler, T. 1999. *Free Trade, Free World: The Advent of GATT*. EH.Net.

CHAPTER 3

Skepticism, Self-Defense/Help and Global Justice

Open world expectations reflect a number of priorities, the most of important of which is the freedom to move valuable goods and services across national or jurisdictional borders, protect 'common areas' that contain resources considered valuable to people around the world regardless of their location from local pressures or regional exploitation, and safely send or acquire information located in another jurisdiction. As demonstrated in Chapter 2, open world expectations also lie along a spectrum, from maximalism (zealotry) and moderate views (criminal skepticism and competitive skepticism) to minimalism (opportunism). The third prong of open world discourse is global justice. If justice, broadly defined, focuses on protecting moral agency, preserving reasonable opportunities and legitimately resolving political conflict in the public/private sector by representatives of the modern state, then global justice aspires to secure the same priorities using norms and institutions that transcend the traditional boundaries of nation-states. Since cross-border disputes are just an extension of on-going disputes between groups located within the same jurisdiction, open world claims tend to be complex, both in their impact and intellectual pedigree. The boundary between the domestic and the global is less important than the boundary between different theories of justice.

A. J. Thomas, *The Open World, Hackbacks and Global Justice*,
https://doi.org/10.1007/978-981-19-8132-6_3

1 Open World Skepticism and Self-Defense

1.1 Ancient and Classical Skepticism

Skepticism about the open world can take many forms. Chapter 2 described two skeptical critiques of the open world: (1) that law enforcement efforts to prevent criminal violations of open world norms are ineffective in terms of capturing identified suspects, punishing convicted violators and deterring potential perpetrators, and (2) that it is very difficult to prevent competitive open world zones from descending into the most dangerous aspects of abuse, ranging from boom/busy cycles, 'race to the bottom' scenarios and the neglect of public and shared resources to rampant intellectual theft, growing economic inequality and the concentration of wealth in the hands of a small number of individuals, firms or governments. Both critiques question some of the underlying 'truths' of open world discourse, in much the same way that ancient skeptics questioned whether individuals could grasp the 'essence' of actual objects.

From its earliest origins, skepticism questioned the nature of truth, the accuracy of human perceptions about reality and the legitimacy of knowledge claims. Ancient Pyrrhonists, for example, avoided making judgments about any object, regardless of its familiarity or apparent concreteness, in order to avoid errors.[1] Aristotle, on the other hand, distinguished between ideal and ordinary knowledge claims.[2] The former exceeded the limits of human perception altogether, while the latter formed the basis of everyday decision-making.[3] Ancient skeptics generally questioned "the very possibility of achieving an objective.. account of reality."[4] Classical

[1] Hannon (2019, 3). Pyrrhonists tried to "free the human soul from worries and reach the spiritual 'autarkeia'." Tinland (2013, 24). See also Susanna Rinard's version of the 'doubting skeptic' who believes that "we ought to suspend judgment on the properties for which, according to others, we lack evidence.. The Doubting skeptic doesn't assert, in a detached academic way, that, in some sense of 'ought,' we ought to suspend judgment. Rather, they urge us to give up our beliefs, as a vegetarian urges us to give up meat." Rinard (2019, 3) and McCormick (2020, 9) ("The skeptical arguments set forth in the Outlines of Pyrrhonism are given as tools to help the reader adopt a way of life free from agitation."); Burnyeat (1983, 140) (Pyrrhonists—ancient skeptics—attempt to tranquility and peace of mind).

[2] Hannon (2019, 22–23).

[3] Id.

[4] McCormick (2020, 21).

skepticism, by contrast, was steeped in the search for rationality, Enlightenment political values and scientific precision.[5] Classical skeptics derived inspiration from Descartes and Hume. Descartes accepted both the limits of human knowledge and the sufficiency of ordinary knowledge,[6] clearly putting him in the category of what we now refer to as 'practical skepticism,' or what others have referred to as 'high-standing skepticism.'[7] Descartes shifted epistemological inquiry from natural philosophy and the essence of objects to the 'science of mind' and the mental aspects of representation.[8] Unlike Descartes, Hume actually identified as a skeptic,[9] but his skepticism was tempered by the recognition that human beings, by their nature, make decisions based on observation and available evidence even if they harbor doubts about the accuracy of those observations or the structure of the 'external world.' "All human life (will) perish"[10] if we don't suspend those doubts. The propensity to believe in spite of our doubts about the external world is fundamental to the human condition,[11] but Hume hoped that our propensity to believe in spite of our doubts would lead us toward tolerance, rather than inaction or prejudice.[12]

[5] Kellner (1985, 3–4).

[6] Hannon (2019, 22).

[7] Hannon (2019, 18) ("the high-standards skeptic is perfectly willing to grant there are differences in degree of justifying that people have ... varying beliefs; she simply denies that we ever reach some ... high-standard required for knowledge").

[8] See, e.g., Rorty (1979, 46) ("Whereas skepticism in the ancient world had been a matter of moral attitude, a style of life, a reaction to the pretensions of the intellectual fashion of the day, skepticism in the manner of Descartes First Meditations was a perfectly definitive, precise 'professional' questions: How do we know that anything which is mental representation anything which not mental?"). See also Williams (2020).

[9] McCormick (2020, 10). ("Like Descartes, Locke wanted to shift inquiry from natural philosophy to the science of mind. Descartes' rationalism is a foundational theory because it purports to discover a non-empirical methodology for defending absolute truth. Locke's empiricism is foundationalist is that it appeals to a class of non-inferentially true, empirical statements upon which to rest everything else.") and Lipton (1990, 817).

[10] Hume (1978).

[11] Hume (1965, 183) ("Nature, by an absolute and uncontrollable necessity has determined us to judge as well as to breathe and feel."). Hume also rejected forms of skepticism that "destroy all assurance and conviction." Hume (1978, 10).

[12] McCormick (2020, 8) (agrees with the view that Rinard's 'pragmatic skeptic,' like the Humean skeptic, is inclined to be more tolerant of opposing views: Hume's tolerance

1.2 Modern and Contemporary Skepticism

After the seventeenth century, the most significant refinement of skepticism took place in the late nineteenth and early twentieth centuries with the pragmatism of Pierce, James and Dewey. Like Hume and Locke, turn of the century pragmatists conceptualized human knowledge and certainty as a product of mind rather than 'natural essences.' As a result, pragmatists perceived an irreconcilable conflict between foundationalism, the basis of ancient and even certain aspects of classical skepticism represented by Descartes, and pragmatism. Both Pierce and Dewey, for example, did not believe that people lacked the capacity to grasp reality or acquire precise knowledge.[13] For Dewey, the best way to allay skeptical doubts about human capacity was to abandon the search for essences (or foundations), encourage people to acquire practical knowledge about the world around them, and feel confident about the coherence and accuracy of that knowledge.[14] Rorty's critique of epistemology in Philosophy and the Mirror of Nature takes direct aim at the foundational sensibilities of both ancient and classical skepticism. Epistemology transformed ancient skepticism into a systematic, disciplinary narrative perfectly suited for Enlightenment minds consumed with the search for rationality.[15]

Rorty's 'interpretive pragmatism'[16] also focused on language, both as a more sustainable basis for modern philosophy than epistemology and/or foundationalism,[17] and as a basis for cross-cultural dialogues and conflict resolution.[18] Interpretivism, historicism (from Hegel and Dewey),[19] and anti-foundationalism inform Rorty's 'liberal ironist,' or a "person who faces up to the contingency of his or her own most central beliefs and desires."[20] But the liberal ironist is not anti-social or racked with doubt

for other views stemmed from his embrace of altruism and sympathy. See Lipton (1990, 857).

[13] Girel (2013, 5).

[14] Id. at 7.

[15] Tinland (2013, 24).

[16] Lipton (1990, 856).

[17] McCormick (2020, 23).

[18] Lipton (1990, 819).

[19] Tinland (2013, 27).

[20] McCormick (2020, 26).

in a world filled with uncertainty and imprecision. On the contrary, liberal ironists have confidence in practical knowledge and welcome the possibility of meaningful social solidarity. Liberal ironists can live with the apparently irreconcilable conflict between 'private irony' (lingering doubts, historical contingency, linguistic fluidity) and the requirements of actual engagement in political and cultural life (pragmatism, belief in the legitimacy of democratic institutions, confidence in share perceptions and consensus).[21] Unfortunately, Rorty's liberal ironism is extremely fuzzy,[22] and in practice, impermissibly tolerant of 'unforced agreement.'[23] Rorty's version of pragmatic skepticism places a great deal of faith in the political and cultural logic of legitimate democratic institutions, particularly with regard to the time-honored, conflict-avoidant practice of balancing.[24] At least morally, it is difficult for Rorty's pragmatic skeptic to 'take a side.'

Ideally, it should be possible, however, for a pragmatic skeptic committed to anti-foundationalism, historicism, interpretivism and socio-political engagement, to not only 'take a side' morally, but also reconcile their moral beliefs with broader, systematic notions of justice. Rawls did not identify as a skeptic and firmly rejected foundationalism as the basis for modern moral philosophy, but he popularized and defended a practice that Rorty implicitly supported. 'Reflective equilibrium' reconciles individual moral beliefs with broader theories of justice, while subjecting those beliefs to a continual process of re-examination and revision.[25] Wide reflective equilibrium is a process undertaken by reasonably free people to continuously re-examine, revise and systematize their beliefs relative to a broader set of moral, political or institutional goals. Individuals "work back and forth …."[26] The end product, though stable, will always impose upon reasonable person 'the burdens of judgment.'[27] The '*strum und drang*' required to yield a stable equilibrium of comparably reasonable sentiments leads to a coherent, internally consistent, systematic theory.

[21] Tinland (2013, 27).

[22] Lipton (1990, 864).

[23] Lipton (1990, 864).

[24] Id.

[25] Rawls (1972, 48). Rawls also referred to reflective equilibrium as 'due reflection.' Rawls (1999a, 32).

[26] Daniels (2000, 128).

[27] Id. at 129.

Wide reflective equilibrium enables theorists to incorporate their own moral intuitions into broader, systematic theories of justice, while continually exposing those intuitions to re-examination and revision. Those intuitions might be shared by skeptics and non-skeptics, but skeptics use wide reflective equilibrium to prioritize intuitions, in spite of their historical contingency and debatable status as universal norms, while satisfying the pragmatic demands of sustained political engagement and democratic legitimacy. Self-identified skeptics have used wide reflective equilibrium to describe the resilience of organized, largely irreconcilable moral clusters (liberal, conservative and radical) in deeply pluralistic, proceduralist and legal systems.[28] Constitutional skeptics have used a version of wide reflective equilibrium to propose morally appealing reforms to complex constitutional systems that are generally perceived to be deeply flawed.[29] Skeptics can also use a version of wide reflective equilibrium to reconcile their intuitions about flaws in the enforcement of open world rules with broader theories of self-defense and global justice.

1.3 Intuitions, Observations and Open World Skepticism

Shared political and social intuitions have always been central to moral inquiry, even when they are subjected to constant revision and re-examination. Shared intuitions are often perceived as deeply emotional, 'pre-rational' or anti-empirical. Consequently, the appeal to some kind of intuitionism seems to reflect a broader critique of theories that rely solely on rationality, pure utility maximization, perfect information or absolute predictability, four qualities that behavioralists rarely find in either public or private life. Researchers have found that regardless of educational level or political belief, individuals are more likely to employ 'bounded rationality' in their everyday lives, combining inferences from whatever information they have with hunches, heuristics, probability calculations and adaptive strategies. Perfect rationality is an illusion, even in a complex, highly bureaucratized institutional environment.

The intuitions of political communities are deeply and inextricably embedded in the fabric of their laws and norms, regardless of whether political decision-makers are prepared to acknowledge it. In terms of

[28] Lipton (1990, 871–78).

[29] Seidman (2014, 7–9).

assessing the significance of intuitions, the relevant inquiry is not whether intuitions are sufficiently rational, or even if political decision-makers regularly rely on intuitions. Instead, we should devote our considerable energies to two broader questions: (1) whether intuitions become relevant when they are organized into recognizable, stable political clusters or ideologies, and (2) whether intuitions become relevant when they become the basis of legitimate, defensible assertions of governmental authority.

The idea that shared intuitions only become relevant or carry some kind of persuasive weight when they are organized into recognizable, stable political clusters is relatively non-controversial. Research has already demonstrated that shared intuitions often map onto settled ideological/political belief systems, and that individuals/groups with organized, culturally coherent views tend to be more successful in changing constitutional norms and rules than individuals/groups with vague or disorganized belief sets. These claims find support in the work of Jonathan Haidt, who has found a correlation between five shared intuitions pairs and recognizable political orientations: harm/care (liberal), fairness/reciprocity (liberal), in-group/loyalty (conservative), authority/respect (conservative) and purity/sanctity (conservative).[30] Justice, and what Haidt refers to as 'related virtues' like fairness and equality, animate contemporary liberals more than contemporary conservatives; the opposite is true for authority, respect, purity, sanctity and loyalty.[31]

The relative value of Haidt's scheme is that it suggests the possibility of an unspoken consensus on the primacy or significance of certain questions. Most Americans, for example, will probably agree with the following statements:

1. checks and balances are crucial to American democracy
2. sovereignty in the United States ultimately resides in the people, not a religious institution
3. managing risk is crucial to securing political and economic freedom in America
4. racism, sexism and class privilege are recurring stains on America's legacy

[30] Haidt (2007, 99).

[31] Id.

5. capitalism, in some form, is crucial to the American socio-political experiment.

Americans may not consciously know why they agree with the aforementioned statements, suggesting that their agreement is more intuitive than deliberative or reflective. What matters, however, is that it is perfectly possible for most Americans to agree with all five statements regardless of their political beliefs, and without knowing how to solve a particular public policy question raised by any statement, or combination of statements.

Popular, shared political intuitions derive their power from a variety of sources, ranging from relative uniform civic education mandates to social media, but the results are comparable to what Rawls described in Political Liberalism as an 'overlapping consensus' on non-comprehensive public values and American constitutional norms. Super-generalized shared intuitions of the variety discussed in this section are also similar to Sunstein's concept of 'incompletely theorized agreements,' repeatedly projected by lawyers and decision-makers into political space. Theorists with different views may disagree about whether either a particular, super-generalized shared intuition should influence the decision-making process of judges, or whether a particular interpretation of a super-generalized shared intuition is sound or practical, but there is very little disagreement about whether super-generalized shared intuitions should be completely excluded from political deliberation.

Broadly speaking, therefore, super-generalized shared intuitions are valuable for two reasons. First, they identify potential points of unanimity or agreement among various actors in the political landscape: decision-makers, judges, lawyers ordinary laypersons, political parties and even social and religious institutions. And second, they aggregate substantive themes and tropes that constrain politics at all levels, ranging from referenda, legislation and elections to Supreme Court decisions.

Both open world discourse and globalization have been around long enough to generate enormous amounts of statistical data, legislation, political commentary, international treaties, anecdotal observations and intuitions. Open world skepticism is based on empirical information, anecdotal information and intuitions, some shared with other individuals who don't identify as open world skeptics, and some shared with fellow skeptics. The core intuition for crime skeptics is that certain 'truths' about open world discourse are simply wrong, such as the perception that decentralization and self-regulation necessarily increase trust and decrease crime. The practical reality, however, is more complex. Banking intermediaries concentrate power, but anecdotal evidence and lay intuition also suggest that intermediaries tend to reduce fraud. Crime skeptics believe that the clash between our intuitions and 'truth-narratives' in open world zones has provided both criminals and ordinary people with new opportunities to engage in fraud and violence, challenge rule of law norms and undermine jurisdictional cooperation agreements between law enforcement agencies in different countries.

Competitive skepticism rests on a different set of observations and intuitions about what people, organizations and nation-states do in hyper-competitive situations. Open world policies definitely encourage competition, and competition can increase efficiency and decrease transaction costs. But the intuition of competitive skeptics is that unchecked open world policies expose the competitive shortcomings or weaknesses of countries, accelerate a 'race to the bottom' that depresses wages and encourages intellectual property theft, and normalizes 'moral ubiquity' problems, or rule of law challenges precipitated by the perception that 'everyone is doing it.'

The intuition shared by crime and competitive skeptics is that there are troubling 'legitimacy gaps' in open world discourse. Table 1 describes three legitimacy gaps, starting with basic conflicts between the ordinary perceptions of open world policies and the actual impact of those policies. There are times, for example, when open world policies appear to accomplish the exact opposite of what was intended, or reward bad faith while punishing good faith. Open world policies may also remain on the books even though the original rationale for having them no longer exists, or the enforcement infrastructure required to enforce them no longer exists. A third legitimacy gap concerns the integrity and credibility enjoyed by self-labeled 'experts' who tout the alleged neutrality of open world policies that clearly benefit, and are really designed to benefit (for efficiency-related, strategic or political reasons), the most affluent, privileged members of society.

2 Justice v. Utility

The claim that legitimacy gaps in open world discourse matter to crime and competitive skeptics is relatively non-controversial. The more controversial claim, however, is that the same skeptics may be sympathetic to

Table 1 Open World Legitimacy Gaps

Conflict between perception and reality	(a) free trade policies for some goods make it more difficult to transport other goods across state lines (b) greater protection for security researchers who hack into computer systems also increases internet harassment and ignores bad faith in the security research community (c) enlarging commons areas increases over-utilization problems
Obsolescence	(a) instead of enabling the poor or unbanked, crypto-currency is horded and manipulated by the rich (b) temporary employment shifts become permanent (c) dramatic increases in academic travel and foreign workers completely overwhelm immigration screening and credentialing processes
Expert community recalcitrance	(a) banking experts develop polices based on 'fundamental' economic principles that systematically benefit the rich (b) 'code is law' experts resist 'hard fork' solutions to crypto-currency crashes

retaliatory self-help, either as a solution to persistent enforcement or incentivization problems in open world zones, or as a mechanism to re-empower 'victims' of open world abuse. This section examines two possible understandings of the plight of vulnerable people in the global environment.

2.1 *Deontological Justice*

2.1.1 Rights Theory

Rights are an extremely important norms category, providing decision-makers with a complex arsenal of arguments, critiques, defense and justifications for policy choices. Rights, in the absolutist sense, consist of core social values that democracies, in particular, try to remove from the '***sturm und drang***' of ordinary political processes. They are, at least in the public imagination, counter-majoritarian 'trump cards' that could be positive (i.e., guaranteeing the authority to do something) or negative (i.e., guaranteeing the authority to be free from something) in character. Rights-talk lends itself to abstract pronouncements about inviolable 'zones of autonomy' from social, political or institutional interference, and rights-theorists tend to focus on procedures capable of ensuring the legitimacy of important distributive priorities. A legitimate administrative or bureaucratic decision-making process, therefore, will not only recognize rights claims, but also provide a mechanism for balancing competing rights claims in an environment dominated by competing political preferences, behavioral perceptions and empirical assertions.

Perhaps most importantly, rights tend to be things that individuals own, and carry with them, if not for the duration of their lives, then at least for the duration of their membership in modern democratic communities. In the classic liberal tradition, for example, ownership plays a central role in the definition of rights and the validity of subsequent rights claims. For Kantian liberals, rights stem from our capacity to exercise reason, whereas for contractarians ranging from Hobbes to Rawls, rights flow from one's capacity to not only exercise reason but also enter into first-order social agreements on core moral norms with similarly situated persons determined to avoid semi-chaotic political arrangements like the state of nature or even an unstable *modus vivendi*.

For modern welfare-state liberals, rights are also crucial components of any societal attempt to ensure equality.[32] For many liberals, the historical record is littered with crisis situations in which both the executive and legislative branches manifestly failed to protect rights and equality norms because of majoritarian pressure, strategic opportunity on the part of legislative or executive actors, or the mistaken belief that rights are always less important than other social goods during crises. The prioritization of procedural fairness in the rights-tradition tends to translate into

[32] According to liberal theory, most social and cultural differences between the ruler and the ruled are irrelevant. Informed consent, rather than hereditary authority, is be the basis of political legitimacy. What this means in practice is that the state, as the principle embodiment of political authority, must remain neutral as much as possible, limiting itself to the enforcement of moral norms and formal principles that all persons, regardless of background, can understand and accept. In order to maintain, at least in principle, a degree of theoretical equality between persons, liberalism prioritizes the 'right' (deontology) over the 'good' (teleology). Deontological moral frameworks start from the assumption that it is always possible to determine if certain actions are right or wrong within the context of a well-functioning society without inquiring too deeply into the ends or distributive consequences of the action itself. Kant's formulation of rights-based liberalism, for example, assumes that freedom is the most important human characteristic, and that in order for persons to grasp the concept of freedom, they have to be able to distinguish between their own goals and the goals of others.

faith that courts should not only provide a robust procedural framework for the resolution of crises either caused by, or the cause of, persistent hierarchical problems and rule of law violations, but also try to minimize the risks posed by arbitrary and capricious rules created by the executive branch or democratically elected bodies. In fact, proceduralism, usually couched in the language of liberal neutrality, limits the consideration of a whole host of moral questions from the public decision-making processes in order to protect the basic rights of citizens to participate in civil society and prevent the government and other powerful interests from gaining undemocratic control over social, economic and cultural instruments capable of undermining political stability[33] and the rule of law.

2.1.2 *Fairness and the Difference Principle*

In the Theory of Justice, Rawls proposed a philosophical justification for the modern liberalism, stripping classical liberalism[34] of its foundational roots. In its place, Rawls proposed a set of assumptions about the basic needs and desires of similarly situated persons, all operating under a veil of ignorance,[35] and a set of assumptions about the minimum distributive requirements of the modern welfare state. 'Justice as fairness' is based on the core assumption that subjectivity is inextricably linked to the desire for freedom and justice,[36] but that desire is constrained by scarcity and social inequality. The 'original position' is a stylized, hypothetical model that captures and filters the preferences of people lacking any prior knowledge of the moral values, strengths and weaknesses of other people. According to Rawls, individuals operating under the original position will develop two principles in order to resolve disputes over the proper distribution of resources and opportunities: 1) the principle of fair equality of opportunity and 2) the difference principle (redistribute resources to benefit the

[33] Hill (1998, 297–309) (discussing investor skittishness in response to political turmoil).

[34] Social contract theorists constructed limited ideal-types in order to establish the minimum conditions required for legitimate political authority. See also Lessnoff (1990).

[35] Rawls (1972, 136–42).

[36] Rawls (1972, 407–504).

least advantaged).[37] The latter distributes the benefits of natural talents across the entire society.[38]

Rawls' model emphasizes the procedural dimensions of justice as fairness, which prioritizes fair equality of opportunity in the distribution of entitlements (opportunities attached to offices and positions open to all)[39] rather than absolute equality of results. Fair opportunity, in other words, constrains the difference principle (all inequalities must be arranged to confer the greatest benefit to the least advantaged), not the other way around. The procedural orientation of justice does not, therefore, provide an exact measure of the range of permissible inequality in the distribution of entitlements. It protects an individuals' right to pursue entitlements and social welfare benefits through fair procedures that ignore irrelevant social differences, but it does not guarantee that an individual will necessarily receive desired benefits, or even that the benefits received will correspond to deeply held notions of moral desert or fitness.[40] Rawls also did not believe that justice as fairness was particularly well-suited for modern societies dominated by welfare rather than property-ownership. The objective of the former is to provide minimum

[37] The difference principle is meaningless without the guarantee of equal opportunity. See, e.g., Rawls (2001, 43) (discussing the lexical priority of equality).

[38] Rawls (1972, 75–83, 101). Neither principle guarantees complete equality of results. The distribution of social resources or life opportunities may not, for example, be enough to equalize the life chances of any effected group, regardless of identity. But the principles of justice go further in trying to preserve the hypothetical equality shared by reasonable persons in the original position than libertarian alternatives, which assign the problem of socio-economic inequality to the vagaries of bargaining and social choice.

[39] Roemer (1996, 283) (entitlements are based on a public system of fair rules).

[40] According to Rawls, the abilities and talents with which one is born are mere accidents of birth. As a result, no one deserves to benefit or gain advantage from their innate talents. Moral desert, in other words, stems from actions, not biological accident. Legitimate expectations should be set by social institutions rather than natural endowments. Nozick, on the other hand, resists transferring control over inborn talents and social assets to the entire community. Nozick (1974, 228–29). And even if, according to Nozick, we agree that people don't deserve (in a moral sense) to benefit from their natural talents, the absence of a real owner of natural talents and assets should create a presumptive in favor of leaving the assets where they fell (with individuals) rather than transferring control to society (or social groups). Rawls' definition of desert and expectation, therefore, makes it difficult, if not impossible, for individuals to control property in themselves, projecting the control of assets and social resources onto an amorphous web of social institutions and customs. Id. at 188–89.

standards of living for individuals accidentally facing misfortune or hardship. Societies faced with massive social welfare transfers might be willing to accept vast inequalities, in violation of the difference principle. In property-owning democracies (of the socialist or liberal varieties), policies are designed to facilitate competition between similarly situated individuals, and as such, fewer inequalities will be permitted by the general population. Justice as fairness, therefore, merely provides a backdrop for the distribution of talents, assets, political positions and entitlements distributed by the government, using criteria that is primarily procedural and structural rather than desert-based, emphasizing individual rights to fair access over moral claims based on intrinsic moral desert, natural law or historical tradition.

2.1.3 Primary Goods

In order to provide a basis for interpersonal comparisons, a foundation for redistributive policies premised on the difference principle, and minimize the suffering of the least well-off, Rawls proposed the concept of primary social goods, or "things that people in a well-order society may be presumed to want, whatever their final ends."[41] Primary social goods are "rights and liberties, opportunities and powers, income and wealth."[42] All rational life plans "require, for their execution, certain primary goods, natural and social."[43] Since life plans are largely incommensurable,[44] primary goods are indexed.[45] The fundamental liberties are weighed equally; other primary goods are expected to "vary in their distribution."[46] In order to protect the least advantaged, Rawls stipulates that "some can accumulate more primary goods if they are acquired in ways which improve the situation of those who have less."[47] Prospects "improve when someone can anticipate a preferred collection of primary

[41] Rawls (1999c, 260).

[42] Rawls (1972, 92).

[43] Id. at 93.

[44] Roemer (1996, 189).

[45] Rawls (1972, 93).

[46] Id.

[47] Id. at 94.

goods."[48] Rawls is only concerned about the distribution of primary goods, not the distribution of individual well-being.[49]

Critiques of Rawls' understanding of primary goods typically fall into three categories. The first is that Rawls' set of primary goods may not reflect changing preferences. Rawls placed a high premium on giving people the freedom to revise their life plans and preferences.[50] Primary goods, on the other hand, are relatively static, implicitly tethered to an almost Marxian/Hegelian understanding of personality and the basic requirements of human flourishing. If there is a clash between primary goods and individual preferences that violate the principles of justice (or even the difference principle in isolation), Rawls does not provide a clear way to resolve the conflict.[51] A second category of critiques contend that a just distribution of Rawls' index of primary goods may not provide high talent people with an incentive to achieve their highest potential.[52] The third category, leveled by Sen and many others, is that Rawlsianism is fetishistic because "Rawls takes primary goods as the embodiment of advantage, rather than taking advantage to be a relationship between persons and goods."[53] Simply possessing a primary good may not improve life prospects, if one lacks the capabilities required to take advantage of primary goods.[54]

[48] Id. at 95.

[49] Kaplow and Shavell (2002, 27, n21).

[50] Rawls (1972, 94).

[51] Fehige (2000, 273–74).

[52] Roemer (1996, 183–85).

[53] Sen (1999b, 326).

[54] Roemer (1996, 189).

2.2 *The Utilitarian Legacy*

2.2.1 *Consequentialism*

Consequentialism[55] is teleological or results-oriented, attaching moral significance to actions rather than the intentions of moral agents, or principles like fairness and justice. Utilitarian formulations of consequentialism emphasize four points: (1) the maximization of general welfare,[56] (2) the identification of desirable social goods, (3) a method for weighing the costs and benefits of specific actions, yielding a sum total that maximizes welfare and (4) a method by which questions of morality can be shifted from purely deontological considerations of right or fairness premised on the moral sensibilities of specific agents (agent-relativity) to a purely teleological, objective determination of moral rightness that remains neutral between agents.[57]

For many consequentialists, rights-prioritization appears to obscure two important social, economic and political trends. The first trend is that during emergencies, the parties responsible for implementing urgent priorities usually enjoy more enforcement power from the majority or the state than parties concerned about the protections of rights. The larger, and arguably more important, question of whether rights ever assume priority over security (or even if genuine security is even possible without robust rights guarantees), is rarely entertained by the majority. Another

[55] For critiques of consequentialism, see Oberdiek (2004, 200) ("consequentialism permits, and indeed requires, a norm of 'lives for convenience'"), Finnis (1998, 513) (aggregation principles flatten human potential). It is more difficult to identify nonconforming individual wrongs under an agent-neutral system because the chain of causation is too attenuated. Agent-neutrality ensures, least in principle, that the consideration of moral questions remains impartial, even if it reduces human flourishing to an intrinsic social goods equal in stature to the maximization of material wealth. But utilitarianism is morally clumsy, demanding more from moral agents used to pursuing notions of the good consistent with common-sense beliefs about the priority of basic social norms and crude approximations of the general welfare than most individuals are prepared to sacrifice. Prioritarian or preference utilitarianism equates the good with the maximization of individual preferences and social priorities, rather than the maximization of pleasure or wealth. It retains its consequentialist sensibility because preferences are considered to be artifacts or consequences of social development. Neal (1997, 35).

Preference or prioritarian utilitarianism is generally more appealing and popular than psychological utilitarianism.

[56] Sen (1979, 468) (welfarism defines the good solely in terms of "the respective collections of individual utilities").

[57] Darwall (2001, 3).

troubling trend, or disturbing socio-psychological consequence, of large-scale emergencies, in particular, is that certain institutions appear to be more competent (or capable) of making rapid decisions during crises than other institutions. These trends demonstrate the dependence of rights claims upon a particular set of distributive preferences considered socially desirable regardless of the rights underpinning them. Even if we assume that rights have priority over security during both ordinary and extraordinary moments, many people believe, along with the consequentialists, that rights are little more than abstract stylizations of underlying welfare (or wealth) maximization norms.

2.2.2 *Welfarism and Wealth*

Rawls' primary goods model, and Dworkin's resourcist model,[58] are 'means-oriented' because both theories assume that in order for individuals to have a reasonable chance to achieve desired ends, they must possess, in the broadest sense, essential objects (such as food, autonomy and rights) as well as opportunities. None of these objects are necessarily considered to be desired ends in and of themselves; the mere possession of these objects does not necessarily lead to happiness. But without them, happiness, self-realization and flourishing are almost impossible to attain. By contrast, proponents of wealth maximization believe that income and/or wealth, as opposed to a laundry list of abstract rights or desired resources, is the key to happiness. For some people, wealth (defined primarily as money or income) is just another indispensable means to well-being, particularly in liberal democratic societies, whereas for others, wealth maximization is the most desirable end for most people. The conflict between primary goods/resourcism and wealth maximization raises a number of important issues, including whether it is possible to define desirable ends in non-economic terms, whether the state can reasonably secure equality of opportunity without engaging in significant income/wealth distribution, or even whether the aggregation of individual utilities can be accomplished without reducing human lives to a monetary or exchange value. For our purposes, however, two additional inquiries are particularly important: (1) whether means-oriented and ends-oriented are interchangeable or at least capable of co-existing

[58] For more on Dworkin's formulation of resourcism, see Dworkin (1981).

within the same framework, and (2) the distinction between attainable and unattainable ends.

Welfarism is based on the assumption that individual well-being is vital to human happiness and flourishing. Classical utilitarianism equated the general welfare with the maximization of pleasure (hedonism) or the minimization of pain,[59] while more modern utilitarians embrace broader notions of social welfare and well-being. But regardless of how general welfare is defined in utilitarian theory, its consequentialist underpinnings require the identification of social goods deemed to be intrinsically valuable to well-being.[60] The good is defined relative to the pursuit of either acts or rules that generate results (or consequences) consistent with the maximization of the sum total of social values or resources prized by all persons, properties with intrinsic value.

Wealth maximization is just one mechanism that theorists have used to measure individual well-being and the overall social welfare. The target indicia used by wealth maximization theory consists of objects, ideas or skills that have value in a market system based on money or currency.[61] In addition to being a function of the perceived monetary value of something, wealth is also a function of what someone is willing to pay,[62] even if the monetary value of a desired object has never been determined before.[63] As a mechanism for determining preference satisfaction, however, wealth maximization is deeply problematic. For many people, wealth maximization does not lead to happiness or even utility.[64] Wealth itself is difficult to determine without a 'natural set of prices.'[65] When social welfare is measured solely in terms of increasing overall welfare,

[59] Bentham (2003, 12), reproduced in Darwall (2001, 12), and Neal (1997, 35) (comparing psychological utilitarianism and prioritarian utilitarianism).

[60] There might be a disagreement over what constitutes hedonism or pleasure-seeking in one context versus the next, for example, but there can be no disagreement amongst persons about the overall desirability of pleasure itself.

[61] Posner (1981, 61).

[62] Kaplow and Shavell (2002, 35), and Plott and Zeiler (2005, 530–32) (defines willingness to pay).

[63] In addition to approximating wealth, willingness to pay is also a rough metric for social utility and well-being because "the greater a person's willingness to pay, the greater level of welfare they receive from the product…." Calandrillo (2001, 981).

[64] Posner (1981, 62–65).

[65] Kaplow and Shavell (2002, 36).

gross disparities in wealth, opportunities and happiness between persons can be tolerated, and distributive justice questions can largely be ignored. Wealth maximization not only ignores the fact that some people are better prepared to take advantage of whatever wealth they do possess than others,[66] but also obscures any consideration of the proper distribution of winners and losers in social policies designed to increase general welfare.

2.2.3 *Efficiency*

Alternative and arguably more popular social welfare models focus on efficiency rather than wealth maximization. Efficiency models foreground the proper distribution of winners and losers, and the calculus of marginal and total utility, in policies that increase the general welfare. The Pareto standard, which measures efficiency in terms of "[making] at least one person better off and no one worse off," focuses on marginal rather than total utility.[67] The most enduring critique of Pareto efficiency is that it discourages redistribution because it can only be achieved by "[situations] in which no one can not be made better without making anyone else worse-off – irrespective of who would be affected or to what extent."[68] This bias against redistribution dramatically increases the likelihood that Pareto efficient policies will leave, largely untouched, a tremendous amount of human suffering and misery.[69] Among its many shortcomings, the Pareto model also assumes perfect competition (i.e., no externalities, entry/exit problems, low transaction costs),[70] that people are overwhelmingly self-interested (ignores moral values, non-utility, altruism),[71] significantly undervalues transactions that occur outside of the ordinary marketplace, such as domestic labor by women),[72] and assumes that most people have enough resources to be self-reliant when bargains in the marketplace are not possible.[73] Contrary to the claims of its supporters, Pareto efficiency

[66] Sen (1999a, 14).

[67] Posner (1981, 54).

[68] Von Staveren (2009, 107).

[69] As Sen once observed, "A society can be Pareto optimal and still perfectly disgusting."

[70] Von Staveren (2009, 109).

[71] Id. For additional behavioral critiques of rationality, see Kahneman (2003).

[72] Von Staveren (2009, 109).

[73] Id. (discusses Sen's critique of 'trade-independent security').

is not neutral because it strongly favors the status quo (no harm, libertarian preference), wrongly assumes that wealth transfers from the most to least advantaged will disincentive the former from pursuing excellence or innovation, and arguably encourages the wasteful use of resources by the most advantaged.[74]

The Kaldor-Hicks (or 'potential Pareto superior') standard, by contrast, measures efficiency in terms of making sure that "any increases in value be sufficiently large that the losers can be fully compensated."[75] In theory, Kaldor-Hicks permits limited redistribution if the winners of market transaction actually compensate the losers (achieving Pareto efficiency), but winners are unlikely to compensate losers due to differences in bargaining power.[76] Since hypothetical compensation is not the same as actual compensation,[77] losers may not be compensated in the real world, leaving winners with a wealth increase and losers with no compensation or increase in marginal utility.[78] Unless the government compensates losers from the gains of winners,[79] or ensures that losers without compensation receive compensation over time from future policies,[80] Kaldor-Hicks can only hold the promise or potential of meeting the Pareto standard.[81] If, however, market losers agree (or consent), *ex ante*, to absorb losses, the absence of actual compensation does not violate Pareto requirements. This scenario effectively reconciles (or harmonizes) Pareto and Kaldor-Hicks.[82]

Unlike Pareto or Kaldor-Hicks, cost-benefit analysis (CBA) defines efficiency in terms of the relatively commonsensical notion "that things are worth doing if the benefits resulting from them outweigh the costs."[83] In theory, CBA can be applied to a much range of social activity than

[74] Id. at 110.

[75] Posner (1981, 91).

[76] Von Staveren (2009, 110).

[77] Adler and Posner (1999, 190).

[78] Posner (1981, 92).

[79] Posner (1981, 92).

[80] Adler and Posner (1999, 190.

[81] Kaldor-Hicks is referred to as 'potential Pareto superiority'. Posner (1981, 91).

[82] Posner (1981, 94).

[83] Sen (2000, 934).

Pareto or Kaldor-Hicks,[84] but in practice, it must be possible to attach an 'explicit valuation' to inputs (policies, preferences, values), in terms of either price or 'willingness to pay.'[85] In spite of its apparent flexibility, however, CBA efficiency is deeply flawed for three reasons. The first is that in many cases, defining one's willingness to pay (through surveys) is a function of existing wealth and social distribution.[86] People with more resources may be willing to pay more for a preference than people with fewer resources and may use that valuation to further signal their affluence. Conversely, people without means may be unwilling to reveal what they would actually be willing to pay because of what that valuation says about their socio-economic status.[87] Second, CBA valuation may ignore the difference between one's willingness to pay X for a preference in one context and one's willingness to pay Y for the same preference in another context.[88] It may also be difficult to factor the cost of contingencies and externalities (foreseeable or unforeseeable) into a willingness to pay disclosure.[89] Third, and perhaps most importantly, there are some highly valuable human preferences (preventing tragedies, religious devotion, family solidarity) that simply elude monetary valuation.[90]

2.2.4 Means vs. Ends Revisited

Normative theories based on utilitarianism are consequentialist or 'ends-oriented.' Wealth maximization is also considered to be ends-oriented because it uses aggregation to calculate total utility, and monetary accumulation to ultimately measure well-being or preference satisfaction. The assumption, in other words, is that wealth is an end that translates, almost immediately for most people, into increased welfare and happiness. Rights, fairness and to a lesser extent primary goods, by contrast, are means rather than ends-oriented. But characterizing wealth maximization as ends-oriented raises three additional issues: (1) whether deontological/consequentialist normative theories can inform well-being analysis,

[84] Sen (2000, 934).

[85] Von Staveren (2009, 110).

[86] Von Staveren (2009, 110) and Sen (2000, 945–6).

[87] Sen (2000, 946) (perception of free-riding).

[88] Id.

[89] Von Staveren (2009, 111).

[90] Id.

(2) whether wealth maximization is either the sole end or just a means to larger, non-economic ends and (3) even if we concede that wealth maximization is a desirable end, the difficulty of achieving that end for some people transforms wealth maximization into little more than 'means pretending to be an end.'[91]

There have been many attempts to develop hybrid models of normative ethics that combine ends-oriented consequentialism and means-oriented Kantianism or Rawlsianism. In all of these projects, the same two questions dominate discussion. The first is whether liberal theory, which purports to be a theory of the right, is really a theory of the good. Clearly any substantive commitment to rules reflects a theory of the good, as does any commitment to rights and procedural norms that enhance autonomy and an individual's capacity to take advantage of opportunities to improve their lives without fear of danger or arbitrary government abuse. The purported neutrality of liberal government, in addition, does not prevent the state from either promoting antidiscrimination norms reflecting egalitarian values (equality is itself a theory of the good), or distributing valuable socio-political or economic resources in order to guarantee a minimum economic 'good life' for the vast majority of citizens. Consequently, even though there may be a considerable amount of disagreement between Kantian or Rawlsian liberals and consequentialists (of the non-utilitarian variety) in theory, in practice, liberal governments have all but given up the pretense they aren't promoting a thin theory of the good in legislation and policy choices, even as they purport to embrace, as a core principle of government, formal neutrality.

The second issue is that there are many policy areas in which ends-oriented thinking actually strengthens the argument for targetted means. Utilitarians, for example, might call for the explicit maximization of psychological states (such as pleasure, hedonism or satisfaction) or preferences that reflect Kantian or Rawlsian moral priorities, effectively encouraging governments to pursue policies that maximize the likelihood of an aggregate increase in the substantive moral and distributive outcomes that

[91] Kaplow also characterized capabilities theory as a means-oriented theory, rather than an ends-oriented theory. Kaplow (2007, 604) ("Interestingly, Sen has criticized Rawls's notion of primary goods because of the difficulty of valuing what are in fact means independent of the ends they serve). See, e.g., Inequality Reexamined, 80. He does not see this criticism as applicable to his own means-based approach, however, because he equates the capabilities he considers with that which is intrinsically good," citing Sen (1992). Capabilities theory is actually a more attainable ends-oriented normative theory.

liberals cherish the most. Utilitarians have also tried to define efficiency and the maximization of social welfare in terms of consent, or more specifically, in terms of what people are willing to sacrifice for the greater good. Conceptualizing one's willingness to abide by distributive justice principles that preserve rights and fairness using a hypothetical social contract, for example, is the centerpiece of Rawls' 'justice as fairness.' Rawls carefully distinguished contract doctrine from utility in Theory of Justice[92] but Kaldor-Hicks contains an implicit contract between winners and losers requiring winners to hypothetically compensate losers.[93] That implicit contract, along with other facts,[94] may account for the political appeal of Kaldor-Hicks,[95] in spite of the very real possibility that it might fail to meet the Pareto standard if compensation is not actually made.[96]

Critiques of the idea that wealth maximization is not a full-fledged ends-oriented normative theory generally fall into two categories. The first consists of those who believe that wealth accumulation is not, for a sizable portion of the general population, the most desirable human end. Very few people believe that wealth is irrelevant to happiness and satisfaction, especially after a minimum threshold of economic self-sufficiency

[92] Rawls (1972, 565).

[93] Adler and Posner (1999) ("Naive contractualism says this: where a governmental official is choosing among options {Ai} at time T, she should choose the option A* such that everyone in the population, given her actual beliefs at time T, would agree to A* as opposed to the other options in {Ai}.In effect, the official's choice must be ex ante Pareto-superior to her other possible choices. This view would routinely create moral dilemmas for governmental officials (since, for many choice situations, no option is Pareto-superior to the others given actual citizen beliefs) and thus contractualist theorists have typically proposed other, more sophisticated variants of contractualism. One variant, in effect, was offered by the economists Kaldor and Hicks (in the days when welfare economists were deeply uncomfortable with interpersonal comparisons and thus advanced views that were closer to contractualism than consequentialism in structure.") Id.

[94] Verkerke (2002, 14) ("Kaldor-Hicks efficiency offers a more realistic assessment of when we might agree that a reassignment would be socially beneficial. A particular match is Kaldor-Hicks efficient if and only if there is no alternative match in which the gainers (whether employers or other workers) could compensate the losers and still come out ahead from the reassignment.").

[95] Calandrillo (2001, 984) ("I, like most, generally opt for the Kaldor-Hicks notion of wealth-maximization (with certain caveats) when I speak of efficiency. It is that concept of maximizing overall social welfare that regulatory policy should strive to achieve, for it intuitively feels closest to serving our nation's best interests by seeking the greatest good for the greatest number.").

[96] *See generally* Feldman (1998, 417).

is reached, but no one, so the argument goes, considers wealth the most desirable end in an of itself. This criticism has been leveled by many theorists, including proponents of primary goods, resourcism and capabilities theory. Sen's critique of wealth maximization and traditional economic development is based, in part, on the belief that wealth accumulation is not the most important human end.[97] It is either just one of many desirable ends or simply a means to more desirable ends. A second category of critiques contend that wealth accumulation is a desirable end, but it is an unattainable end because the accumulation of wealth is very difficult to a sizable portion of the general population. If the central advantage of ends-oriented theories over means-oriented theories is that the former provides a more compelling description of the conditions that people actually require in order to flourish, then the elusiveness of wealth maximization (i.e., the aggregation of individual wealth accumulation) for the average person is as problematic as the abstractness of primary goods.

3 Self-Defense vs. Self-Help

3.1 Self-Defense

3.1.1 Threshold Distinctions

In the popular imagination, retaliatory self-help measures are broadly reminiscent of *lex talionis*, or "an eye for an eye, and a tooth for a tooth."[98] The Books of Exodus and Leviticus in the Old Testament endorse *lex talionis*, but early scholars such as Maimonides conceptualized it as a proportional punishment mechanism grounded in money compensation.[99] Eventually, *lex talionis* gained prominence as the cornerstone of retributive justice theories, not in the simple sense of 'an eye for an eye,' but as a justification for severe punishments like the death penalty.[100] Even in its vestigial form,[101] *lex talionis* also serves as a reminder of

[97] Sen (1999a, 14).

[98] See, e.g., Tables. 449BC. Twelve Tables, $8.2. See also Miller (2006, 30) (*lex talionis* emphasizes "equivalence, balance and precision…").

[99] See Hekmat (1994, 28).

[100] Finkelstein (2002, 13–14).

[101] In its vestigial form, *lex talionis* is simply revenge. See Sherwin (2003, 1406–13) (concluding that retributive actions may be justifiable, but critiquing acts of private revenge). That said, revenge also has its proponents. See, e.g., French (2001, 89–91)

the importance of 'making individuals whole' in the meting out of legal punishments.

But retaliatory self-defense is a deceptively complex concept. There are at least five important threshold distinctions to consider in any case or controversy involving self-defense. The first is the fundamental distinction between actions undertaken in response to a perceived or actual violation of oneself, and actions undertaken in response to a perceived or actual violation of another. Another important distinction concerns the tendency to conflate self-defense with self-help. The latter may involve nothing more than defensive measures that do not, in any way, affect the operation or function of the perceived violators company, property or personhood. Victims simply take steps to increase their resilience or vulnerability to attacks ('hardening their defenses'), without directly striking back at the perpetrator.

Self-defense cases are generally divided into two broad categories distinguishable by causal principles: reactive self-defense (actions undertaken in response to an actual or perceived violation) and anticipatory self-defense (similar actions undertaken prior to a perceived or actual violation).[102] Since self-defense is usually a response to an actual or perceived emergency, self-defense discourse often turns on the issue of necessity. Both reactive anticipatory self-defense make two crucial assumptions about human affairs: (1) that actions arguably undertaken out of necessity are extraordinary and unusual (i.e., behavior that the average person wouldn't do under normal circumstances), and (2) that an individual acting out of necessity did not provoke another person to engage in behavior that the first person later found threatening. The first assumption distinguishes ordinary from extraordinary circumstances, while the second assumption seemingly ignores the possibility that in many cases, legally relevant effects precede causes.[103]

and Murphy (2003, 17–26) (suggesting vindictive behavior is not always irrational or immoral), Zipursky (1998, 85) (noting that, under a civil recourse approach, "our society ... avoids the mayhem and crudeness of vengeful private retribution, but without the unfairness of leaving individuals powerless against invasions of their rights") and Zipursky (1998, 82–86).

[102] Lindner (1939, 253) (defines anticipation as a "transitional mode of apprehending" that combines "a more primitive aspect of organic function (and) primitive searching" with a "prophetic form of imagining which forecasts the remote and independent future").

[103] Fumerton and Kress (2001, 93) (causation requires temporal direction but legal necessity does not).

In a factually complex case involving overlapping causal sequences, supervening events and mixed motives, the reactive/anticipatory distinction matters less than the perception of impossibility, or the "absence of possible alternatives",[104] a perception that must be both reasonable and credible. But the reasonableness of this perception is itself the product of framing effects. It is not uncommon, for example, for policy-makers to frame on-going cultural or ideological conflicts in necessitarian terms as a pretext for military action. By limiting their options to purely strategic or military solutions, policy-makers not only avoid any searching inquiry into the moral implications of government policy, but they also leave the impression that any decisions culled from such an inquiry could have only a negligible effect on the realization of the desired outcome. And when a clear causal link between what appears to be the only reasonable but otherwise unlawful alternative and a desirable outcome can be established,[105] the compelling logic of necessitarian reasoning can be superseded by broader policy objectives. Attenuated[106] causal chains also affect certainty. Both reactive and anticipatory self-defense are framed responses to risk, but the chain of causation is less complicated in reactive self-defense cases, thereby increasing the certainty that a particular risk assessment will ultimately turn out to be correct.

3.1.2 *Agent-Relativity v. Agent-Neutrality*

Theorists have also developed both agent-relative and agent-neutral formulations of self-defense. Agent-neutral formulations make general assertions about the moral appeal of certain kinds of self-defense claims regardless of the circumstances faced either by victims or perpetrators. It is, therefore, irrelevant if the victim's perception of an imminent threat was reasonable (as opposed to perceptions based on super-sensitivity, paranoia or ignorance); that the perpetrator's actions might have been involuntary, forced or accidental; that the victim's self-defense might have been a response to a violation of property rights rather than bodily integrity; or that the victim's retaliation might have caused much more

[104] Braine and Clark (1972, 142–145) (describes factually, judgmentally, and cognitively prior conditions that foreclose future possibilities).

[105] Randall (1976) (medical necessity defense from glaucoma patient permitted because no alternative was available).

[106] Fumerton and Kress (2001, 87) (over-attenuated causal choices prevent the assignment of legal responsibility and damages).

harm than the perpetrator's initial violation. Agent-relative accounts, by contrast, focus on the very specifics that agent-neutral theorists try to avoid, such the difference between defenses of property and defense of life, or whether the victim goaded the perpetrator into threatening the victims' life or goods.

Almost by definition, agent-neutral accounts of self-defense are more 'collectivist' than agent-relative accounts. As a result, utilitarians who define the maximization of general welfare in terms of the aggregation of individual welfare determinations generally prefer agent-neutral accounts. Proponents of agent-relativity accounts, on the other hand, assign priority to rights or deontological rules that distinguish persons who might claim self-defense. But agent-relative formulations of self-defense lie along a spectrum. At one end of the scale, Kantian models prioritize deontology, Hohfeldians and Nozickians emphasize the co-relative right to self-preservation, and Locke emphasized individualized claims of self-protection based partially on natural law. On the other end of the scale, however, there are those who believe that state assertions of force to protect individuals based on notions of fairness or justice are agent-neutral, even though those same notions derive their legitimacy and coherence from underlying, agent-specific rights claims.

3.1.3 Imminence

The fear of imminent harm is central to self-defense, regardless of whether the crisis is evaluated under domestic or international law. Courts have held that imminence is a function of the amount of time victims expect to pass before harm actually occurs, but that fear is a direct reflection of the victim's experiences. A direct confrontation between the defendant and the victim, for example, may lead to the perception that the attacker intends to kill the victim immediately,[107] but courts have also permitted self-defense claims for battered or abused victims who were repeatedly threatened with death by an attacker.[108] Courts have also considered, and largely rejected, self-defense claims from individuals who hired a third

[107] Janes (1993, 241–42) ("a threat, or its equivalent, can supply self-defense when there is reasonable belief that the threat will be carried out").

[108] Robinson (1992, 91) (permits self-defense by battering victim when "torture appears interminable and escape impossible").

party to harm or kill serial batterers.[109] For some courts, planning is not necessarily inconsistent with the concept of imminence.[110] Nor does self-defense necessarily require an immediate spur-of-the-moment, reaction to a direct confrontation with the attacker.[111]

American formulations of imminence under international law can be traced back to provisions in the US Constitution[112] and the Caroline incident, when the British seized, in US territory, an American steamer filled with supplies destined for insurgents battling the British in Canada and sent the ship toward Niagara Falls, ultimately killing two US citizens on board.[113] Proportionality also limited British action against the steamer to physically stopping, boarding, emptying and then disposing of the ship altogether, quite possibly at the risk of endangering the lives of individuals on board to refused to leave.[114] But the Caroline incident illustrated the limits of imminence in self-defense operations: deterrence may demand anticipatory action, even though no military strike or invasion has taken place, and an attack is not imminent. When military forces abandon reaction as a justification for self-defense, imminence itself becomes speculative.[115] According to Daniel Webster, Caroline stood for the proposition that military force requires "a necessity of self-defence, instant, overwhelming, leaving no choice of means, and no moment for

[109] Walker (1986) (self-defense should be permitted for victims who hire third parties) Walker (1986).

[110] Ha (1999, 191) ("inevitable harm' is not the same as imminent harm. A reasonable fear of future harm does not authorize a person to hunt down and kill an enemy").

[111] See MPC (2005, Section 3.04(1)): 'immediately necessary' is more flexible than strict imminence.

[112] See Constitution (1776a, I.10) (prohibits military forces from "engaging" in war "unless actually invaded, or in such imminent Danger as will not admit of delay"). Constitution (1776a, I.10).

[113] See Yoo (2003, 572) ("the Caroline test has been distilled into two principal requirements. First, the use of force must be necessary because the threat is imminent and, thus, pursuing peaceful alternatives was not an option. Second, the response must be proportional to the threat"), citing Dinstein (1992, 208–12, 219–20).

[114] Ashburton (1858).

[115] See, e.g., Bobbitt (2008, 435) ("As a matter of law, it is certainly true that any right to self-defense implies a right to act while action is still possible. The test for imminence is applied to the harm, not to the event sought to be prevented.").

deliberation."[116] Modern American presidents, however, take a variety of factors into consideration before authorizing the use of military force, and imminence is just one of those factors.[117]

Imminence also constrains presidential authority to use military force in self-defense. Presidents have traditionally enjoyed considerable authority to respond to imminent or actual attacks without additional Congressional authorization.[118] Both the Commander-In-Chief Clause[119] and settled legal precedent[120] provide the president with inherent power to initiate military action in self-defense, and that power has been used repeatedly by presidents since the beginning of the republic.[121] But the president's broad authority to initiate military action in self-defense[122] depends on the nature and duration of combat.[123] Congress retains the historical authority[124] to declare war, and in the public imagination, wars are large in scope and long in duration. But

[116] Miller (1934, 449) (analyzes the text of the Caroline Correspondence of Mr. Webster, Secretary of State, and Lord Ashburton, British plen).

[117] Framework (2014, 9) citing Bethlehem (2012) (those factors include "the nature and immediacy of the threat; the probability of an attack; whether the anticipated attack is part of a concerted pattern of continuing armed activity; the likely scale of the attack and the injury, loss, or damage likely to result therefrom in the absence of mitigating action; and the likelihood that there will be other opportunities to undertake effective action in self-defense that may be expected to cause less serious collateral injury, loss, or damage").

[118] *See* Egan (1988, 530) ("Unless Congress specifically has provided otherwise, courts traditionally have been reluctant to intrude upon the authority of the Executive in military and national security affairs."); Wormuth and Firmage (1986, 133–34, 142–49) ("in the case of executive wars, none of the conditions for the establishment of constitutional power by usage is present").

[119] Hamdi (2004, 579–85) (characterizes the President as "Commander in Chief of the Armed Forces").

[120] Prize (1863, 668) ("If a war be made by invasion of a foreign nation, the President is not only authorized but bound to resist force by force. He does not initiate the war, but is bound to accept the challenge without waiting for any special legislative authority.").

[121] Corn (1998, 210) (Chief Justice Marshall endorsed "a broad scope of inherent presidential power to order military conduct absent any congressional authorization" in Barreme (1804).

[122] For more on the President's narrow domestic emergency power, see. Constitution (1776a, art. I, § 8, cl. 10) and Constitution (1776b, art. I, § 8, cl. 15).

[123] Stromseth (1996, 872–86) (historical support for broad Presidential authority to order deployment of troops).

[124] Turner (1994, 914) (presidential power to initiate or respond to force is has always been limited in scope and duration).

the War Powers Resolution,[125] a controversial piece of legislation[126] that was passed over President Nixon's veto in 1973,[127] places constraints on the president's authority to use military force in self-defense.

The War Powers Resolution contains a provision that limits presidential authority to use military force to sustained, rather than intermittent, military engagements. The former implies that American military force is more appropriate when used in response to an imminent threat. The sustained combat debate dates back to the passage of the War Powers Resolution in 1973.[128] Members of both the Ford and Carter Administrations believed that neither the hostilities nor imminent hostilities language of section 4 (a)(1) applied to "irregular or infrequent violence in a particular area."[129] President Reagan embraced a similar reading of section (a)(1) when he failed to file a report after Marines operating in a peacekeeping capacity were killed during an attack in Lebanon. The Senate, however, rejected Reagan's view that the attack on the Marines was merely an "isolated or infrequent act of violence" that did not trigger the War Powers Resolution. They concluded that the attack that killed the soldiers was a hostile act falling under section (a)(1) that required a presidential report.[130] The Clinton Administration adopted a similar view when it authorized military strikes to support forces defending UN

[125] WPR (1973). Section 1541(c) states that "The constitutional powers of the President as Commander-in-Chief to introduce United States Armed Forces into hostilities, or into situations where imminent involvement in hostilities is clearly indicated by the circumstances, are exercised only pursuant to (1) a declaration of war, (2) specific statutory authorization, or (3) a national emergency created by attack upon the United States, its territories or possessions, or its armed forces." WPR (1973).

[126] Turner (2013). *See also* Turner (1991), Fisher and Adler (1998, 1–6, 10–12) ("the Resolution... grants to the President unbridled discretion to go to war as he deems necessary against anyone, anytime, anywhere, for at least ninety days"), Glennon (1995, 28–31), Ford (1995, 700–78), Ides (1984, 631–52), and Rushkoff (1984, 1332–33) (defends War Powers Resolution).

[127] Nixon (1973).

[128] Fisher (1995) (Article II-based justifications for the use of military force from Ford to Clinton).

[129] Leigh (1975).

[130] SenateReport (1983).

peacekeepers in Somalia.[131] According to the Administration, the sixty-day termination requirement did not apply to "intermittent engagements regardless of whether they constitute actual hostilities."[132] President Obama followed suit after authorizing air-strikes against Libya in 2011 because the air-strikes campaign[133] did not "involve sustained fighting or active exchanges of fire with hostile forces, nor do they involve US ground troops."[134] Since the War Powers Act was passed, therefore, successive presidents have tried to weaken the link between imminence and the legitimate use of military force.

3.1.4 Deterrence

Self-defense is a deterrent mechanism that may signal the presence, or absence, of law. When an individual, for example, approaches an empty intersection and refuses to cross, it is more likely that the law, in the form of a red signal along with the possibility of being caught and even fined by an observant police officer, deterred them from crossing more than the fear of being hit by a passing car. Moral adherence and the fear of public shaming can explain obedience to the law in many situations, but most people obey the law because of the fear of punishment or penalty. Deterrence regimes are also negative in that they prevent someone from taking an action, but the failure to act can produce either positive or negative results. A fine, for example, could very easily provide a pharmaceutical company with an incentive to produce a safer or more effective vaccine, whereas a poorly implemented penalty or regulatory restraint could provide an individual with an incentive to act recklessly, free-ride on taxpayers insurance payments during a personal health emergency or even destroy land in a commons area. Deterrence is particularly sensitive

[131] On March 24, 1999, Clinton announced the commencement of NATO strikes against Yugoslav targets, and on April 28, the Clinton submitted a report to Congress, "consistent with the War Powers Resolution." *See* Campbell (2000).

[132] Gilman (1992).

[133] Savage (2011) (US military's activities in the NATO-led air war in Libya amounted to "hostilities").

[134] Savage and Landler (2011). *See also* Ackerman (2011).

to the reputational capital of the deterrer. If a deterrer has not demonstrated either the capacity or the will to deter others in the past, a current deterrent threat is unlikely to be effective in the present.[135]

3.1.5 *Degrees of Intrusion*

The final component of self-defense concerns the degree of intrusion. As I mentioned before, self-defense measures lie along a spectrum, from least intrusive to most intrusive. Let us consider two scenarios that might trigger retaliation: dramatic over-use of an easement and hacking into someone's computer system. On the one end of the self-defense scale, over-use of an easement might trigger minimally intrusive measures like warning the easement holder, recording instances of over-use and contacting local authorities. Least intrusive responses to a cyber hack might be limited to merely entering into the suspected perpetrators system in order to learn ascertain his identity/location, as well as ascertain whether the hacker acquired data from an external source. Since hacks can be channeled/controlled through a third-party system, least intrusive active measures may require access to a third computer as well. Moderately intrusive responses to easement over-use might consist of putting fences on the boundaries of the easement or placing sleeping policeman on the easement. In the case of a hack, moderate measures might consist of placing repositories of fake information (honey pots)[136] on a computer system with the expectation that an unauthorized user will try to remove the repository. Once the hack has occurred, the target will place a beacon in the hacker's computer, and modify the hacker's system to retrieve the stolen data. The most intrusive self-defense response to easement over-use might consist of physically altering the land or property of the easement holder. The most intrusive responses to hacks might shut down the hacker's system, either partially or completely. All of these hackbacks assume

[135] Haas (2001, 245) (discusses the significance of credibility to any deterrent action), and Levy (1996, 191) ("when states issue deterrent threats, it is often in response to prior threats by a potential initiator").

[136] Honeypots (2012) (honeypots are " a computing resource, whose sole task is to be probed, attacked, compromised, used or accessed in any unauthorized way").

that traceback techniques can determine the identity and address of the hacker with precision.[137]

3.2 *Self-Help*

3.2.1 Consent and Violence

Self-help also lies along a spectrum that includes self-defense. At one end, there are claims involving the use of deadly force to protect one's own life or another person's life, followed by the use of deadly force to protect an object, and the use of non-deadly force to protect an object. Traditional self-defense (including 'Stand Your Ground' and Castle Doctrine)[138] and national defense claims enjoy significant public support if the claim is reasonable and made in good faith, even if the case involves extreme violence or resulted in considerable harm to attackers or third parties. States never permit the use of deadly force to protect property, but they do provide ample protection for those who use non-deadly self-help tactics to protect both personal and real property. More controversial authorizations of self-help that don't include the use of deadly force to protect human life, on other hand, depend on four factors: (1) actual or implied consent, (2) minimization of violence, (3) the efficiency or private conflict resolution mechanisms and (4) mitigation.

Since legal authorizations of self-help are usually based on private law, courts require either actual or implied consent from the individuals against whom self-help is administered. Ideally, that consent should not be obtained through trickery, misrepresentation or deceit.[139] Consent is required for repossession of dwellings under Article 9 of the UCC[140] and repossession of chattel under Article 2.[141] Individuals can agree to permit a creditor to enter the individuals premises and remove specific

[137] Tsagourias (2012, 229–32) (proposes attribution threshold sufficient to trigger national self-defense), and Aghaei-Forushman and Zineir-Heywood (2013, 128–29) (distinguishes between logging and marking traceback strategies).

[138] Hoffman and Nyikos (2008, 18–19).

[139] Cohen (1998, 2007).

[140] Cohen (1998, 2004) (self-help upon debtor defaults) and Sharkey (2013, 682) (Article 9 repossession).

[141] Smith (2005, 85–86) Repossession under Article 9 does not trigger due process. Id. at 86, n. 58.

items via contract,[142] and courts routinely permit creditors to remove items 'in plain view' if the possessor offers no resistance. In some cases, failure to pursue self-help measures authorized by law does not prevent creditors from alternative avenues of legal relief, including a civil suit for breach of contract or failure of specific performance. Requiring consent in self-help cases performs three functions. First, it appears to preserve an arena of deregulated self-governance over many commercial transactions. Second, it discourages the public perception that self-help measures permit behavior that would otherwise be considered a trespass.

Third, and perhaps most importantly, consent minimizes the possibility of violence or 'breaches of the peace' during self-help operations. The fear of violence overshadows almost every consideration of self-help tactics. Both Article 2 and 9 of the Uniform Commercial Code, for example, incorporate the breach of the peace standard,[143] and most courts don't permit parties to waive the breach of the peace requirement via contract.[144] Courts have split on whether verbal threats alone (as opposed to threats combined with violence) violate the peace, whether it is possible to breach the peace solely through electronic communication,[145] and whether breach of the peace violations also necessarily violate personal and/or intimate privacy guaranteed by the Fourth Amendment.[146] Once again, violence during self-help operations increases the likelihood of chaos, instability and the erosion of public trust in law enforcement, and even state sovereignty.[147]

3.2.2 *Efficiency and Mitigation*

Proponents of private law and deregulation more generally also believe that self-help, properly structured and administrated, provides an efficient mechanism for private conflict resolution.

In a society that extends long-term protection to a subset of the most valuable ideas, for example, efficient laws will encourage inventors to take extra precautions to protect those ideas, rather than protect every idea or

[142] Cohen (1998, 2011).

[143] Cohen (1998, 2004).

[144] Cohen (1998, 2005, n.51).

[145] Cohen (1998, 2003).

[146] Cohen (1998, 2004).

[147] Hoffman and Nyikos (2008, 4–5) and Sharkey (2013, 682).

concept they produce. Self-help provisions in trademark law that require inventors to signal their intention to protect their most valuable ideas, therefore, are more efficient than laws that require inventors to protect everything.[148] Without the signaling self-help provisions in trademark law, inventors might engage in extraordinarily wasteful efforts to protect their most valuable ideas, ranging from distributing portions of a larger product across a wide swath of the labor force in order to prevent anyone from understanding complete projects[149] to engaging in a costly arms race in order to monitor the activities of competitors.[150] If transaction costs are low,[151] self-help may provide aggrieved parties with an efficient resolution.[152]

The final overarching constraint on self-help is mitigation, or the expectation that plaintiffs will take steps to protect themselves from attacks or violations perpetrated by the defendant. The most obvious, and frequently employed, mitigation strategy consists of exclusion practices that prevent either authorized or unauthorized, parties from accessing property, services or ideas.[153] Manifesting an intent to exclude could be a s simple as putting up a sign, erecting a fence,[154] labeling certain information as private[155] proprietary or even classified, locating valuable in a building with blackened windows or under a tarp,[156] installing locks[157] or physical resistance to repossession[158] to complex algorithms or elaborate obstructive schemes involving third parties. In some situations, exclusion

[148] Lichtman (2004, 19).

[149] Id. at 24.

[150] Id. at 26–27.

[151] Sharkey (2013, 680).

[152] Epstein (2005, 26) ("The dominant feature of any self-help strategy is that it pairs a quick, cheap and reliable remedy with incomplete relief, that is, relief which by definition and design does not leave the aggrieved party as well as he would have been if the other party had faithfully performed its obligations in the first place.").

[153] Plaintiffs who cross a boundary to abate a nuisance, on the other hand, are less likely to receive judicial deference than plaintiffs who erect boundaries to exclude potential or actual trespassers. Smith (2005, 90).

[154] Lichtman (2004, 18, n.31) and Sharkey (2013, 685, n.37).

[155] Sharkey (2013, 699).

[156] Lichtman (2004, 27).

[157] Smith (2005, 81).

[158] Smith (2005, 85).

by a potential defendant is little more than a signal to the plaintiff that formal legal process is necessary because unilateral self-help might lead to violence. Under trademark law, however, failure to employ reasonable precautions in the protection of intellectual property precludes a remedy for trademark theft,[159] essentially transforming exclusion into a mitigative burden that forecloses formal legal redress.

Generally speaking, mitigative burdens are a double-edged sword in self-help cases. On the one hand, imposing a mitigative burden on plaintiffs seems efficient because victims might be the 'cheapest cost avoider.'[160] Some proponents of self-help in tort and criminal law, however, reject mitigative burdens because such burdens appear to weaponize the vulnerabilities of victims.[161] In many ways, the problematic nature of mitigative burdens illustrates the dilemma of forcing the general public to choose between primary goods (i.e., mitigative burdens are appropriate because everyone has a technical right to mitigative tools) and wealth maximization (i.e., mitigative burdens are appropriate even though they ignore the prohibitive costs of effective hardening).

3.3 Rights-Based Defenses of Self-Defense and Self-Help

Rights-based justifications for self-defense are extremely familiar in liberal democratic societies with a demonstrated commitment to core fundamental rights. No expectation is more fundamental than self-preservation, regardless of whether the expectation is conceptualized as a right or the product of natural law. For Locke and Grotius, individuals who attempt to extinguish the life of another are little more than beasts, utterly incapable of asserting any claims superior to the victim's basic claim to self-defense. Similarly, culpable (non-innocent) attackers may forfeit their right to self-preservation if they violate a law subject to capital punishment. But grounding self-defense, and by extension retaliatory self-help, in rights theory is deeply problematic for three reasons.

The first criticism applies to all rights-based claims, not just self-defense or retaliatory self-help. Rights-based theories suffer from two circularity problems: (1) rights are invoked as a solution to majoritarian tyranny even

[159] Lichtman (2004, 18).

[160] Sharkey (2013, 683).

[161] Sharkey (2013, 683, n 33).

though they reflect the legitimate preferences of the majority of citizens in liberal democratic societies, and (2) rights are invoked as a solution to natural law or divine authority even though the most appealing rights claims, including the right to self-defense, appear to be pre-political and fundamental. Both objections lead to the same intuition, which is that rights are a form of expression, not justification. The rhetorical power of 'rights-talk,' both as an anti-majoritarian trump card and as an implicitly pro-majoritarian heuristic device, merely postpones consideration of the ultimate justification for moral claims; it does not end the inquiry. The second problem is that in the absence of a firm moral foundation for rights, such as consent of the governed, deontological rules or social contract—rights-based defenses are more like excuses than justifications. Conceptualizing self-defense as an excuse allows civilized societies to violate the general prohibition against taking a life, while leaving room for the acceptance of 'lesser rights' excused by emergency circumstances. The third problem relates to the concept of forfeiture. As a preliminary matter, it is very difficult to reconcile the view that culpable aggressors forfeit the right to self-preservation with the general prohibition against taking someone's life. Even if we get around the forfeiture problem for capital punishment and wartime decisions, we are still left with the question of whether forfeiture should be temporary or permanent, or even if forfeiture should only operate in favor of specific victims, extend to third parties, or even the entire world. If self-defense is defined primarily as a correlative right, then forfeiture should only operate in favor of a specific victim and only as long as the threat posed by an aggressor actually exists.

3.4 *Utilitarian Defenses of Self-Defense*

Utilitarian formulations of self-defense start from the assumption that permitting someone to use deadly force in order to preserve their own life increases the general welfare. Since utilitarian theory requires the aggregation of individual welfare assessments, the baseline scenario doesn't require a searching inquiry into the circumstances surrounding the aggressor's threat to the victim's life. Using a 'lesser harmful results' calculus, the general harm caused by permitting someone to use force to defend their lives from an aggressor will be less damaging than the harm caused by dutiful adherence to the general prohibition against taking another life, especially if it's the life of someone who wants to take your life. Modified, act-utilitarian versions of the basic scenario automatically discount the interests of unjustified aggressors relative to the interests of the victim, incorporating the responsibility of a special class of victims into the aggregation of welfare assessments. Other versions point to the socially beneficial affects of letting people defend themselves if their lives are threatened, regardless of the moral status of aggressors or victims, particularly in terms of reducing social anxiety and deterring further violence. 'Lesser harmful results' (or 'lesser evil') formulations are usually utilitarian, although there are many non-utilitarian versions of 'lesser evil' theory.

Critiques of utilitarian formulations of self-defense generally fall into four categories. The first category suggests that there is a fundamental disconnect between the aggregative underpinnings of utilitarianism and the exceptional circumstances that usually arise in the context of self-defense. Rights theorists, in particular, argue that the right to self-defense always trumps other considerations, regardless of whether it increases the general welfare or not. Another category of critiques, closely related to the first category, is that utilitarian accounts of self-defense can lead to majoritarian tyranny, particularly if courts, operating as act-utilitarian institutional checks, lack the power to constrain or overturn oppressive legislative restrictions on self-defense that target vulnerable populations. A third category of critiques questions whether 'lesser harm' (or lesser evil) theories should ever become the basis of self-defense models, both utilitarian and non-utilitarian, because they are so open-ended and malleable. The fourth category questions the general utilitarian aversion to questions of fairness and responsibility that are so central to self-defense claims, both in terms of how these claims are evaluated by the general public and in term of how they are treated by courts.

4 Force and Meaningful Freedom

Many critics of 'lesser harmful results' theory (i.e., the harm caused by a defense of one's own life is less dangerous than the harm caused by allowing an attacked person to succumb if they cannot defend themselves), or rights theory (threatened individuals have a right not to be killed), have turned to a version of 'forced choice' theory. The central premise of forced choice theory in the context of self-defense is that when an aggressor threatens someone's life, they effectively deprive victims of the freedom to choose any option except the use of deadly force to preserve their own life.[162] The absence of meaningful choice transforms self-defense into a form of excuse, rather than the 'right thing to do' (justification).[163] Necessity follows a similar logic because it is invoked to excuse an attacked party's necessary use of deadly force to preserve their own life or the life of another.[164]

Forced choice theory assumes that victims, faced with the prospect of imminent death or severe harm, are 'forced' into using deadly force against an attacker. Their decision-making is not voluntary; whatever freedom they enjoy after a potential threat became clear is not meaningful. Forced choice theorists also recognize that the element required to establish the culpability of aggressors directly impacts the scope of meaningful freedom that victims possess. Non-culpable parties against whom self-defense claims might be brought, including innocent bystanders and non-agent aggressors acting on behalf of others, for example, do not constrain the meaningful freedom of victims as much as culpable aggressors.

Forced choice theory also characterizes self-defense as an excuse rather than justification, and that distinction is extremely important for two reasons. First, excuse-based models, like necessity, almost attribute the victim's action to diminished capacity. The absence of meaningful decision-making power forces the victim down a path that can be understood by almost everyone. It also explains why the victim had to deviate from the general prohibition against taking another life. Forced choice theory also treats the conduct of non-culpable aggressors as excusable.[165] In other words, when an innocent bystander or non-agent aggressor contributes, either directly or indirectly, to a situation that endangers the victim's life, this action is excused even if it falls below normal thresholds of foreseeability.[166]

Second, conceptualizing self-defense as the unfortunate by-product of diminished capacity—a negative claim—often lacks the moral appeal of a positive assertion of a right to preserve one's life, which is the strength of rights-based theories. In many quarters, someone should not have

[162] Wallerstein (2005, 1006).

[163] Wallerstein (2005, 1006).

[164] Wallerstein (2005, 1005, n.25).

[165] Id. at 1013.

[166] Id.

to apologize for saving their own life; they should be applauded. One can mitigate this problem, however, by linking the diminished decision-making capacity of the victim to broader distributive and global norms that establish a positive basis for political and institutional protection.

The bridge between forced choice theory and broader perspectives on distributive and global justice is a framework that I call the 'force-multiplier principle.' Under the force-multiplier principle, the moral and legal legitimacy of a necessity claim is tied to both the quality and quantity of reasonable options available to a claimant during a crisis. In its simplest form, the force-multiplier principle is relatively non-controversial if we also assume that most people react to the perception of an imminent (or on-going) crisis in an unusual or extraordinary manner not because they completely reject rule of law norms, are already inclined toward socially disruptive behavior or are deeply psychologically damaged, but because they feel, quite reasonably, that they have no choice.

In assigning priority to the reasonableness of the necessity claimant's perceptions of an imminent or on-going crisis, and the potential alternatives available to him either before or during a crisis, the force-multiplier principle assigns partial responsibility to the claimant if his perceptions are patently unreasonable or entirely implausible given the circumstances in which he finds himself and his psychological state at the time. In this sense, it rewards reasonable good faith interpretations of facts relevant to a perceived crisis and punishes claimant responses to crises that could be characterized as purely voluntary or strategic.

As a general theory of moral desert, the force-multiplier principle encourages decision-makers to not only link the credibility of necessity claims to the quality and quantity of reasonable options available to the claimant (based on their own objectively or subjectively determined understanding of reason), but also favor necessity claims brought by demonstrably good faith claimants (i.e., claimants whose perceptions of a crisis fall well within the bounds of ordinary behavior) over extremely reckless claimants because it avoids unjust enrichment and punishes crisis-generating parties whose actions are likely to produce harmful spillover effects and unforeseeable cascades. In other words, the force-multiplier principle defends, for both moral and practical reasons, on policies that reward arguably good faith claims and punish arguably reckless claimants.

References

Ackerman, B. 2011. Legal Acrobatics, Illegal War. *N.Y. Times*, June 21.

Adler, M., and R. Posner. 1999. Re-thinking Cost-Benefit Analysis. *Yale Law Journal* 109: 165–247.

Aghaei-Forushman, V., and A. Zineir-Heywood. 2013. On Evaluating IP Trace-back Schemes: A Practical Approach. *IEEE Security and Privacy Workshop*: 127–134.

Ashburton, L. 1858. Letter from Lord Ashburton to Daniel Webster, US Secy of State (July 28, 1842). British and Foreign State Papers 1841-2. http://www.yale.edu/lawweb/avalon/diplomacy/britain/br-1842d.htm.

Barreme. 1804. Little v. Barreme, 6 U.S. (2 Cranch) 170 (1804).

Bentham, J. 2003. *An Introduction to the Principle of Morals and Legislation* (reproduced in Darwall, 2001).

Bethlehem, D. 2012. Self-Defense Against an Imminent or Actual Armed Attack by Nonstate Actors. *American Journal of International Law* 106: 769–803.

Bobbitt, P. 2008. *Terror and Consent: The Wars for the Twenty-First Century*. New York: Alfred A. Knopf.

Braine, D., and M. Clark. 1972. Varieties of Necessity. *Proceedings of the Aristotelian Society* (Supplementary Volumes) 46: 139–187.

Burnyeat, M. 1983. Can the Skeptic Live His Skepticism? In *The Skeptical Tradition*, ed. Myles F. Burnyeat. Berkeley.

Calandrillo, P. 2001. Responsible Regulation: A Sensible Cost-Benefit, Risk Versus Risk Approach to Federal Health and Safety Regulation. *Boston University Law Review* 81: 957–992.

Campbell. 2000. Campbell v. Clinton, 203 F.3d 19, 20 (D.C. Cir. 2000).

Cohen, J. 1998. Copyright and the Jurisprudence of Self-Help. *Berkeley Technology Law Journal* 13: 1089–2020.

Constitution. 1776a. U.S. CONST, art. I, § 8, cl. 10.

Constitution. 1776b. U.S. CONST, art. I, § 8, cl. 15.

Corn, G.S. 1998. Presidential War Power: Do the Courts Offer Any Answers? *Military Law Review* 157: 180, 210.

Daniels, N. 2000. Reflective Equilibrium and Justice as Political. In *The Idea of a Political Liberalism: Essays on Rawls*, ed. Victoria Davion and Clark Wolf. Lanham: Rowman and Littlefield.

Darwall, S. 2001. *Consequentialism*. Oxford: Blackwell Publishing.

Dinstein, Y. 1992. *War, Aggression and Self-Defense*, 3rd ed. Princeton: Princeton University Press.

Dworkin, R. 1981. What is Equality? Part 2 : Equality of Resources. *Philosophy and Public Affairs* 10 (4): 283–345.

Egan. 1988. Dep't of the Navy v. Egan, 484 U.S. 518, 530.

Epstein, R. 2005. The Theory and Practice of Self-Help. *Journal of Economics and Policy* 1: 1–26.

Fehige, C. 2000. Justice Beyond Desires Norman Daniels, Reflective Equilibrium and Justice as Political. In *The Idea of a Political Liberalism: Essays on Rawls*, ed. Victoria Davion and Clark Wolf. Lanham: Rowman and Littlefield.

Feldman, A. 1998. Kaldor-Hicks Compensation. In *The New Palgrave Dictionary of Economics and the Law*, ed. Peter Newman. London: Springer.

Finkelstein, C. 2002. Death and Retribution. *Criminal Justice Ethics* 21 (2): 12–21.

Finnis, J. 1998. On The Practical Meaning of Secularism. *Notre Dame Law Review* 73: 491–526.

Fisher, L. 1995. *Presidential War Power*. Cambridge: Harvard University Press.

Fisher, L., and D. Adler. 1998. The War Powers Resolution: Time to Say Goodbye. *Political Science Quarterly* 113: 1–14.

Ford, C. 1995. War Powers as We Live Them: Congressional-Executive Bargaining Under the Shadow of the War Powers Resolution. *Journal of Law and Policy* 11: 609–652.

French, P. 2001. *The Virtues of Vengeance*. Oxford: Oxford University Press.

Fumerton, R., and K. Kress. 2001. Causation and the Law: Pre-emption, Lawful Sufficiency and Causal Sufficiency. *Law and Contemporary Problems* 64 (4): 83–90.

Gilman, B. 1992. *Response to Rep Benjamin Gilman and Senator Jesse Helms from Assistant Secretary of State Wendy Sherman*, July 21.

Girel, M. 2013. From Doubt to Its Social Articulation: Pragmatist Insights. In *Pragmatism and the Social Dimension of Doubt: Fresh Perspectives*, vol. 2.

Glennon, M. 1995. Too Far Apart: Repeal the War Powers Resolution. *University of Miami Law Review* 50: 17–35.

Ha. 1999. Ha v State, 892 p.2d 174, 191 (Alaska App.).

Haas, M. 2001. Prospect Theory and the Cuban Missile Crisis. *International Studies Quarterly* 45: 241–254.

Haidt, J. 2007. When Morality Opposes Justice. *Social Justice Research* 20 (1): 98–108.

Hamdi. 2004. Hamdi v. Rumsfeld, 542 U.S. 507, 579–585.

Hannon, M. 2019. Skepticism: Impractical, Therefore Implausible. *Philosophical Issues*.

Hekmat, F. 1994. The Origins and Evolution of Lex Talionis in the Abrahamic Faiths.

Hicks, J. 1939. The Foundations of Welfare Economics. *Economic Journal* 49: 696–724.

Hill, C. 1998. How Investors React to Political Risk. *Duke Journal of Comparative and International Law* 8: 283–312.

Hoffman, W., and S. Nyikos. 2008. Governing Private Sector Self-Help in Cyberspace: Analogies from the Physical World. *Carnegie Endowment for International Peace*.

Honeypots. 2012. European Union Agency for Network and Info. Sec., Proactive Detection of Security Incidents Ii: Honeypots 17.

Hume, D. 1965. *A Treatise on Human Nature*, ed. L.A. Selby-Bigge. Oxford: Clarendon Press.

Hume, D. 1978. *An Enquiry Concerning Human Understanding*, ed. P.H. Nidditch. Oxford: Oxford University Press.

Ides, A. 1984. Congress, Constitutional Responsibility and the War Power. *Loyola Law Review* 17: 599–652.

Janes. 1993. State v Janes, 121 Wash. 2d 220, 241-2 (1993).

Kaldor, N. 1939. Welfare Propositions in Economics and Interpersonal Comparisons of Utility. *Economic Journal* 49 (195): 549–552.

Kaplow, L. 2007. Louis Kaplow, Primary Goods, Capabilities ... or Well-Being. *Philosophical Review* 116 (4): 603–632.

Kaplow, L., and S. Shavell. 2002. *Fairness Versus Welfare*. Cambridge: Harvard University Press.

Kellner, D. 1985. Critical Thought, Marxism and Modernity. *Studies in Soviet Thought* 44 (2): 144–148.

Leigh, M. 1975. *Letter from State Department Legal Advisor Monroe Leigh and Dod General Counsel Martin Hoffman*, June 3. To House Foreign Affairs Committee Chair Clement Zablocki, and Office of Legal Counsel (Carter), 4a Op. O.L.C.

Lessnoff, M. 1990. *Social Contract Theory*, ed. Michael Lessnoff. New York: New York University Press.

Levy, J. 1996. Loss Aversion, Framing and Bargaining: The Implication of Prospect Theory for International Conflict. *International Political Science Review* 17 (9): 179–195.

Lichtman, D. 2004. How the Law Responds to Self-Help (John M. Olin Program in Law and Economics). Working Paper No. 232.

Lindner, R. 1939. An Experimental Study in Anticipation. *American Journal of Psychology* 51 (2): 253–260.

Lipton, R. 1987. Altruism and Sympathy in Human Ethics. *Austrakasian Journal of Philosophy* 65 (1): 18–32.

Lipton, R. 1990. Beyond Skepticism, Foundationalism and the New Fuzziness: The Role of Wide Reflective Equilibrium in Legal Theory. *Cornell Law Review* 75 (4): 810–877.

McCormick, M. 2020. Value Beyond Truth-Value: A Practical Response to Skepticism. *Synthese* 198 (9): 8601–8619.

Miller, H. 1934. *Treaties and Other International Acts of the United States of America*, vol. 4, ed. Hunter Miller. Government Printing Office.

Miller, W. 2006. *Eye for an Eye*. New York: Cambridge University Press.

MPC. 2005. *Use of Force in Self-Protection*. Model Penal Code (Thomson/West, 2005).

Murphy, J. 2003. *Getting Even: Forgiveness and Its Limits*. New York: Oxford University Press.
Neal, P. 1997. *Liberalism and Its Discontents*. New York: New York University Press.
Nixon, R. 1973. Veto of the War Powers Resolution. *Public Papers* 5: 893 (October 24).
Nozick, R. 1974. *Anarchy, State and Utopia*. New York: Basic Books.
Oberdiek, J. 2004. The Ethics of Risk Regulation: Towards A Contractualist Re-orientation. *Rutgers Law Journal* 36: 199–205.
Plott, C., and K. Zeiler. 2005. The Willingness to Pay-Willingness to Accept Gap and the Endowment Effect: Subject Misconceptions, and Experimental Procedures for Eliciting Valuations: Reply. *American Economic Review* 101 (2): 1012–1028.
Posner, R. 1981. *The Economics of Justice*. Cambridge: Harvard University Press.
Prize. 1863. The Prize Cases, 67 U.S. (2 Black) 635, 668 (1863).
Randall. 1976. US v. Randall, 104 Daily Wash. L. Rep. 2249, DC Superior Court.
Rawls, J. 1972. *A Theory of Justice*. London: Oxford University Press.
Rawls, J. 1999a. *The Law of Peoples*. Cambridge: Harvard University Press.
Rawls, J. 1999b. Reply to Alexander and Musgrave. In *John Rawls, Collected Papers*, ed. Samuel Freeman. Cambridge: Harvard University Press.
Rawls, J. 1999c. A Kantian Conception of Equality. In *John Rawls, Collected Papers*, ed. Samuel Freeman. Cambridge: Harvard University Press.
Rawls, J. 2001. *Justice as Fairness: A Restatement*, ed. Erin Kelly. Cambridge: Harvard University Press.
Rinard, S. 2019. Pragmatic Skepticism, Philosophy and Phenomenological Research. *Rutgers Epistemology Conference*.
Robinson. 1992. Robinson v State, 417 S.E.2d 88, 91 (S.C. 1992).
Roemer, J. 1996. *Theories of Distributive Justice*. Cambridge: Harvard University Press.
Rorty, R. 1979. *Philosophy and the Mirror of Nature*. Princeton: Princeton University Press.
Rushkoff, B. 1984. Note, A Defense of the War Powers Resolution. *Yale Law Journal* 93 (7): 1330–1354.
Savage, C. 2011. 2 Top Lawyers Lose Argument on War Power. *N.Y. Times*, June 18.
Savage, C., and M. Landler. 2011. War Powers Act Doesn't Apply for Libya, Obama Says. *N.Y. Times*, June 15.
Seidman, L. 2014. Constitutional Skepticism: A Recovery and Preliminary Evaluation. ssrn.com/abstract=2387031.
Sen, A. 1979. Utilitarianism and Welfarism. *Journal of Philosophy* 76 (9): 463–489.

Sen, A. 1992. *Inequality Reexamined*. Cambridge: Harvard University Press.

Sen, A. 1999a. *Development and Freedom*. Oxford: Oxford University Press.

Sen, A. 1999b. Equality of What? In *Equal Freedom*, ed. Stephen Darwall. Ann Arbor: University of Michigan Press.

SenateReport. 1983. S. Rep. 98-242 at 7, and Pub. L. No. 98-119, section 2 (b), 57 Stat. 805 (1983).

Sharkey, C. 2013. Trespass Torts and Self-Help for an Electronic Age. *Tulsa Law Review* 44 (4): 677–706.

Sherwin, E. 2003. Compensation and Revenge. *San Diego Law Review* 40: 1387–1413.

Smith, H. 2005. Self-Help and the Nature of Property. *Journal of Law, Economics and Policy* 1 (1): 69–107.

Stromseth, J. 1996. Understanding Constitutional War Powers Today: Why Methodology Matters. *Yale Law Journal* 106: 845–915.

Tables. 449BC. Twelve Tables, $8.2. Available at http://avalon.law.yale.edu/ancient/hammint.asp.

Tinland, O. 2013. The Pragmatist Skepsis as a Social Practice: Skepticism, Irony and Cultural Politics in Rorty's Philosophy, in Pragmatism and the Social Dimension of Doubt: Fresh Perspectives. *European Journal of Pragmatism and American Philosophy* 2: 21–30. Matthias Girel (dir.).

Tsagourias, N. 2012. Cyber Attacks, Self Defense and the Problem of Attribution. *Journal of Conflict and Security Law* 17: 224–244.

Turner, R. 1991. *Repealing the War Powers Resolution: Restoring the Rule of Law in U.S. Foreign Policy*. Pavilion.

Turner, R. 1994. War and the Forgotten Executive Power Clause of the Constitution: A Review Essay of John Hart Ely's War and Responsibility. *Virginia Journal of International Law* 34: 903–951.

Turner, R. 2013. The War Powers Resolution at 40: Still an Unconstitutional, Unnecessary, and Unwise Fraud That Contributed Directly to the 9/11 Attacks. *Case Western Reserve Journal of International Law* 45 (1): 109–143.

Von Staveren, I. 2009. Efficiency. In *The Handbook of Economics and Ethics*, ed. Jan Pell and Irene Van Staveren. Edward Elgar Publishing.

Verkerke, J. 2002. Is the ADA Efficient? Working Paper No. 02-9, May 13. http://ssrn.com/abstract_Id=311888, at 14.

Walker, L. 1986. A Response to Elizabeth M Schneider's Describing and Changing Women's Self-Defense Work and the Problem of Expert Testimony on Battering. *Women's Rights Reporter* 9: 223–236.

Wallerstein, S. 2005. Justifying the Right to Self-Defense, a Theory of Forced Consequences. *Virginia Law Review* 91: 999–1073.

Williams, M. 2020. Descartes' Transformation of the Skeptical Tradition. In *The Cambridge Companion to Ancient Skepticism*, ed. R. Bett. Cambridge: Cambridge University Press.

Wormuth, F., and E. Firmage. 1986. *To Chain the Dog of War: The War Power of Congress in History and Law*. Dallas: Southern Methodist University Press.
WPR. 1973. The War Powers Resolution, Pub. L. No. 93-248, 87 Stat. 555 (1973) (codified at 50 U.S.C. § § 1541–1548) (1994).
Yoo, J. 2003. International Law and the War in Iraq. *The American Journal of International Law* 97 (3): 563–576.
Zipursky, B. 1998. Rights, Wrongs, and Recourse in the Law of Torts. *Vanderbilt Law Review* 51: 1–29.

CHAPTER 4

Capabilities, Entitlements and VRN

1 Capabilities Theory

1.1 Needs and Functionings

The capabilities approach, principally developed by Amartya Sen and Martha Nussbaum, attempts to close the gap between the abstract identification of desirable social goods considered necessary for socio-political flourishing and the actual conditions of peoples' lives. For Nussbaum, basic capabilities necessary for flourishing include, among other things, control over one's immediate environment, body health and integrity, relative freedom from constant anxiety, fear or pain, freedom of imagination and freedom to develop social affiliations.[1] Sen's formulation of capabilities, by contrast, focuses primarily on an individual's relative freedom to take advantage of available opportunities and 'functionings.'[2] Both Nussbaum and Sen identify risk thresholds, or structural/political/environmental barriers to the actualization of opportunities (which could be earned or purely fortuitous), capabilities and natural talents, sufficient to justify either intervention or significant socio-economic redistribution by stakeholders, states or enforcement regimes.

[1] Nussbaum (2000, 78–80).

[2] Sen (1999a, 75-76).

A. J. Thomas, *The Open World, Hackbacks and Global Justice*,
https://doi.org/10.1007/978-981-19-8132-6_4

According to Sen, the primary goods model is flawed because it treats primary goods as a proxy for advantage rather than recognizing that advantage is a function of the relationship between persons and goods. Since simply possessing primary goods (or having ready access to such goods) does not, in and of itself, guarantee individual (or even collective) well-being, the concept of primary goods shifts attention away from the way that people actually live their lives.[3] Sen does not suggest that individuals do not desire primary goods, or even that it is impossible to organize shared socio-political expectations considered essential for flourishing in most societies into a prioritized list or index. His argument, instead, is that even if we can identify these primary goods (either objectively or subjectively), individual capacities to take advantage of primary goods differ so widely, and the motivation to value primary goods differs so significantly from person to person, that the practical effect of primary goods, as a basis for distributive justice (of the incremental or sweeping variety) is almost negligible.[4]

The capabilities approach is equally critical of utilitarianism, even though it shares, with utilitarianism, the tendency to equate well-being with the relationship between persons and goods. Where they differ is on the relevance of mental states to distributive questions, which is the tendency to equate the maximization of well-being (or welfare) with desire satisfaction or pleasure. Aggregate increases in well-being are the only real objective in utilitarian analysis. Individual notions of well-being are merely factored into the overall aggregate measure of well-being, and each individual contribution to the aggregate well-being is measured in the exact same way. Like the primary goods approach, therefore, utilitarianism largely ignores individual differences, not because it blithely assumes that the mere possession of primary goods will automatically improve the well-being of everyone, regardless of their capabilities or life challenges, but because it uses an indicia or well-being—desire satisfaction—that downplays individual life patterns that deviate from aggregate models of well-being. The inadequacy of this approach is particularly

[3] Id. (the primary goods model focuses on "income rather than one what incomes does, on the social basis of self-respect rather than on self-respect itself…").

[4] Sen (2002, 83) (suggesting that we judge "individual advantage in terms of the respective capabilities, which the person has, to live the way he or she has reason to value. This approach focuses on the substantive freedoms that people have, rather than only on the particular outcomes with which they end up.").

problematic for people who suffer from (or under) capability constraints that limit options and lower expectations. Such individuals tend to develop desires that reflect neither their true preferences nor their true long-term best interests.

The central advantage of the capabilities approach over its two main rivals—primary goods (resourcism) and utilitarianism—is its sensitivity to the complexities of individual decision-making practices. It identifies aspects of social life—the basic need for health, sustenance, self-esteem and control over one's environment and interpersonal relations—that all persons arguably need but it also recognizes that if those needs are not satisfied, individuals will find it difficult to either take advantage of primary goods or pursue desires that reflect their true preferences unconstrained by diminished expectations. Inadequate need satisfaction undermines the meaningful realization of individual well-being, regardless of how it is defined. However, one is still faced with the seemingly imponderable task of ranking or assigning priority to a relatively large list of capabilities, only some of which will be available to specific individuals at a particular time.

Capabilities theorists don't believe that individual freedom is solely dependent on the acquisition of primary goods or income. Nor is it the victim of diminished expectations caused by insurmountable socio-political obstacles, or at the mercy of any aggregate increase in well-being. Instead, meaningful expressions of individual freedom require the creation of a space in which needs, functionings and capabilities can be adequately satisfied, and unless satisfaction occurs, it is almost impossible to suggest that someone failed to take advantage of available opportunities.

The concept of need satisfaction in Sen's thought hinges on three major questions. The first is whether the satisfaction of needs consists of merely giving individuals, occupying a space in which capabilities and/or functionings can be realized, an opportunity to make an informed choice, or both the opportunity to make a meaningful choice and the achievement of a desired outcome in the form of a life plan or agency goal. Since the bulk of Sen's critique of Rawls is that the primary goods paradigm fails to bridge the actual gap between the mere possession of core moral goods like rights and the ways in which people actually live their lives, one would expect Sen to insist, if not upon outcome equalization, than some weaker version that enables people to "escape morbidity, to be adequately nourished, to have mobility, to achieve self-respect, to take part in the life of their community...," and to experience pleasure, regardless of whether it contributes to the fulfillment of an important life goal or not.[5] This interim position, which Cohen dubbed 'midfare' (halfway between primary goods and welfare),[6] not only links capabilities and functionings to actual human needs (and/or life conditions) short of the realization of long-term goals, but it also demands the relative equalization of functioning 'life-sets'[7] across all persons.[8] So that regardless of the situation, the universal satisfaction of need requirements, in the form of the relative parity of functioning life-sets across persons, can become the basis for meaningful interpersonal comparisons, subsequent distributive justice claims and the possible redistribution of socio-economic resources.

1.2 Defining Tastes and Preferences

By identifying baseline requirements for need satisfaction, we can begin to not only imagine attributes shared by all individuals, regardless of location or status, but also understand why these attributes might be something that all societies want to protect in both their own citizens and individuals from other parts of the world. Identifying baseline requirements for need satisfaction across societies, in other words, can become the basis for a theory of global justice. Global justice models must consider the kinds of decisions that extremely oppressed individuals make, especially when those choices potentially undermine norms specifically put in place to help the worst-off members of society.

Sen's discussion of tastes is particularly revealing on this point. Most economists define tastes as preferences[9] framed by social biases,[10] socio-psychological conditioning or cognitive limitations. Even though tastes are formed by many factors, it is not uncommon for tastes to deviate

[5] Roemer (1996, 188 n7).

[6] Roemer (1996, 189).

[7] Id at 190.

[8] Id.

[9] Becker (1996, 3–4) (equating preferences with tastes).

[10] Cooter (2000, 129–34) (shared tastes for local public goods tend to generate neighborhood clustering effects).

from personal or moral values.[11] Nor is it uncommon for tastes to conflict with commonly held notions of personal well-being (either objectively or subjectively determined), or policies that increase the aggregate welfare of a particular community. Since values are generally 'stickier,' or more resilient, than tastes, the latter are particularly sensitive to the "framing of choices, the context in which choices are made, and the method by which choices are elicited."[12] One could very easily develop, for example, expensive tastes (preferences completely out of sync with reasonable expectations), or even offensive tastes (preferences that involve harming someone else).[13]

Given the fluidity and complexity of tastes, no one should be surprised if force or coercion plays a direct part in the formation of expensive, offensive or other preferences that people tend to develop, but the use of force or coercion is particularly important for the formation of 'cheap tastes,' or preferences developed by individuals in an extremely disproportionate power relationship with another, such as "the battered slave, the tamed housewife, the broken unemployed, the hopeless destitute."[14] The cheap taste phenomenon is extremely problematic because an individual with cheap taste is likely to undershoot rather than overshoot (as one would expect with expensive and offensive tastes), any reasonable projected life path or agency plan going forward because their needs have not been satisfied, all reasonable functioning paths have effectively been blocked and capabilities are practically non-existent. And yet, these people are not irrational. In fact, the decision to remain in an extremely dysfunctional home with a violent spouse who beats you periodically is completely rational when the alternative is likely spousal abuse toward children. The same is true for someone who trades away the right to vote permanently in order to guarantee shelter and sustenance. Entrenched cultural expectations may provide additional reasons to embrace, and even defend, cheap tastes.

The cheap taste phenomenon provides a perfect illustration of what happens when choices are made without securing basic capabilities and functionings. Individuals who adopt cheap tastes also approximate the constrained decision-making conditions of people in crises. How do we

[11] Kuran (2013, 241) (describes the evolutionary basis for a deviance between values and preferences).

[12] Id. at 232.

[13] Roemer (1996, 188).

[14] Sen (1987, 11).

alleviate the suffering of these people and enhance equality? The solution is a theory of social responsibility.

1.3 Responsibility

Both Nussbaum and Sen believe that individuals can be held responsible for the choices they make if they were sufficiently 'free' when they made them. This belief relies on a whole host of other considerations, such as the meaning of free choice in particular situations, whether these choices can be irretrievably blocked by bad luck,[15] whether it is possible to reason backward from the quality of those choices (especially choices motivated by 'cheap tastes') to the kind of freedom that agents experience, and ultimately whether the agent's basic needs, capabilities and functionings were satisfied before the choice was made. Nussbaum and Sen also leave open the possibility that responsibility can be used, not just as a way to justify state power to punish people who make bad choices even when their needs have been satisfied and their capabilities relatively equalized with others, but also as a way to justify state power to reward people who make good choices (or encourage/nudge people to make good choices) even if their needs haven't been sufficiently satisfied.

According to Rawls, the state is responsible "for maintaining certain basic liberties and opportunities and for providing a fair share of primary goods .... (l)eaving it to individuals and groups to forms and revise their aims and preferences accordingly."[16] The stated responsibility operates (within well-ordered societies) in concert with broader social norms ensuring that individuals "take responsibility for dealing justly with one another on the basis of a public measure of (generalized) needs."[17]Nussbaum and Sen believe that primary goods are too far removed from the more basic needs that govern the way that people actually live their lives, but that the state is responsible for equalizing capabilities (to compensate for natural born differences, handicaps or even bad luck) The obligation to intervene extends beyond formal legal guarantees, but falls short of full outcome equalization.

[15] Roemer (1996, 270).

[16] Rawls (1999b, 241).

[17] Id. at 242.

Nussbaum and Sen want to reward good choices, punish bad choices and improve the life prospects of people who make responsible choices, using a version of prioritarianism, which requires the state to prioritize policies that minimize the suffering of the worst-off[18] when it dispenses justice or distributes socio-economic resources. Nussbaum and Sen also want to prioritize specific capability-enhancing strategies that improve the condition of the worst-off who act without fault (broadly defined), encourage responsible behavior and discourage irresponsible behavior by the badly off, and quite possibly, other members of society as well.

Responsibility-based prioritarianism establishes a credible link between capability-enhancing policies, the intrinsic moral value of improving the lots of worse-off (using an objective rather the strict egalitarian or sufficientarian standard), and merit-based criteria for distributive justice (using responsibility or the reasonable absence of fault). But responsibility-based prioritarianism requires a list of attributes or norms that we could rank along a list from most responsible to least responsible (or, in the alternative, a much longer, but non-indexed, list of responsible and irresponsible norms). Without that list, either defined generally or linked to a range of behavioral and/or moral responses typically demonstrated in specific situations, it is difficult to identify which capability-enhancing policies will generate the responsibility norms most deserving of state recognition in a prioritarian scheme designed to minimize the suffering of the worst-off.

Both Nussbaum and Sen provide some guidance on this issue. Nussbaum, for example, supplied a capabilities list, starting with physical attributes/priorities (life, bodily health, bodily integrity, sense and emotions), proceeds to moral and/or socio-cultural factors (practical reason, affiliation, relations with other species and play), and concludes with spatial factors (control over one's environment).[19] At first glance, Nussbaum's capabilities list has little to do with responsibility. Instead, Nussbaum's prioritized list seems to proceed from the least controversial, irreversible impacts on human flourishing (physical and mental health) to more contested, debatable adverse impacts (the freedom to play or control over physical space). One could also argue that the list proceeds from the most serious potential crises to the person (as opposed to the community, institution or nation-state), to the least serious, or from

[18] Parfit (2000, 347).

[19] Nussbaum (2009, 76–78).

nearly universally accepted (and arguably objective) attributes/capabilities to more culturally driven capabilities, to arguably particularistic or even extreme capabilities valued by a relatively small number of unique individuals.[20] Nussbaum, in other words, seems to prioritize maxim in egalitarian policies that enhance the capabilities of those falling below a given non-welfarist threshold, and even suggests that unspecified "large items on the list" are more important than unidentified "more specific subsections",[21] but she does not provide a sorting or prioritizing mechanism for below threshold conflicts, and ignores luck egalitarian or prioritarian solutions to above threshold conflicts.[22] Sen is also frustratingly oblique and non-committal about the ideal ordering of capabilities or even development priorities[23] beyond the baseline of need satisfaction. But regardless of how they rank capabilities, both Sen and Nussbaum believe that the realization of each listed capability is necessary in order to ensure a measure of meaningful freedom. With that freedom comes the power to not only make reasonable choices, but also make choices for which one expects to be held responsible.[24] Every cumulative or incremental increase in capabilities enhances the prospects for individual responsibility and future accountability.

1.4 Equalizing Opportunities

Sen also imposes a duty upon the state to equalize opportunities (through capability and functioning sets) rather than results for both practical and philosophical reasons. As a practical matter, Sen recognizes the inherent danger of granting the state power to equalize results regardless of the long-term consequences, including the loss of rights. Many lives have been destroyed and freedoms crushed by authoritarian (and/or fascistic)

[20] Nussbaum's distributive model also contains aspects of sufficientarianism (everyone falling below a particular non-welfarist threshold should receive 'enough' benefits, in the form of capability-enhancing assistance, to meet minimum requirements) and egalitarianism, without considering the appeal of other factors, such as how cost/benefit analysis or intuitive moral norms like responsibility might severely limit or even eliminate the state's obligation to enhance the capabilities of the worst-off. Nussbaum (2005, 216).

[21] Id. at 194.

[22] Stein (2009, 506–9).

[23] Sen (1999a, 286) (resists ordering of preferences or development priorities).

[24] Id. at 284 (without freedom, neither choice nor responsibility are defensible).

states determined, quite possibly in a genuine attempt to alleviate massive, and potentially destabilizing, class disparities, to equalize the realistic life prospects of individual citizens. But Sen's main philosophical objection to the equalization of results stems from his definition of freedom, which is "the substantive opportunity ... to pursue ... our objectives and goals."[25] Freedom, in other words, is the meaningful opportunity to pursue things that one "has reason to value,"[26] not the guarantee that one will actually attain all of one's desires. Equalizing capability and functioning life-sets is a way to render choices more meaningful but it cannot, and should not if history is any judge, empower the state to guarantee either that two people with comparable capabilities will pursue the same long-term goals, or that the same two people will actually achieve the same long-term goals or life plans. Both courses, or justifications for state power, are inadvisable because both require the state to ignore relevant differences between persons as it distributes scarce socio-economic resources, such as the fact that one person might have failed to advance life goals because of enormous, random, unavoidable or inexplicable bad luck (which makes him more worthy of assistance), or that another person might fail to achieve life goals, in spite of achieving basic need satisfaction, because of personal irresponsibility (which makes him less worthy of assistance).

Sen, therefore, believes that the state should only equalize capabilities that satisfy basic needs and provide meaningful opportunities rather than guarantee individual outcomes or long-term life goals. The relative freedom and informed decision-making that flows from the equalization of capabilities and opportunities also provides the state with an opportunity to distinguish between persons on distributive justice questions (i.e., the distribution of scarce socio-economic resources). The basis for the distinction, however, is not utility (the aggregation of individual well-beings) or a non-egalitarian welfarist calculation[27] "because this depends significantly on how (one) chooses to live and the uses to which one puts the opportunities and resources and liberties made available to them."[28] It is, instead, personal responsibility on the part of decision-makers or individual agents. Like Rawls and Dworkin, Sen believes that the state

[25] Sen (2002, 10).

[26] Id. at 10.

[27] Roemer (1996, 164).

[28] Sen (2002, 12).

can assign priority to rights claimants (speaking loosely) who act responsibly, a burden of action that is, of course, only possible if adequate need satisfaction and capability equalization has been achieved. It is only then that a person can be said to act relatively 'freely,' all things being equal, and it is only then that the state can, with complete moral and political justification, either punish or reward responsible agents for the choices they make.

On the question of whether the state should guarantee equal outcomes or merely equal opportunities (capabilities), Sen rejects both the utilitarian/welfarist view that is geared towards the maximization of aggregate utility and overall welfare at the general expense of egalitarian concerns and any consideration of personal responsibility as a second-order justification for differential distribution of scare socio-economic goods, as well as resourcist views that balance the acquisition of primary goods against personal responsibility in the calculus of guaranteeing equal opportunities. The critical nexus for distributive justice, in Sen's view, is the relationship between need satisfaction sufficient to relatively equalize capabilities and personal responsibility on the part of the moral agent. The former provides individuals with the means to make life choices and decisions that they feel good enough about to take responsibility for their actions, and the latter proves the state with a way to distinguish, for purposes of dispensing distributive justice, between reasonable claims worthy of serious consideration (and even emulation), and relatively unreasonable claims that should be ignored, discouraged or punished.

The central challenge for Sen is that even though it makes perfect sense to link meaningful freedom of choice with basic need satisfaction and the reasonable equalization of capability, and equally good sense to use personal responsibility to encourage both good behavior and constrain the possibly uncontrollable growth of state power (which would increase if the state were given full authority to equalize both opportunities and outcomes), it is not entirely clear which behavior Sen wants to explicitly encourage at the policy level once need satisfaction and capability equalization have occurred. It is clear, for example, that Sen wants to discourage the acquisition of cheap tastes, which are themselves the best sign that massive capabilities inequalities exist, but is not clear, at the policy level, which behavior Sen wants to encourage. Nor is it particularly clear, from his indexed list of capabilities and functionings, whether the state or some other enforcement agency should prioritize the reasonable equalization of specific capabilities over other capabilities in order to more effectively encourage behaviors and choices considered to be more desirable than other possible behaviors.[29]

Addressing the dual questions of exactly which behaviors should be encouraged and which capabilities should be prioritized at the policy level does not force Sen to commit to one particular distributive justice model applicable to all situations regardless of the circumstances or socio-economic conditions, but it does require him to do more than simply place capability equalization (with no discussion of which capabilities matter the most) in the collection of things that the state must consider when it attempts to equalize opportunities, or personal responsibility generally (without referring to which acts of responsibility are more desirable than others) as a legitimate way to distinguish morally praiseworthy and problematic rights claims.

2 The Limits of Sen and Nussbaum

2.1 Lists, Temporality and Choice

Capabilities theory has generated its fair share of criticism, only some of which is relevant for a capabilities-based defense of retaliatory hackbacks. One of the most important differences between Sen and Nussbaum, for example, is that Sen refused to propose a list of 'core' capabilities based on a philosophical (as opposed to empirical) rendering of personhood. Nussbaum, on the other hand, did provide an extensive list of capabilities, and grounded those capabilities in an Aristotelean understanding of personhood, flourishing and functioning. Sen's refusal to ground his theory of functioning and capability in Marxianism or Aristoteleanism is not a serious point of disagreement between Sen and Nussbaum because Sen simply claimed that his theory did not require such grounding. He did not say, however, that he disagreed with Nussbaum's decision to ground her theory in Aristotle.

The provision of a capabilities list is a larger point of contention between Sen and Nussbaum. Nussbaum's decision to provide such a list generated three additional critiques of capabilities theory: (1) whether any capabilities list is exclusive or merely illustrative, (2) the contents of any capabilities list may reflect asymmetric power relations and (3) whether Nussbaum's list is similar to Rawls' list of primary goods or 'core' human

[29] Roemer (1996, 192).

rights lists. The first two critiques suggest that Nussbaum's capabilities list is under-inclusive. Allison Jaggar, a prominent feminist theorist, attributes the under-inclusiveness of Nussbaum's list to Nussbaum's liberal proceduralist sensibilities, which tend to privilege men over women, as well as the Global North over the Global South.[30] Rawls shares the same characteristics with Nussbaum, which at least partially explains the striking similarity between Rawls' primary goods list and Nussbaum's capabilities list. Samuel Freeman has also argued that "capability and justice as fairness are much closer than Nussbaum appreciates",[31] particularly with regard to people with severe disabilities who may possess primary goods, but completely lack the capability required to participate in ordinary society.[32]

Sen, on the other hand, is frequently subjected to three additional critiques: (1) that capabilities theory is merely an "instrumental means to an end" rather than an end unto itself,[33] (2) Sen under-theorizes the relationship between capability and choice and (3) Sen fails to adequately 'temporalize' capabilities. Both the choice and temporal critiques were proposed by Bernard Williams. The choice critique challenges Sen's oft-repeated assertion that capabilities exist independent of choice.[34] If, someone, for example, is 'capable' of fixing a car or voting, that capability continues to exist whether they choose to act or not. That capability did not also come into existence merely because someone 'chose' to be capable, according to Sen. The capability came from circumstances and/or innate ability, not personal choice.[35] Consequently, people born with advanced mathematical ability are automatically more likely to be successful as math professors; football players with strong arms are automatically more likely to be successful as quarterbacks; married women with strong economic and emotional from their biological family or friends are automatically more likely to leave their husbands; people with a longer life expectancy are automatically more likely to secure a decent standard of living.[36]

[30] See e.g. Jaggar (2006); See also IEPCapabilities (2022).

[31] D'Amato (2014, 6) citing Freeman (2006, 412).

[32] Freeman (2006, 412), cited by Amato (2014, 4).

[33] D'Amato (2014, 5).

[34] See e.g. Williams (1987, 97–99).

[35] Id. at 99.

[36] Id. at 100.

According to Williams, however, capability is inextricably linked to choice. Consequently, football players with strong arms are only more likely to be successful as quarterbacks if they choose to be quarterbacks, and those with longer life expectancies are only likely to attain a decent standard of living if they choose to spend their golden years on economic sustenance rather than asceticism or deep meditation. Without the choice to act on, or take advantage, of a capability, capability itself is nothing more than a probability assessment. Williams' second critique focuses on the crucial relationship between capability and the passage of time. According to Sen, capability to do X means that the capability to do X exists right now. Consequently, when a football player acquires the ability to throw long balls, he is capable of acting on that capability immediately. When an abused woman attains the capability to punish her abusive husband, she is able to act on that capability immediately, whether action takes the form of calling the police, physically hitting back or draining his bank account. According to Williams, however, it is perfectly possible to attain the capability to do X without having the immediate ability to act on that capability in a preferred fashion. Someone, for example, may acquire a license to practice medicine in state X, but decide to remain in state Y for six months because of a family emergency. The costs of immediately moving to state X—economic, emotional, familial, reputational—may be too high to justify moving immediately, even though the person is technically capable of practicing in state X.[37]

2.2 *The Means-Ends Problem*

Another critique of Sen, arguably more damaging than Williams' critique of choice and temporality, is that capabilities analysis is a theory of means rather than a theory of means. In its weak form, the 'means-ends' critique is that there is a significant overlap between many capabilities and resources. If we treat these resources as instrumental means rather than ends, then we must also treat the capabilities with which they overlap as instrumental means as well.[38] The strength of the weak means-ends critique depends on which capability deficiency one selects because capability deficiencies lie along a spectrum. At one end of the spectrum, there

[37] Id.; See also IEPCapabilities (2022).

[38] D'Amato (2014, 5).

are people who lack capability because of disabilities or cognitive difficulties. The idea that people with severe disabilities or cognitive difficulties fail to take advantage of opportunities solely because they lack primary goods or personal wealth is laughable. The capability deficits of people at this end of the capability scale are profound. At the other end of the scale, there are capabilities that overlap with important resources. In the development arena, Sen himself acknowledges that food and shelter can be secured with resources.[39] But the relationship between these kinds of capabilities and resources is much more complex than it appears to be. On the one hand, both resources/primary goods and capabilities (food/shelter) appear to be instrumental means to a better life, but on the other hand, resources appear to be an instrumental means to the attainment of food and shelter, capabilities that could be considered ends in themselves.[40] Even if we temporarily suspend, for the sake of argument, the 'means-ends' problem, it is impossible to reject the argument, made by Thomas Pogge, that certain kinds of capabilities "[require] institutional distributions that take into account the capability of people to convert resources into opportunities."[41] At the center of the spectrum, however, there are intractable, irrefutable, situational capability problems that don't overlap cleanly with resources or primary goods. 'Poor' overlap, in this context, means that it is not accurate to characterize resources as a necessary means to the attainment of situational capability characterized as an end. Nor is it accurate to treat both resources and situational capability as instrumental means necessary for the attainment of other ends.

The strong 'means-ends' critique is that Sen's belief that capabilities are intrinsically valuable seems to be based on a theory of well-being comparable to that of the utilitarians and wealth maximizers.[42] For critics, Sen's frequent assertions that capabilities have intrinsic value always seemed more strategic than substantive. In terms of the latter, Sen has always been somewhat vague. Sen never subscribed, for example, to Nussbaum's view that capabilities are based on Aristotelean values, not because Sen disagreed with Nussbaum on this question, but because he

[39] D'Amato (2014, 5), citing Sen (1988, 162–64).

[40] D'Amato (2014, 5).

[41] Id. See also Pogge (2002, 169–174).

[42] Kaplow (2007, 2).

believed that his theory "didn't need it." The strategic significance of assessing the intrinsic value of capabilities, however, has never been vague in Sen's universe. In response to the argument that there is a tremendous amount of practical overlap between resources/primary goods and specific capabilities, Sen's defenders responded by claiming that capabilities had intrinsic value, and that it is wrong to simply "assign an equivalent, merely instrumental value for capability."[43]

A similar strategy was deployed by Sen's defenders against utilitarians and wealth maximizers. In Inequality Re-Examined,[44] Sen argued that capability theory "[makes] room for a variety of doings and beings as important in themselves (not just because they may yield utility, not just to the extent that they yield utility)."[45] Without a strong substantive justification for the intrinsic value of capabilities, according to Shavell, Sen's model seems to focus on well-being and welfare,[46] albeit a 'thick' theory of well-being/welfare that considers a wide variety of factors, not just the accumulation of wealth. Defending the latter seems to be the focus of Sen's writings on development.[47] Since, according to Shavell, Rawls also seems to be pre-occupied with well-being, wealth maximizers and welfarists focus on well-being, and both are means-oriented, capability theory appears to be just another means-oriented framework. The real challenge for Shavell is not whether capability is a pure or even partial ends-oriented theory compared to its rivals. On the contrary, the real challenge is deciding which theory provides the best proxy or surrogate for well-being beyond income and clumsy inter-personal comparisons. It is here that Rawls and Sen become important.[48]

2.3 *The Overlap Problem*

Any capabilities-based defense of retaliatory hackbacks has to address potential overlaps and means-ends problems. The former is a concern because a capability-based defense has to establish, first and foremost,

[43] D'Amato (2014, 4).

[44] Sen (2006).

[45] Id. at 80.

[46] Kaplow (2007, 4).

[47] Kaplow (2007, 17–18).

[48] Kaplow (2007, 18).

that it is not only analytically and conceptually distinguishable from its primary competition, but that the political, moral and legal defense of a capability model is not dependent on concepts or policies developed by either Rawlsians or utilitarian/wealth maximizers. Conceptually, it is not difficult to understand why someone would fail to take advantage of opportunities even if they have primary goods/rights or wealth. It is also not difficult to understand many policies designed to alleviate the suffering of chronically oppressed people may not be successful if those policies consist entirely of wealth accumulation, resource transfer or the establishment of rights.

But self-defense and self-help are more than just the failure to take advantage of available opportunities. An abused married woman with means and legal rights who refuses to leave her home because she is afraid of what her husband will do to her children is not, in the public imagination, the same as a similar individual who chooses to defend or help herself in this situation, not by calling the police, hiring a private attorney or lock herself in a safe room inside of her own home, but by either paying someone to kill her husband, asking a friend to kill her husband or killing the husband herself. There is also, in the public imagination, a fundamental difference between a repeatedly hacked medium-sized advertising firm that refuses to contact the police, initiate a private suit or rely on hardening because it is afraid that someone will discover major vulnerabilities in its network software, and an identical firm that either convinces one of its employees, or hires an outside programmer, to retrieve stolen files from a suspected hackers computer system. In both cases, capability-deficient victims 'failed' to take advantage of available opportunities for redress even though they have rights, resources, primary good and wealth. But when you add self-defense/self-help to the aforementioned scenarios, one question emerges: if the abused, affluent mother can summon the courage to kill her abusive husband or enlist the assistance of other to do the same, or the repeatedly hacked mid-size advertising firm can summon the courage to convince one of its employees or an outside blackhat programmer to retrieve stolen files from the suspected hackers computer, then both parties appear to have more capability than counterparts who did not resort to self-defense or self-help.

Our intuition, in other words, is that capability-deficient people who resort to self-defense or self-help have more 'capabilities' than capability-deficient people who don't resort to self-defense or self-help. Imagine, for example, three possible responses to a perceived or actual crisis, all involving self-defense or self-help. In scenario 1, the victim is hopeless and weak (ie. convinced that there are no available solutions to their suffering). In scenario 2, the victim resorts to self-defense or self-help out of weakness because more traditional rule of law avenues and hardening are unlikely to be effective. In scenario 3, the victim resorts to self-defense/self-help out of strength because he believes that he can achieve redress or justice faster and cheaper on their own, even if traditional rule of law avenues and hardening are available and likely to be reasonably effective. The victim in scenario 3 appears to be more 'capable'—less repressed, more optimistic about their own circumstances or the effectiveness of law enforcement—than the victims in scenarios 1 and 2.

But capability is more than just an intuitive notion. It is a complex framework, the central focus of a philosophical project that stands in stark contrast to primary good/rights and utility/wealth maximization. Earlier, I mentioned three kinds of capability-deficient people: (1) those with disabilities or cognitive difficulties, (2) situationally capability-deficient people, and (3) those with existential constraints like food or shelter. In theory, one could imagine people in all three categories using self-defense or self-help, particularly if Williams argument about linking capabilities to temporal constraints is factored into the equation. Cognitive and disability issues, as well as food and shelter issues, could dissipate over time, increasing capabilities. But realistically, situationally capability-deficient are more likely to use self-defense or self-help, even if their adaptive preferences are entrenched. These victims, in particular, may be able to summon more strength or courage to use self-defense or self-help than other capability-deficient people, and to do so without undermining the central claim that capability-deficient people often choose to retaliate even though they have rights/primary goods or wealth. But the overlap problem creates an existential dilemma, not because situationally capability-deficient people, like food/security deficient people, occupy a point on the capabilities spectrum likely to overlap with resources, goods and wealth, but because there is empirical evidence suggesting that self-defense and self-help are more likely to be effective (as a deterrent or immediately retributive mechanism) if the victim either has wealth or enough expertise to navigate rules designed to assist people who resort to self-defense or self-help legally. Consequently, a capabilities-based defense of retaliatory hackbacks that focuses on situationally capability-deficient entities has to walk a fine line between victims who resort to self-defense/self-help because they lack the capability to take advantage of traditional avenues of redress even though they have wealth, primary goods or rights, and victims who are capable enough to use their wealth, resources or mastery of the law to increase the effectiveness of their retaliatory efforts.

Means-ends critiques of Sen and Nussbaum raise an entirely different set of issues once we identify a class of victims with enough capability to employ self-defense or self-help. Beyond the concerns raised by Pogge (both resourcism and capability theory are instrumental means to an end rather than ends in and of themselves) and Shavell (the intrinsic value that presumably distinguishes capability theory from resourcism is just a broader version of the same objective that pre-occupies both resourcists and welfarists: well-being),[49] capability theorists have to defend a claim more ambitious than the general assertion that capability is more end-oriented than utility or primary goods. It must also demonstrate that capability theory is more likely to secure attainable ends than is competitors. In order to make his argument, Sen and Nussbaum much prove that capabilities theory accurately describes the constraints that individuals face in their daily lives, constraints that they cannot completely remove, either in theory or practice, solely through primary goods or wealth. Capability theory must explain why policies that focus exclusively on wealth transfers or the establishment of rights often fail; or why people with wealth or legal rights consistently fail to take advantage of opportunities or underperform. It must show that cognitive, situational or existential constraints on individual or collective freedom are a better predictor of future success than the implementation of policies designed to increase general well-being based on either the aggregation of individual utilities (using wealth of efficiency as a metric) or the endowment of primary goods or the establishment of formal legal rights.[50]

Capability-based defenses of retaliatory self-defense and self-help must secure 'attainable ends'. It should be possible to secure these attainable ends relatively quickly, presumably faster than one could secure the same ends by contacting the police, initiating a private lawsuit or hardening. In order to be truly attainable, the ends secured should not completely

[49] Kaplow (2007, 2).

[50] Sen (1999a, 90).

exceed the reasonably foreseeable outer boundaries of an individuals' capability-deficiencies. Securing attainable ends should not expose anyone to more political or legal liability than necessary or reasonable given their capability-deficiencies. And it should assign responsibility to all persons involved in the event or violation that lead to the use of self-defense or self-help, as well as all persons involved in or affected by retaliation. Both Sen and Nussbaum have written extensively about the relationship between capabilities and responsibility. In Development As Freedom, for example, Sen stated that

> Responsible adults must be in charge of their own well-being; it is for them to decide how to use their capability. But the capability that a person does actually have ... depends the nature of social arrangements, which can be crucial for individual freedoms. And the state and society cannot escape responsibility.[51]

Like Sen, Nussbaum has also analyzed the relationsip between responsibility and capabilities, both in terms of individual conduct (such as the responsibilities of married women in abusive relationships) and in terms of enforcing the constitutional and human rights of capability-deficient people. But neither Sen nor Nussbaum have addressed the complex task of apportioning legal responsibility in self-defense or self-help cases when the retaliatory acts of the victim caused damages. Those damages might be the 'attainable ends' sought by the victim, but the manner in which the victim retaliated may warrant splitting the cost of damages between the victim and the aggressor. The next section considers this question as part of a broad, capabilities-based defense of retaliatory self-defense and self-help applicable to situationally capability-deficient individuals, organizations and nation-states in public and private law cases.

Thus far, we have considered potential overlaps between capability theory and fairness (primary goods, rights), as well as capability theory and wealth maximization. A third category concerns potential overlaps between capability theory and efficiency. Efficiency is usually measured in terms of whether a particular 'change,' ranging from personal preferences to entire policy regimes, makes people 'better-off' or 'worse-off,' both considered to be relatively accurate measures of utility or human satisfaction. Efficiency skews toward the narrow question of "whether

[51] Sen (1999a, 288).

an economy produces at its possibility frontier, that is, generating maximum possible market output with given resources."[52] Traditional economics also skews toward positive (efficiency) rather than normative (equity, liberty, rights), questions.[53] Since cost savings are considered a desirable goal from the standpoint of traditional economics, policy regimes that encourage people to pursue the 'least expensive' solution to social conflicts (such as striking voluntary agreements that don't require state intervention, or mitigation strategies that minimize the extent of damages) are considered to be 'efficient.' Alternative efficiency rationales like innovation are often considered desirable not for their own sake, but because they lead to more streamlined, faster and cost-effective results. As a result, disentangling cost, market mechanisms, wealth maximization and social goals ultimately reducible to cost saving from conventional definitions of efficiency is quite difficult.

According to Sen, non-market factors should be considered in virtually every theory, including theories that purport to measure efficiency. Among the non-market values Sen considers to be the most important for human flourishing is freedom. Relatively early in his career, Sen surmised that it would be impossible to achieve a robust threshold of personal freedom under a Pareto-optimal regime.[54] Later, he concluded that in order to preserve the importance of non-market factors like freedom in any theory of human happiness, theorists had to turn their attention to the 'opportunity aspect of freedom.' Eventually, Sen developed an alternative efficiency model better equipped to evaluate the socio-economic and distributive effects of capability deficiencies. That model starts with the concept of 'opportunity-freedom.' 'Opportunity-freedom' draws inspiration from Isaiah Berlin's famous distinction between negative and positive liberty. The former is freedom from something, such as government interference, threats to bodily integrity, intrusions onto land, or the violation of interests. The latter, on the other hand, is more participatory.[55] While opportunity-freedom is closely related to positive liberty, its counter-narrative, 'process-freedom,' concerns "the process by which … [choices]

[52] Von Staveren (2007, 1).

[53] Id.

[54] Sen (1970).

[55] Berlin (1961) (distinguishes between positive and negative liberty).

are made and the number and characteristics of unchosen alternatives."[56] The essential difference between process and opportunity freedom, therefore, is that the former focuses on "freedom of action and decisions," whereas the latter focuses on "actual opportunities that people have, given their personal and social circumstances."[57] Even though process freedom is deeply vulnerable to manipulation, it is the preferred liberty model of classical economists,[58] libertarians (who tend to "[confine their] attention only to appropriate procedures"),[59] and consequentialists (who tend to focus only on "adequate opportunities").[60] Even though the process and opportunity aspects of freedom are clearly distinguishable, robust models of freedom, according to Sen, must account for both aspects.[61]

Opportunity freedom also links responsibility to equality. For most theorists, equality is an aspirational ideal, both in terms of equality of opportunity and equality of results (outcomes). Sen rejected equality formulations based on utility,[62] welfarism[63] or primary goods,[64] in favor of 'basic capability equality,'[65] but he also belongs to philosophical tradition that prioritizes equalizing opportunities for people who exercise a measure of personal responsibility given their life circumstances.[66] Equality capability calibrated to reflect opportunity sets and responsible assessments of available opportunities based on life circumstances is a priority for Sen, but his equalization formula is fuzzy, according to his critics, creating the distinct possibility that someone could be perceived as

[56] Foster (2010, 21).

[57] Sen (2000b, 17).

[58] Id.

[59] Id. Libertarians also prefer negative over positive liberty. Sen (2002, 514).

[60] Sen (2000b, 17).

[61] Id. at 291 ("... freedom is concerned with processes of decision-making as well as opportunity to achieve valued outcomes...").

[62] Sen (1995, 308 n.15).

[63] Id. at 315-323.

[64] Id. 323-6. See also 328-9 ("the focus on basic capabilities can be seen as a natural extension of Rawls concern with primary goods, shifting attention away from goods to what goods do to human beings.").

[65] Id. 327-9.

[66] Kanbur (2016, 8). (circumstances should include life conditions "outside of the individuals control.")

"not responsible for their opportunities... [but also] responsible for their choice of functioning vectors and agency goals."[67] Sen conceded that "only a partial set of functioning vectors and capabilities ... is possible" in terms of securing an equality of opportunities,[68] but he still struggles to "separate opportunity from outcome... and circumstances from efforts."[69]

Finally, opportunity-freedom refines the concept of efficiency. In Rationality and Freedom, Sen analyzed four versions of the "basic criterion of opportunity freedom comparison of different achievement sets."[70] Under Axiom R, achievement set A and achievement set B are equal (i.e., provide the same amount of opportunity-freedom) if there is a "one-to-one correspondence between" a subset of A and a subset of B, and "every element of [subset] A is regarded as no worse than a corresponding element of [subset] B."[71] Under Axiom P, set A produces more opportunity freedom than set B if every element in a subset of set A is strictly preferred to a corresponding element in a subset of set B.[72] Axiom P* modifies Axiom P in that only "some element in [subset A] is preferable to every element of [subset B]."[73] Under Axiom O, at least one element of set A must be preferred to all elements of set B in order to ensure that set A "offers more opportunity freedom than [set] B."[74]

Axiom O can generate two distinctly different possibilities. The first, an actual increase in freedom, "requires the presence of a more preferred alternative, but the presence of a more preferred alternative does not necessarily guarantee an increase in freedom."[75] The second possibility, "at least as much freedom,"[76] "requires some element of [set] A to be better than every element of [set] B," but it does not require some

[67] Roemer (1996, 192).

[68] Id. at 193. See also Sen (2002, 315) ("opportunity freedom must frequently take the form of incomplete orderings.").

[69] Kanbur (2016, 10).

[70] Sen (2002, 515).

[71] Id.

[72] Id. at 516.

[73] Id.

[74] Id.

[75] Id. at 517.

[76] Id.

element of [set] B to be better than every element of [set] A."[77] Without that symmetry, the second possibility can only produce a partial or incomplete picture of any increase in freedom. Both possibilities reflect the realization that "we cannot be sure that there is an expansion of opportunity-freedom unless there is an opportunity of getting a better alternative,"[78] and even if there were a better alternative, someone could reasonably argue that the better alternative did not strictly expand their opportunities.[79] The "variability of reasonable judgments" on the best way to increase opportunity freedom though the availability of better alternatives is simply unavoidable.[80] As a result, Axiom O can only produce a 'weak efficiency' of opportunity-freedom,' which Sen defined as the following:

> [T]he state of affairs is weakly efficient in terms of opportunity-freedom if there is no alternative feasible state in which everyone's opportunity-freedom is surely unworsened and at least one person's opportunity-freedom is surely expanded.[81]

Sen's model, just one of many attempts to combine positive and normative economics, turns the concept of efficiency on its head. Pareto usually focuses on allocation preferences, but preference arrangements also contain information about available opportunities. The former ranks consumption bundles, whereas the latter ranks opportunity sets. With regard to allocation preferences, an arrangement is Pareto optimal if it allocates in a manner that makes at least one person better-off without making anyone else worse-off. With regard to opportunities, however, an arrangement is Pareto optimal if it creates a situation in which "it is impossible to improve one agent's opportunities without reducing another agent's opportunities."[82]

According to some critics, Sen's model is best understood as the first welfare theorem filtered through the lens of 'no resentment,' such

[77] Id.

[78] Id. at 518.

[79] Id.

[80] Id.

[81] Id.

[82] Tadenuma (2012, 108).

that "the optimum of opportunity freedom can be attained in perfectly competitive markets when agents do no regret the choices, among feasible alternatives, that they make."[83] Regret theory focuses on the ways in which people challenge the wisdom of past decision-making. In the popular imagination, people 'regret' decisions that fell short of expectations, whether those decisions lead to actions (commission) or inaction (omission). For a wide variety of reasons, people are generally averse to making decisions they are likely to regret later. The general study of 'regret aversion' began with papers by Sugden, Loomes and Bell in the 1980s that examined the affect of anticipated disappointment on decision-making.[84] If, for example, the expected utility of a decision was 50, and the actual utility generated by the decision was only 30, an individual will experience 'regret' at having made the decision. Decisions that yield more than the expected utility, on the other hand, lead to feelings of 'elation.'[85] Factors that can increase or decrease the likelihood of post-decisional regret include "outcomes received by others," "outcome narrowly missed" and "alternative outcomes that can readily be imagined."[86] Regrets about omissions tend to be stronger than regrets about commissions,[87] and people tend to regret making decisions that did not reflect the way they actually felt when the decision was made.[88] Lost opportunities, or opportunities available to others but not oneself, are a prime source of regret. The immediate solution to 'opportunity-regret' is offering people as many reasonable opportunities as possible, but studies have shown that in some situations, offering people a wide range of options actually lengthens the decision-making process and reduces welfare.[89] Feelings of regret can also be produced by 'event-splitting' events.[90]

[83] Von Staveren (2007, 13) citing Qizilbash (2005).

[84] Connolly and Butler (2006, 142).

[85] Id. at 144.

[86] Id. at 145.

[87] Connolly and Zeelenberg (2002, 213).

[88] Id. at 215.

[89] Irons and Hepburn (2007, 192)

[90] Bleichrodt and Wakker (2015, 502) See also Starmer and Sugden (1993) (splitting expected outcomes into multiple events can produces feeling of regret) and Humphrey (1995).

For other critics, Sen's model merely extends Pareto efficiency, which usually applies to 'commodity spaces' and preference fulfillment, into the "space of actual functionings" (non-market and market) and capability fulfillment without resolving the central shortcoming of the Pareto approach: tolerance for low marginal subjective well-being.[91] Regardless of its shortcomings, Sen's model does address efficiency and equity simultaneously.[92] Unlike market-driven efficiency models that protect 'process-freedoms' well,[93] and CBA models that fail to capture the value of many kinds of opportunity expansion,[94] opportunity-freedom efficiency "appears better able to assess incommensurable outcomes and to value these in the project participants' own terms."[95] It is also well-equipped to improve "well-being outcomes for the population as a whole."[96]

3 Reconceptualizing the Melamed/Calabresi Matrix

Apportioning legal and political responsibility for self-defense and self-help acts is a crucial part of a capabilities-based defense of retaliatory hack-backs. This section proposes an apportionment framework that reflects the strengths of capability called the 'violation / retaliation' nexus (VRN) (Table 1). VRN resembles the Melamed/Calabresi entitlement matrix (MCEM) (Table 2), a famous model that analyzes remedies available to plaintiffs and defendants, but there are important differences between the frameworks, starting with the meaning of entitlements.

[91] Sen (2002, 522). See also Von Staveren (2007, 13-14). (Sen's model appears to "[copy] the strong no-harm principle of the Paretian efficiency criterion" that "does not allow redistribution from low to high marginal subjective well-being.")

[92] Sen (2000, 120) ("[E]quity-motivated interference with the working of the market mechanism can weaken efficiency achievements even as it promotes equity.")

[93] Sen (2002, 513).

[94] Von Staveren (2009, 112).

[95] Id.

[96] Id. at 113.

Table 1 The Violation / Retaliation Nexus

Level of Protection > 'Entitlement Holder'	*Presumptive rule*	*Desert rule*
Plaintiff	- Impermissible attack - Permissible retaliation (zone 1)	-Permissible attack - Permissible retaliation (zone 2)
Defendant	- Permissible attack - Impermissible retaliation (zone 3)	- Impermissible attack - Impermissible retaliation (zone 4)

Table 2 Melamed-Calabresi Entitlement Matrix

Level of Protection > 'Entitlement Holder'	*Property Rule*	*Liability Rule*
Plaintiff	Rule 1 : Plaintiff gets injunction against Defendant's pollution of trespass	Rule 2 : Defendants' Trespass / Pollution is allowed, but damages but damages are paid to Plaintiff
Defendant	Rule 3 : Defendant's Trespass / Pollution is allowed (ie. injunction is denied)	Rule 4 : Plaintiff gets injunction against Defendants' Trespass or Pollution, but must pay damages to compensate defendant

3.1 Entitlements

The Melamed/Calabresi matrix describes how courts resolve disputes over entitlements without resorting to the simple logic of 'might makes right.' Every modern society expects its courts to "decide which of the conflicting parties will prevail. The entitlement to make noise versus the entitlement to have silence. The entitlement to pollute versus the entitlement to forbid them – these are the first order of legal decisions."[97] In

[97] Calabresi and Melamed (1972, 1095).

many ways, the bare bones concept of an entitlement is the least controversial part of the MCEM. Melamed and Calabresi devote much more energy on the categorization of entitlements and the overarching legal sensibilities (or rules) that protect them than the underlying assumptions behind the entitlement concept, at least in part because the term entitlement appears to be nothing more than a right created by existing legal and political institutions. But incorporating Melamed and Calabresi's understanding of entitlements into a capability-based model is more complex than it appears for three reasons.

First, the MCEM is part of the founding canon of law and economics.[98] Even though the bare-bones concept of 'entitlement as a right' under MCEM is roughly comparable to the Hohfeldian concept of rights as a correlative concepts,[99] entitlements are conceptually entangled with wealth and utility. Certain kinds of entitlements, for example, cannot be taken away from someone even if someone else is prepared to pay an enormous sum of money for the entitlement. Such an entitlement is immune from utility calculations.[100] Capability claims also partially overlap with wealth maximizing practices, not because someone with wealth can never do anything to alleviate the suffering of victims with significant capability deficits but because people with capability deficits may be entitled to things that can't be reconciled with wealth maximizing norms, efficiency or increases in general welfare. By extension, capability-deficient people may be 'entitled' to things that modern societies have not yet recognized or codified as formal rights. In fact, capability-based claims may be conceptualized as requests, expectations or even naked assertions of sovereignty rather than formal rights/entitlements as they are defined by the MCEM.

Blindly incorporating the MCEM's concept of entitlements into a capability-based model is also problematic because Sen wrote extensively about capability-based entitlements that cannot be reduced to wealth maximization or primary goods, and yet may still have a place in modern legal systems. In Development and Freedom, Sen discussed two kinds of entitlements. The first consists of entitlements derived from his list of

[98] Krauss (1999, 1).

[99] Hohfeld (1913) (describes reciprocal rights and entitlements).

[100] Krauss (1999, 2).

instrumental freedoms, economic facilities and political freedoms. Political freedoms and civil rights, for example, refer to opportunities people have to "determine who should govern," and "include the political entitlements associated with democracies."[101] Economic facilities, on the other hand, "refer to the opportunities that individuals respectively enjoy to utilize resources for the purposes of consumption, or productivity, or exchange. The economic entitlement that a person has will depend on the resources owned or available for use as well as o conditions of exchange."[102] This narrow, traditional definition of entitlements largely conforms to the definition of entitlements in the MCEM although it is perfectly possible to imagine entitlements in Sen's model that are not, as yet, formally recognized by law. It is also perfectly possible to someone to possess generally recognized 'entitlements' and still lack the capability to feed themselves.[103] This leads to a second, broader understanding of entitlement, one that rejects calculating a countries general welfare in terms of the "total food supply."[104] Entitlements also refer to things "over which.. [one] can establish ownership and command."[105] For food security, the object to be controlled is obviously the production of food, but Sen leaves room for conceptualizing entitlements as anything "over which one has established ownership or command." Calculating an entitlement, in turn, depends on one's endowment, production possibilities and exchange conduct.[106] In the context of food, "entitlement failures" can lead to famines.[107]

Sen's narrow (or freedom-based) and broad (or command-based) definitions of entitlement are relatively consistent, particularly in the context of food security. As one commentator observed, "a person who is force-fed must be said to not have 'established command' over food, and therefore not to possess a corresponding food entitlement."[108] But even beyond food security, we can imagine someone having an 'entitlement'

[101] Sen (1999a, 38).

[102] Id. at 38–39.

[103] Id. at 66.

[104] Id. at 162.

[105] Id. at 162.

[106] Id. at 162–63.

[107] Id. at 164.

[108] Reddy and Daoud (2018, 6).

to do something when they have sufficient freedom and command over certain aspects of their immediate environment, even if that entitlement has not yet been enshrined as a formal legal right or secured by wealth. Consequently, a capability-based model is not bound by the MCEM's definition of entitlement as formally recognized rights, entitlements immune to state interference because they reflect a hypothetical, voluntary bargain, entitlements constrained by private law expectations, or even entitlements made by nation-states based on international law and sovereign reciprocity rules.[109]

3.2 *Property Rules, Liability Rules and Other Rules*

In addition to challenging the very definition of entitlements under the MCEM, capability theorists are also likely to challenge the decisional legal rules that protect entitlements. The MCEM describes three types of entitlements, the first of which is an entitlement protected by the 'property rule.' The property rule leaves undisturbed a hypothertical bargain struck between two parties over the value of an entitlement. In theory, entitlements protected by the property rule require no further state intervention, and they enjoy a presumption of legitimacy if the bargain was voluntary. Entitlements protected by the 'liability rule' are entitlements in which "someone may destroy the initial entitlement if he is willing to pay an objectively determined value for it."[110] State intervention is required but the value of the entitlement may be determined by a third party.[111] Perhaps most importantly, entitlements protected by the liability rule are contested, rather than presumptively legitimate. The third form of entitlement is an inalienable entitlement that cannot be sold or purchased.

Standing alone, the property rule is useful because it provides a decisional pivot point to specific remedies, and a policy shortcut for the presumptive legitimacy of an entitlement claimed by either plaintiffs or defendants, but it also implicitly links the resolution of entitlement disputes to wealth maximization, one of the principal targets of capability theory. The possession of wealth sufficient to strike a reasonable bargain, for example, is necessary in order to receive protection from

[109] D'Amato (1984, 1301).

[110] Calabresi and Melamed (1972, 1092).

[111] Id.

the property rule. The voluntariness of that hypothetical bargain is also questionable if one or both parties lack fundamental capabilities, even if they possess wealth, formal rights or primary goods. As Melamed and Calabresi observed, "distributions of wealth may effect a society's choice of entitlement."[112] People at every point along the capability-deficiency spectrum often lack the freedom to strike a fair bargain, and yet that lack of capability should not exclude them from property rule protection. Capability theorists, therefore, are likely to look for some other metric besides voluntariness or the background assumption of sufficient wealth to strike a reasonable bargain as a marker of the presumptive legitimacy of entitlements protected by the property rule, and largely immune from state intervention.

Unlike entitlements protected by the property rule, entitlements protected by the liability rule apportion responsibility between the parties when either or both parties cannot strike a bargain, at least in part because one of both parties might be at fault. Liability rule protections are appropriate when there are high transaction costs (which could include insurmountable capability-deficiencies), and once again the assumption is that parties possess enough wealth to pay damages depending on the circumstances. But VRN requires the consideration of two other factors before apportioning responsibility in retaliatory self-defense and self-help cases: (1) forcing the aggressor to assume the total cost of all damages caused by the victim's retaliation if the aggressor deliberately tried to exploit the capability-deficiencies of the victim while engaging in a particularly 'dangerous' attack, and (2) forcing the victim to split the cost of any damages caused by the victim's retaliatory acts if the victim deliberately failed to exercise due care, within his/her capabilities, while executing a retaliatory response to the aggressor's attack.

3.3 *Efficiency, Distributional Concerns and 'Other Justice Reasons'*

Under the MCEM, the resolution of entitlement disputes serves three purposes. The first is administrative and/or economic efficiency. Entitlements protected by the property rule will produce Pareto-optimal results if transaction costs are relatively low and the defendant assumes costs as the cheapest cost avoider. Entitlements are also determined by the

[112] Id. at 1096.

distributional goals of a society, and the MCEM once again assumes that wealth distribution will be one of the highest priorities.[113] MCEM also recognizes that the choice of entitlements based on wealth distribution may coincide with the minimum endowment of 'merit goods' such as education, clothes or bodily integrity, an implied overlap between wealth maximization (MCEM incentivizes people to accumulate wealth) and Rawls' primary goods' model.

The most promising opportunity for capability theorists, however, is MCEM's discussion of "other justice reasons." Melamed and Calabresi' focused almost exclusively on entitlement results based on wealth maximization/efficiency and traditional distributional policies designed to promote economic, political and social equality, even though they conceded that "explaining entitlements solely in terms of efficiency and distribution in even their broadest form, does not seem wholly satisfactory."[114] 'Other justice' models were characterized as idiosyncratic, so lacking in definitional precision and detail that "the bipolar efficiency locution is all that is needed."[115] Commentators have speculated about Calabresi's true intentions when he referred to 'other justice reasons,' including, most recently, the possibility that Calabresi was referring to transformative justice or some kind of 'community policing' model that transcends efficiency or distribution.[116] The former focuses primarily on communal accountability and forgiveness.[117] Even though scholars differ on the question of whether capabilities theory is a full-fledged theory of justice,[118] capability theory is a legitimate 'other justice' candidate that explains constraints that people face in their actual lives, constraints that cannot be completely explained—subject to overlap problems—by the possession of primary goods/rights or wealth. Capability-based defenses of retaliatory self-defense and self-help are a perfect test case for Calabresi's 'other justice' idea because there is evidence that people retaliate after crises (as opposed to turning to law enforcement, private suits or hardening) even if they possess rights, primary goods and wealth, and even

[113] Id. at 1099.

[114] Id. at 1104.

[115] Id. at 1105.

[116] Davidson (2021, 1627).

[117] Id. at 1630.

[118] Reddy and Daoud (2018, 1).

more evidence that retaliatory self-defense and self-help cases actually fall into four categories, similar to Rules 1–4 of the MCEM.

3.4 Rules and Zones

3.4.1 Rule 1/Zone 1

Rule 1 of the MCEM extends property rule protection to the entitlements of the plaintiff, resulting in the plaintiff receiving an injunction against the defendant's pollution or trespass. Applying Rule 1 to retaliatory self-defense or self-help requires two additional steps. First, there is an additional action beyond the defendant's trespass or violation: the plaintiff's retaliatory Second, retaliation self-defense or self-help is a remedy available to the plaintiff regardless of whether the plaintiff is entitled to an injunction or damages. Standing alone, self-defense premised on the fear of imminent death or destruction is considered an intrinsic moral right, regardless of whether it is invoked by individuals or the state. Jurisdictions may differ on the question of whether self-defense should be defined as an excuse or a justification, but the fundamental right to assert self-defense is universally recognized. Retaliatory self-help, on the other hand, retains a stronger link to formal law, making it more like a formal injunction or court-order damages. Retaliatory self-defense and self-help are also, in the context of cyber discourse, considered to be an alternative legal 'remedy' to criminal law enforcement, private suits and 'hardening'.

In order to account for all of these concerns, Rule 1 under VRN contains two results: 'impermissible attack' and 'permissible retaliation.' The plaintiff's right to an injunction translates into the idea that the defendant's initial attack is impermissible. Under MCEM, property rule protections of the plaintiff's entitlements are efficient (i.e. Pareto-optimal) if the hypothetical bargain struck over the entitlement is voluntary, there are few transaction costs, and the defendant is the 'cheapest cost avoider.' Under VRN retaliatory attacks are permissible if they are conducted with sufficient attribution, proportionally and out of necessity. There must also be evidence that the plaintiff faced considerable constraints in the actual lives after the attack that prevented them from taking advantage of law enforcement, private suits or hardening, all of which require a significant endowment of primary goods, rights or wealth. The defendant's initial attack is impermissible because the defendant intentionally chose a method of attack or violation that exploited the capability deficiencies of the plaintiff, and the plaintiff did not provoke the attack. The decision to define certain attacks as impermissible does not impose a positive duty on victims to block access, or manifest an intent to exclude, because such a requirement reinforces the 'blame the victim' ethos so central to open world zealotry, and under-estimates the complexities of capability deficiency. The 'presumptive rule' essentially protects the entitlements of good faith plaintiffs with capability deficiencies in 'zone 1' situations.

Zone 1 of VRN protects the entitlements of capability-deficient individuals, groups or nation-states that hack-back in good faith (ie. without provocation), and execute hackbacks with due diligence (i.e., 'reasonable' necessity, attribution and proportionality). That protection can be conceptualized in two ways. The weak version of 'zone 1' protection only requires a showing that the victim was capability-deficient, whereas the strong version of 'zone 1' protection requires an additional showing that the perpetrator was aware of the victim's capability-deficiency and deliberately engineered a cyber attack capable of exploiting the victim's capability-deficiency. 'Weak zone 1' protection is easier to achieve than 'strong zone 1' protection, but the moral argument for VRN protection is undoubtably more compelling in 'strong zone 1' cases.

3.4.2 Rule 3/Zone 3

Under the MCEM, Rule 3 extends property rule protection to the entitlements of defendants, resulting in a determination that the defendant's trespass/pollution should be allowed, and the plaintiff's injunctions should be denied. VRN's presumptive rule also protects the entitlements of defendants in zone 3 situations if there is evidence that the attack/violation was either not designed to exploit the victim's known capability deficiencies, or unjustifiably provoked by a victim with sufficient capability. If either of these conditions are satisfied, there should be a determination that any subsequent retaliation by the plaintiff was impermissible, and that the initial attack/violation was permissible.

3.4.3 Rule 2/Zone 2

Under the MCEM, Rule 2 extends liability protection to the entitlements of plaintiffs, resulting in the defendant's trespass/pollution being allowed, but damages must be paid to the plaintiff. Instead of the liability rule, zone 2 of VRN uses the desert rule to protect the entitlements of the plaintiff. The permissibility of the defendant's attack/violation is questionable or deeply contested, not because the defendant did not intentionally engineer an attack specifically designed to exploit or exacerbate the plaintiff's capability deficits, but because the defendant's justification for doing so could have been in the public interest, or with limited official authorization. But the manner in which the attack took place—high probability of widespread damage, repeat offenders, persistent rule of law violation, attacks on critical infrastructure—can be characterized as particularly dangerous, demanding an enhanced penalty (similar to punitive damages). Retaliation by capability-deficient entities is permissible if the plaintiff believes that law enforcement, private suits and hardening will not deter a particularly dangerous attacker from repeated the same attack in the future, and the plaintiff exercised due care during the retaliation. Defendants must absorb the total cost of any damages caused by the plaintiff's retaliation.

3.4.4 Rule 4/Zone 4

Under the MCEM, Rule 4 extends liability rule protection to defendants, resulting in the plaintiff receiving an injunction against the defendant's trespass or pollution, but the plaintiff must pay damages to compensate the defendant. Zone 4 of VRN, operating under the desert rule rather than the liability rule, also protects the entitlements of the defendant. The defendant's attack is characterized as impermissible, but the plaintiff's retaliation is also characterized as impermissible because it was undertaken without sufficient attribution or proportionality, or it was unnecessary, considering the plaintiff's capability deficiencies. As a result, the plaintiff and defendant must split the cost of any damages caused by the plaintiff's retaliatory action.

4 VRN: Synthesis

VRN reflects a skeptical view of open world discourse in which the transfer of goods, services, idea and people across borders is encouraged, but carefully monitored to prevent criminality and competitive abuse. Central to the monitoring process is the belief, held by many, that individuals, organizations and nation-states that have been attacked in open world zones should not be limited to litigation, hardening or retreat. Placing such a limit on the victims of open world abuse obscures the potentially beneficial effects of retaliatory self-defense/help on open world discourse and deprives victims of important protections demanded by global justice and often provided by international institutions.

VRN uses capability theory to establish the presumptive moral legitimacy of retaliatory self-defense/self-help claims brought by capability-deficient victims, and apportions responsibility for any damages caused by retaliation between the victim/retaliator and the aggressor. By design, VRN is broad and transnational in scope: it applies to retaliatory criminal self-defense claims and civil self-help claims; claims that could be brought in American and foreign courts; and nation-to-nation conflicts in which effective cross-jurisdictional law enforcement of either injunctions or damages is unlikely.

As a 'stand-alone' theory, VRN is always available, even if either injunctive relief (in a civil proceeding for self-help), or a conviction (in a criminal proceeding for self-defense) are likely in the long term. The strategic 'trigger' for presumptively legitimate retaliatory claims under VRN is not whether an injunction or conviction are likely; it is whether formal legal processes or effective hardening are attainable ends given the victim's capabilities. In essence, VRN rewards responsible victims with significant capability deficits, and punishes irresponsible aggressors with significant capability advantages.

References

Amadae, S. 2004. Amartya Sen's Rationality and Freedom. *Economics and Philosophy*. 20 (2), 381–389.

Becker, G. 1996. *Accounting for Tastes*. Cambridge: Harvard University Press.

Berlin, I. 1961. *Two Concepts Of Liberty*. New York: Clarendon Press.

Bleichrodt, H., and Wakker, P. 2015. Regret Theory : A Bold Alternative to the Alternatives. *The Economic Journal*. 125(3), 493–532.

Calabresi, G., and A. Melamed. 1972. Property Rules, Liability Rules and Inalienability: One View of the Cathedral. *Harvard Law Review* 85: 1089–1097.

Cooter, R. 2000. *The Strategic Constitution*, 129–134. Princeton: Princeton University Press.

Connolly, T., and Zeelenberg, M. 2002. Regret in Decision Making. *Current Direction in Psychological Science*. 11(6), 212–216.

Connolly, T., and Butler, D.J. 2006. Regret in Economic and Psychological Theories of Choice. *Journal of Behavioral Decision-making*. 19(2), 139–154.

D'Amato, A. 1984. Is International Law Really Law? *Northwestern University Law Review* 79: 1293–1314. Faculty Working Paper 103.

D'Amato, C. 2014. Rawls and Capabilities. *Virginia Tech. Papers*. vtechworks.edu.lib.vt.edu.

Davidson, A. 2021. Guido Calabresi's 'Other Justice Reasons.' *University of Chicago Law Review* 88: 1625–1631.

Foster, J. 2010. Freedom, Opportunity and Well-Being. *OPHI Working Paper No. 175.* Oxford: Policy and Human Development Initiatives, April.

Freeman, S. 2006. Frontiers of Justice: The Capabilities Approach vs Contractarianism. *Texas Law Review* 85 (2): 385–430.

Hohfeld, W. 1913. Some Fundamental Legal Conceptions as Applied in Judicial Reasoning. *Yale Law Journal* 23: 16–38.

Humphrey, S. 1995. Regret Aversion or Event-Splitting Effects? More Evidence Under Risk and Uncertainty. *Journal of Risk and Uncertainty.* 11: 263–274.

IEPCapabilities. 2022. *Global Ethics: Capabilities Approach. Internet Encyclopedia of Philosophy.* ISSN 2161-0002. Last visited, 11/22/2022. https://iep.utm.edu/ge-capab/.

Irons, B. and Hepburn, C. 2007. Regret Theory and the tyranny of Choice. *The Economic Record. 83* (261): 191–203.

Jaggar, A. 2006. Reasoning About Well-Being: Nussbaum's Methods of Justifying the Capabilities. *The Journal of Political Philosophy* 14 (3): 301–322.

Kanbur, R. 2016. Capability, Opportunity, Outcome … And Equality. *Working Paper.* Charles H. Dyson School of Applied Economic and Management, Cornell University. WP 2016—05. March 2016.

Kaplow, L. 2007. Louis Kaplow, Primary Goods, Capabilities … or Well-Being. *Philosophical Review* 116 (4): 603–632.

Krauss, M. 1999. Property Rules vs. Liability Rules, Paper 3800 (GMU Law School), JEL Series. *International Encyclopedia of Law and Economics.*

Kuran, T. 2013. Moral Overload and Its Alleviation. In *Economics, Values and Organization*, ed. Avner Ben-Her and Louis Putterman. New York: Cambridge University Press.

Nussbaum, M. 2000. *Women and Human Development: The Capabilities Approach.* Cambridge: Cambridge University Press.

Nussbaum, M. 2009. *Frontiers of Justice: Disability, Nationality and Species Membership.* Cambridge: Harvard University Press, Belknap Press.

Parfit, D. 2000. Equality or Priority. In *The Ideal of Equality*, ed. M. Clayton and A. William Basingstoke. Basingstoke: Macmillan Press.

Pogge, T. 2002. Can Capability Theory Be Justified? *Philosophical Topics* 30 (2): 167–228.

Qizilbash, M. 2005. Incommensurability and the First Fundamental Welfare Theorem. *Oxford Economic Papers.* 57, 664–673.

Rawls, J. 1999a. *The Law of Peoples.* Cambridge: Harvard University Press.

Rawls, J. 1999b. Reply to Alexander and Musgrave. In *John Rawls, Collected Papers*, ed. Samuel Freeman. Cambridge: Harvard University Press.

Rawls, J. 1999c. A Kantian Conception of Equality. In *John Rawls, Collected Papers*, ed. Samuel Freeman. Cambridge: Harvard University Press.

Reddy, A., and A. Daoud. 2018. Entitlements and Capabilities. *SSRN Electronic Journal*. https://doi.org/10.21391.ssrn.320924.
Roemer, J. 1996. *Theories of Distributive Justice*. Cambridge: Harvard University Press.
Sen, A. 1970. The Impossibility of a Paretian Liberal. *Journal of Political Economy*. 78 : 152–57.
Sen, A. 1987. *The Standard of Living*. Cambridge: Cambridge University Press.
Sen, A. 1988. The Concept of Development. In *Handbook of Development Economics*, vol. 1, ed. H. Chenery and T.N. Srinivasan. Amsterdam: Elsevier.
Sen, A. 1999a. *Development and Freedom*. Oxford: Oxford University Press.
Sen, A. 1999b. Equality of What? In *Equal Freedom*, ed. Stephen Darwall. Ann Arbor: University of Michigan Press.
Sen, A. 2002. *Rationality and Freedom*. Cambridge: Harvard University Press.
Sen, A. 2006. *Inequality Re-Examined*. New York: Oxford University Press.
Starmer, C. and Sugden, R. 1993. Testing for Juxtaposition and Event-Splitting effects. *Journal of Risk and Uncertainty*. 6 : 235–254.
Stein, M. 2009. Nussbaum: A Utilitarian Critique. *Boston College Law Review* 50 (2): 489–531.
Tadenuma, K., and Xu, Y. 2012. Extensions of the Fundamental Welfare Theorems in a Non-Welfaristic Framework. *Hitotsubashi Journal of Economics*. 53(1) : 107–120.
Von Staveren, I. 2007. The Ethics of Efficiency. *SCEME Working Paper No. 18*. http://www.sceme.org.uk
Von Staveren, I. 2009. Efficiency, in *The Handbook of Economics and Ethics* (ed. by Jan Pell and Irene Van Staveren) (Edward Elgar Publishing).
Williams, B. 1987. The Standard of Living: Interests and Capabilities. In *The Standard of Living*, ed. Geoffrey Hawthorn. Cambridge: Cambridge University Press.

CHAPTER 5

Aggravating Factors, Freedom and Retaliation

Properly conceived, self-defense and self-help claims can preserve the autonomy of attacked individuals during actual crises, as well as the autonomy of aggressors, both *ex ante* (preventing disproportionate retaliation by victims during attacks) and *ex post* (via legal determinations of culpability and appropriate punishment after an attack). In Chapter 3, I argued that VRN conceptualizes the clash between victims and aggressors during crises that lead to retaliation as competing entitlements and competing conceptions of meaningful freedom in the global world. The intellectual, historical and philosophical foundation for VRN combines pragmatic open world skepticism, with capabilities theory. All of these considerations are part of the 'presumption prong' of VRN because they establish the elements necessary make a presumptively legitimate retaliatory self-defense or self-help claim. Such a claim is roughly comparable, but not identical, to an entitlement claim made by an aggrieved plaintiff that is strong enough to warrant injunctive relief under 'rule 1' of the Melamed/Calabresi entitlement matrix. This chapter examines the 'desert prong' of VRN, or factors that would either require victims to pay a portion of the costs for retaliatory action against an attacker, or require an attacker to absorb the total cost of retaliatory action undertaken by the victim.

A. J. Thomas, *The Open World, Hackbacks and Global Justice*,
https://doi.org/10.1007/978-981-19-8132-6_5

1 Constraints on Self-Help/Defense

1.1 Necessity

1.1.1 Modernity and Necessity

For most theorists, the central sociological and institutional markers of modernity—rationality, the movement from traditional or kinship ties to more contract-oriented affiliations, technological advances and the spread of democracy authorized by informed consent—fetishize control. Modernist control from the bottom takes the form of political authority anchored by informed consent, and from the top, it takes the form of political, institutional and bureaucratic structures informed by procedural rationality, neutral decision-making procedures, transparency and the careful manipulation and regulation of non-conforming persons, uses and norms. In the process of taking control over the course of individual lives, the environment, political organization, technological progress and the inevitable growth of bureaucracy, the modern mind has expanded its understanding of both man-made and natural risk.

Necessity is intimately connected to the control of risks. Institutionally, legal necessity frames the imperative to control potential risks initially as an individuals' overarching right to protect his/her person, and extends the right outwards to property, loved ones, associational autonomy and governmental stability. When combined with public authority, the control ethos takes the form of an overlapping consensus or the common good, all thought to stem either from majority preferences channeled through elected representatives (in the case of legislation) or dispassionate, learned judges obligated to protect the public interest. In this form, necessity becomes a collective political judgment rather than a personal decision, a collective judgment usually based on procedural rationality.

Legal necessity is packaged as an exceptional extension of the power of an otherwise well-functioning, modern bureaucratic state, an extension driven by extenuating, extraordinary circumstances and actors who may be profoundly illiberal or pre-modern in their political sensibilities. When democratic states invoke necessity, in other words, they confront the limits of modern political consciousness, with its emphasis on universal norms and the moral legitimacy of ordinary democratic process. Necessity, therefore, is a powerful ideological and symbolic tool, capable of not only enhancing the power of any public or private entity that wields it successfully, but also constructing an image of the modern world. If the central premise of modernity is control, than the image it yields is a necessity claimant fully capable of assessing the threat that confronts him in a crisis or emergency situation.

Necessity operates from the 'bottom-up' during crises when minority factions obtain enough power to force majorities to abandon traditional habits of political obedience. It also operates from the top-down, as part of a public justification strategy. Governments frequently invoke necessity as an extension of their police power authority to protect the public health, safety and welfare, authority ostensibly granted to them—at least originally—by the people. Necessity, therefore, is just one of many means by which modern civil society mediates majoritarian impulses. Majoritarian bodies perform the same function, mediating bottom-up democratic forces as channeled through popular initiatives or the legislature, and top-down institutional forces through judicial review.

Necessity also resolves conflicts between worldviews. The significance of these conflicts operates on two levels, the first of which is descriptive, mechanical or granular. The initial challenge is to identify the exact manner in which the conflict occurs, and a comparable framework for any possible resolution. 'Weak' resolutions between different necessitarian worldviews, for example, merely consist of a simple overlapping consensus on the descriptive or analytical relevance of specific key terms like the historical definition of a crisis, the psychological markers of heightened stress in emergency situations or even the presumptive legitimacy of rule of law norms in all but the most extreme circumstances. Such a weak consensus model may explain the appeal of legal necessity cases and crisis-oriented legal regimes couched in the language of political and/or moral minimalism,[1] which resolves many incommensurability problems

[1] Minimalism is also a response to the counter-majoritarian character of judicial review, limiting the power of courts to undermine legislative will. Minimalist courts practice what can only be described as 'strategic avoidance.' By striking down vague laws, using the non-delegation doctrine to limit Congressional authority, insisting on clear legislative statements, striking down old laws that lack popular support, basing decisions on actual facts rather than hypotheticals, and insisting on public justifications, minimalist courts manage to issue narrow rulings limited to the facts before them, and avoid controversial speculations about future problems. Minimalism also facilitates democratic dialogue and legislative decision-making. 'Void for vagueness' rulings, for example, force legislatures to deliberate before coming up with more precise statutory language. And narrow substantive rulings that do not speculate about future possibilities or hypothetical situations also leave room for democratic dialogue.

and cultural disputes generally attributable to the 'fact of pluralism' through procedural norms and restrained administrative decision-making.

An intermediate model might consist of a shared recognition that complex crisis-related regimes either cannot (due to limitations within all theories) or should not (for political reasons) be resolved without using insights and strategies from more than one major necessitarian narratives. Such a perspective informs, either implicitly or explicitly, many complex crisis-related legal regimes ad convergence proposals. In fact, it is not uncommon for policy-makers with entrenched opposed macro-theoretical sensibilities to embrace the same regulatory scheme because it contains a cluster of ideas that all stakeholders find attractive, either for purely strategic reasons or for deeply normative reasons. Typically, the 'hard cases' lie within this cluster (or convergence point). The central challenge is whether policy-makers can balance (or accommodate) seemingly divergent macro-theoretical perspectives within one framework without lapsing into a defense of the status quo.

A third, or strong, point of conflict between different necessitarian worldviews consists of a fundamental disagreement on core moral tenets, assumptions and priorities in legal necessity cases, a disagreement that will inevitably lead to different policy proposals. The strong model, therefore, generates profound and largely irresolvable conflicts at the level of both theory and practice.

What are the philosophical and distributive implications of weak, intermediate and strong necessitarian models? The principal implication is that the task of finding an overlapping substantive consensus, or functional convergence, between different worldviews is relatively easy if we adopt the weak model, but the task grows more difficult as we move from the intermediate to the strong model. The burden of justification for the end product of any comprehensive framework that attempts to reconcile different necessitarian worldviews also increases, a burden that draws its strength from ongoing pluralist critiques of moral universalism and philosophical foundationalism. These critiques question whether the broad sweeping moral principles that ground a particular necessitarian worldview (such as liberalism or utilitarianism) are more granular than they appear to be at first glance, distinguishable not only in terms of thick (perfectionism) versus thin (proceduralist) values, but also in terms of functional versus purely aspirational values. They also question whether it is not only possible to distinguish the universalistic ambitions of broad moral frameworks from the merely contingent, political preferences of a particular community, but also whether it is still possible to justify universalistic distributive preferences about the proper way to resolve crises in an increasingly globalized world.

1.1.2 The Impact of Reasonableness

All emergencies involve at least three categories of persons: (1) those who cause, react to or are affected by (stakeholders) the emergency; (2) those in a position to decide what the emergency is, how long it persists and how it ends; and (3) those in a position to resolve the crisis. Regardless of how a legal dispute develops, decision-makers will not only have to decide which category a particular necessity claimant belongs to (and obviously they can belong to more than one), but also decide if the claimant's action during a crisis, actions driven by the claimants' motivations, cognitive limitations, pre-dispositions, perceptions of possible harm, cultural background, respect for the presumptive validity of rule of law norms and moral preferences, should in any way be relevant to the broader question of whether the claimant's action should be either excused or punished (if a legal violation did in fact occur). Each decision implicates the concept of reason, defined in this case not in terms of rationality per se (using the narrow rational choice definition), but rather as a combination of commonsensical explanation and partial justification. According to this broad definition of necessitarian reason, the decision-maker considers whether a claimant can give reasons for his actions, whether a similarly situated (or even completely disinterested) person could give reasons for the claimant's actions, whether the decision-maker can provide good reasons for either excusing or punishing the claimant and whether the capacity (or existence) of reasons for the claimant's actions are even relevant in the determination of guilt, the assignment of responsibility, the resolution of the crisis or the reward of damages if appropriate.

At every stage of a legal necessity controversy, reason, broadly defined, exerts an influence on the direction of events. At the beginning of the controversy, for example, almost everyone, including people who reacted instinctively or reflexively to the perception of an imminent emergency, or to the perception that an ongoing emergency was going to get worse, will attempt to show that his/her reaction to a perceived crisis was 'reasonable' under the circumstances. At the end of the controversy, when legal or political responsibilities are being assigned, the decision-maker will give reasons for their decisions, reasons that will typically reflect either majoritarian preferences or well-established socio-political or procedural norms.

Reason, in other words, will do everything from explaining the claimant's behavior to justifying the courts' disposition, imperceptibly re-aligning minoritarian perceptions with majoritarian expectations.

Even though there is general agreement that reason, broadly defined, plays an important part in the framing and disposition of legal necessity claims, theorists differ on two key aspects of this question: (1) the ideal redistributive medium (or criteria) for the disposition of necessity claims and (2) the identity of an ideal decision-maker in crisis cases. It is not enough, in other words, for claimants to give reasons for their actions or institutional entities to give reason for their decisions. These reasons must reinforce a desirable moral ideal, advance a particular distributive objective or either directly or indirectly enhance the authority of a collective decision-making body presumably well equipped to resolve crises.

The need to identify the ideal redistributive medium and decision-making body dominates necessitarian thought, producing two dominant intellectual traditions. The first tradition designates rights as the ideal redistributive medium (or conceptual arbiter). An imbalance of rights tied to equality and liberty produces crises according to this view. The second tradition designates the maximization the overall social welfare as the ideal redistributive medium for crisis resolution.

Necessity under VRN reflects a capability-based worldview. It assigns priority to the reasonableness of the necessity claimant's perceptions of an imminent or ongoing crisis, and the potential alternatives available to him either before or during a crisis. It also assigns partial responsibility to the claimant if his perceptions are patently unreasonable or entirely implausible. In this sense, VRN rewards reasonable good faith responses of crises, and punishes bad faith responses to crises. Most importantly, however, VRN recognizes that too many victims of attacks in open world zones must choose between ineffective rights-based solutions to crises, and unattainable utility-based solutions to crises. The pressing burden of 'need satisfaction' that many capability-deficient victims face every day is a crisis that must be resolved before they can take advantage of rights or income.

1.2 *Proportionality*

1.2.1 Common Law Antecedents

The concept of proportionality suggests that there is a strategic, institutional, cultural or even intuitive link between the severity of an offense and the severity of punishment.[2] Complex philosophical discussions of proportionality, which could range from various attempts to perfect punishment, and reconcile the severity of punishment with core liberal principles, or justify the proper ratio of harm to punishment, only become truly relevant when they form the basis for public justification of government sentencing policies. In the modern democratic world, therefore, proportionality principles are both inward-looking and outward-looking, reconciling notions of retribution, just dessert and reasonable suffering with the demands of public reason, institutional accountability and bureaucratic rationality.

Proportionality principles assume that we can identify, through a complex political, institutional and moral calculus, the best ratio of harm to punishment, as well as rank offenses from least harmful to most harmful. In most cases, the latter is relatively trivial. No one doubts, for example, that first-degree murder is more harmful and dangerous to the collective consciousness than breaking and entering or jay-walking, but commonsensical calibrations of harmful offense can also generate unbelievably nonsensical results that seem to ignore any sense of proportionality.

Decision-makers must balance intuitive notions of severity, moral outrage and calibrations of harm to punishment against the facts of a specific case that could push the boundaries of just dessert and appropriate punishment. And this balancing act requires a careful consideration of how common notions of proportionality are filtered through the justice system (process) and how proportionality colors popular perceptions of guilt and sentencing. Decision-makers always consider three factors as they search for a proportional resolution of a crime, and these factors appear repeatedly in the common law, constitutional discourse and international legal norms. The first is the identity of the accused. On the surface, the concept of proportionality may appear to be 'agent-neutral.' For committed civil libertarians who believe that the death penalty is intrinsically immoraland should always be considered cruel and unusual,

[2] Proportionality almost becomes irrelevant if the entity responsible for imposing punishment does not have to account for the severity of punishment imposed. In a totalitarian society, for example, the government does not have to explain or justify its actions or punishments to the public, or justify the imposition of punishment for reasons unrelated to any consideration of the difference between the severity of offense and the severity of punishment.

the identity of someone accused of committing a brutal, horrific crime on death row is largely irrelevant. But in practice, the identity of persons likely to receive extremely severe punishments is very important.[3] Very few people are surprised, for example, when serial rapists are put to death, or even extremely dangerous murderers who repeatedly profess their intention to kill another inmate or a corrective officer are sentenced to life in prison.

Another consideration for decision-makers is the broad problem of jurisdictional asymmetry. Jurisdictional asymmetry consists of identical crimes committed in different geographical areas that lead to dramatically different sentences. The war on drugs provided many examples of proportionality analysis weighed down by strong jurisdictional asymmetry problems. In a time when two states have already legalized the possession and use of marijuana, for example, it is easy to forget that many states impose stiff fines for the possession and/or sale of harder drugs like cocaine, crack or heroin. The constitutionality of such a harsh sentence was the central issue in Harmelin v Michigan,[4] a case challenging the sentence of mandatory life in prison without the possibility of parole following a conviction of possessing 672 grams (or 1.5 pounds) of cocaine.

In upholding the conviction, Justice Scalia, writing for the majority, made two important points relevant to definition of proportionality. The first was that even though there are "relatively clear historical guidelines and accepted practices that enable judges to determine which modes of punishment are 'cruel and unusual'",[5] the Eighth Amendment itself does not contain a proportionality guarantee. The second point spoke directly to strong jurisdictional asymmetry. Scalia left open the possibility that there could be enormous variations between different jurisdictions. Differences would be constrained by "extreme examples that no rational

[3] Even though decision-makers have frequently used inappropriate markers of dangerousness like race or economic class during sentencing, the idea that the identity of someone should be relevant in defining proportionality is relatively well established, even though we recognize that the likelihood of inappropriately imposing extremely harsh sentences to individuals deemed to be member of a dangerous class is considerable. See Bentham (1843, 399–402) (Decision-makers should always be concerned about over-punishment because "anti-pathy, or a want of compassion for individuals who are represented as dangerous and vile, pushes them onward to an undue severity.").

[4] Harmelin (1991).

[5] Id.

person, in no time or place, could accept,"[6] but that wouldn't prevent jurisdictions from adopting extremely severe penalties for a category of 'grave offenses' that they themselves designate. The content of that category will depend entirely on "how odious and socially threatening"[7] decision-makers consider certain offenses to be, using whatever indicia of social harm or dangerousness that they choose. Except for extreme cases, proportionality does not prohibit radically different sentences in different jurisdictions. Nor does it prevent jurisdictions from considering or ignoring sentencing decisions in other jurisdictions.

In his concurrence, Justice Kennedy found a narrow proportionality principle in the Eighth Amendment, but concluded that it was not violated by the Michigan legislature when it imposed a mandatory life sentence for the possession of 1.5 pounds of cocaine because it was reasonable for the Michigan legislature to conclude that the possession of such a "large … amount of cocaine in terms of violence, crime and social displacement"[8] is dangerous enough to warrant a life sentence. But Kennedy, like Scalia, recognized that the jurisdiction could reasonably adopt radically different sentencing guidelines based on different penological theories, whether the jurisdiction is federal or state, and potentially the political climate in the legislature, which courts should largely defer to on sentencing questions as long as the punishment is not, as Scalia suggested, patently extreme or cruel.

Scalia's 'no-proportionality' and Kennedy's 'narrow proportionality' reading of the Eighth Amendment not only leave plenty of room for latitude between jurisdictions, but also suggest that proportionality is more robust vertically than horizontally. It provides a framework for organizing our intuitions about the proper ratio of punishment to harm in specific cases, but it provides relatively little comparative guidance across substantive areas or jurisdictions.

1.2.2 Limiting and Enabling Proportionality

The debate between Scalia and Kennedy in Harmelin foreshadowed a broader distinction between proportionality as a limiting principle and proportionality as an enabling principle. Limiting proportionality is based

[6] Id. at 963.

[7] Id.

[8] Id.

on the recognition that political communities may wish to inflict punishments far in excess of the original offense, either because of a cultural propensity to over-punish, popular pressure or bureaucratic over-reach. Enabling proportionality, on the other hand, provides governments with an excuse to impose penalties for offenses that the body politic may be reluctant to consider without additional justification. The underlying assumption, not surprisingly, is that a society is either relatively benevolent and forgiving or deeply concerned about the long-term costs of carefully calibrating the harm caused by an offense to a specific form of punishment. A useful analogy could be the President aggressively protecting his unilateral authority to make strategic decisions during wartime, or even our sovereign right to possess nuclear weapons, without actually using those powers. The same is true when states fight strenuously to preserve their unilateral, sovereign right to respond to natural disasters, while taking advantage of FEMA assistance. In all of these cases, the government possesses a power that it refuses to assert. Enabling proportionality provides governments with the technical legal authority to impose a life sentence for drug possession, or even the death sentence for homicides falling short of first-degree murder, even though the state may choose not to exercise its authority in most cases.

Distinguishing between proportionality as a limiting principle and proportionality as an enabling principle highlights the fact that proportionality norms can be perceived as an obligation in one environment, and as a choice in other environments. Certain segments of the American public, for example, may believe that it is perfectly appropriate to chop the hands off of a thief or draw and quarter simple burglars, but these punishments clearly violate the Eighth Amendment. States are obligated to conform to the text and meaning of the cruel and unusual clause. If proportionality is the means by which states conform to the Eighth Amendment, the proportionality operates both as a limiting principle and as a source of constitutional obligation. But consider an alternative: a state decides to impose only a life sentence or a determinate long prison sentence for heinous crimes like first-degree murder, even though the Supreme Court has decided that the death penalty is not a disproportionate sentence for grave offenses within the meaning of the Eighth Amendment. In this situation, proportionality operates as an enabling principle, providing the state with the authority to impose the death penalty even if they choose not to. Proportionality also protects the choice that states have between various punishments (death penalty, life sentence) that do not violate the Eighth Amendment. Depending on the circumstances, therefore, states may decide to characterize adherence to proportionality norms as a strategic choice, rather than a moral or legal obligation.

Similar issues emerge when the state tries to deter foreign powers or sovereign entities from engaging in problematic conduct using proportionality. Proportionality rules designed to deter modern democratic states heavily invested in the rule of law and determined to preserve their reputational capital may not effectively deter rogue or outlier states with few incentives to curb their behavior beyond the fear of economic collapse or military invasion. Regional tensions also create significant jurisdictional asymmetry problems. Pakistan, for example, may be willing to adhere to proportionality principles during a military conflict with India because of India's proximity and possession of nuclear weapons, but it may not adhere to proportionality principles if it decides to invade the Philippines or Brazil. Finally, there is the question of what happens when a country decides to withdraw from international agreements that ban disproportionate or excessive uses of force because it believes that withdrawal is in its own national interest or that withdrawal is consistent with its own sense of proportionality. If so, then one countries' understanding of proportionality may operate as an enabling rationale for abandoning international deterrence efforts, while international agreements banning disproportionate or excessive uses of force may operate as a limiting principle in the same context. During geo-strategic and military conflicts, in other words, proportionality may be an enabling principle for some countries and a limiting principle for other countries. The former animates American efforts to deter the use of disproportionate force by military officers under the law of armed conflict model while the latter animates UN-sponsored efforts to deter the use of disproportionate force by military officers under the international human rights paradigm.

1.2.3 Acts of War

The use of extreme force is almost unavoidable during warfare, the unfortunate but entirely foreseeable byproduct of practical necessityand prolonged psychological duress. But the use of extreme force during warfare is governed by a complex set of laws and norms[9] that sovereign

[9] See e.g. Andropopolis (1994).

nations, concerned first and foremost with their own security needs, may choose to either follow or ignore depending on the circumstances and overarching ideological preferences. The outer boundaries of the complex penumbra of rules and norms governing the use of extreme force (*jus in bello*) are defined by the law of armed conflict (LOAC). Prohibitions on inappropriate extreme force during wartime are also constrained by various provisions of the Geneva Convention and the 'cruel and unusual' section of the US Constitution. Both domestic and foreign provisions, therefore, constrain governmental authorizations of extreme force during all phases of a military or national security crisis, as well as broader socio-political norms that predate any existing constitutional, bilateral or multilateral limitations to which the American government is currently obligated. Ignoring these norms, either temporarily or permanently carries consequences, the most important of which is that a country is then more likely to mistakenly use disproportionate force that eventually undermines, and may even prolong, the war effort itself.

During wartime (either real or imagined), arguably the most important constraint on the use of extreme force comes from formal and informal 'laws of war' (*jus in bello*) provisions that most sovereign states subscribe to even in the absence of formal treaties provisions. Even before we reach the crucial question of what constitutes extreme force during wartime, the laws of armed conflict make certain assumptions about what constitutes armed conflict,[10] the identity and capacities of aggressors and when it is even appropriate to use force in response to aggression (*jus ad bellum*), either in self-defense after an attack or in anticipation of a future attack. Once the threshold of armed conflict has been established (either as a matter of domestic or international law), extreme force is expected to be both proportionate and discriminating. The former constrains government to use only as much force as is necessary to reasonably achieve a desirable or legitimate military or strategic end,[11] and the latter requires that governments only use force against parties (or stakeholders) either directly or reasonably responsible for hostilities.[12] Proportionality constrains the use of force during an armed conflict as well as the manner of response after an attack or potentially in anticipation of

[10] Bothe (2003, 229) (distinguishes between armed affairs and mere threats).

[11] Protocol_I. (1977a).

[12] Protocol_I. (1977b).

a future attack. In federal civil suits against the US government, the question of whether a military officer used disproportionate extreme force is considered a non-justiciable political question, but courts can consider whether military officers used disproportionate extreme force tantamount to a war crime,[13] in federal criminal cases.

1.3 Due Diligence, Attribution and Distinction

Every well-functioning society requires some degree of due diligence from governmental bodies and private entities seeking official protection, recognition, access to valuable goods and services, or public legitimacy. In the popular imagination, due diligence is identical to administrative precision. When companies and governments enter into contracts with other parties, they are expected to determine, with reasonable precision, any facts that might compromise, encumber or nullify the execution of those contracts. Corporate board members are expected to exercise due diligence in their interactions with other board members, customers and, in most cases, employees. That due diligence is essentially a fiduciary responsibility. It is undertaken, in other words, both because it ensures accountability, government recognition and civil corporate relations, and because it protects the corporation itself. Failure to exercise due diligence is an invitation for mischief and, quite possibly, illegality. Pharmacies, for example, that refused to run cursory background checks on doctors who prescribed massive amounts of Oxycodone were eventually shut down. People who contributed money to questionable charitable organization without conducting any background checks soon found themselves under indictment for providing material support to terrorist organizations.[14]

When legitimate nation-states employ deadly force, they also operate under a set of due diligence guidelines for targeting. Discriminatory (or distinction) principles defined by international law require that the

[13] WarCrimesB (1977) (prohibits war crimes). See also WarCrimesA (1907).

[14] Crimm (2005, 218) ("Caution suggests that the broader and deeper the due diligence exercised, the more likely a donor, domestic publicly-supported charity, or private foundation is to avoid the devastating effects of accusations of transgression of the federal anti-terrorism measures."). For specific provisions that criminalize providing material support to terrorists or terrorist organizations, see Material (2006) (terrorist organizations).

government "distinguish between combatants and non-combatants, military objectives and protected people / protected places."[15] The principle of distinction demands that both governments and non-state actors limit the use of force to belligerents alone, but countries at war devote an enormous amount of time identifying two types of individuals: lawful enemy combatants and unlawful enemy combatants. The former typically consists of soldiers on the battlefield, while the latter often refers to civilians who participated, either directly or directly, in hostile activities.[16]

Due diligence guidelines also constrain the selection of potential targets after a cyber attack. If there are significant attribution problems after a cyber attack (i.e. the perpetrator is a weak state that successfully eludes detection or a private, non-state actor operating either on its own or with the clandestine support of a national state), determining how much force to use in self-defense, and against whom will be extremely difficult. In some situations, proof of overall control over the apparatus used in a cyber attack may be sufficient to prove attribution. It might be even more difficult to fashion a proportional response in self-defense if a cyber attack takes the form of a widespread denial of service hack that exploits multiple computers at different locations Given these difficulties, it is not surprising that retaliatory responses to cyber attacks can combine kinetic and non-kinetic actions, as long as the proportionality principle is satisfied.

Very few people in the 1980's realized that a super cyber attack orchestrated by a criminal network, and conducted with the full knowledge and quiet support of a nation-state, could provoke a military response from the attacked nation, and quite possibly even a formal declaration of war. But attribution technologies have significantly improved since the dawn of the computer age. In the same way that Al Qaeda could not hide behind the Taliban in Afghanistan, cyber attackers residing in a foreign country that manage to fatally compromise vital US military computer assets in the United States will not be able to hide behind its host state given recent advances in attribution and de-anonymization.[17] Many victims of cyber attacks can take advantage of attribution technologies—Since

[15] OLH (2001, 31).

[16] Protocol_1a (1977).

[17] For more discussion of traceback technologies and attribution strategies, see Chapter 6, section 2.2.

some attribution and traceback techniques are more reliable than others, political decision-makers will have to evaluate the relative accuracy of a specific attribution technique, as well as determine the victim's capacity to comprehend and select a particular attribution technique, before deciding if 'reasonable attribution' was achieved in a particular case.[18]

2 Enhanced Penalties and Retaliation Costs

Decision-makers have always been sensitive to factors that seem to demand an extra measure of caution, or an extra dose of punishment. The fact that convicted criminals can be imprisoned or executed, and unsuccessful civil litigants can't be punished in the same manner, suggests that there are certain acts so dangerous or deserving of widespread moral opprobrium that society feels the need to physically separate the guilty party from the general population. We also attack extra penalties to crimes involving guns or motivated by bias. VRN attaches 'extra penalties' to certain acts, but in this case, the penalty consists of forcing the attacker to completely absorb the cost of damages caused by the victim's retaliation.

Chapter 4 described the elements required to establish a presumptively legitimate claim under VRN. Section 1 of this chapter analyzed three aggravating factors—failure of necessity, attribution or proportionality—that attackers can use to rebut the presumption that victims should not have to compensate attackers for damages caused by retaliation. This section, by contrast, actually reinforces the intuition behind the presumption for a subset of attacks that lead to retaliation, as well as a subset of culpable aggressors. VRN is based on the partial intuition that many victims lack the necessary freedom and capabilities to effectively protect themselves against attacks in the open world. The absence of timely law enforcement, affordable litigation options or effective hardening only amplifies their capability deficiencies, leaving them vulnerable to a wide variety of attacks unless they have access to retaliation. This scenario, along with a determination that the attacker is a culpable aggressor, is enough to establish the presumptive legitimacy of the victim's retaliation under VRN, but it is not enough to persuasively argue that the attacker must absorb most, if not all, of the costs of retaliation.

[18] For more on reasonable attribution, see Schmitt (2014, 727).

But proof that the attackers' actions fall into the 'extremely dangerous' category tips the balance in favor of the victim/retaliator. This proof is compelling enough to warrant the imposition of an enhanced penalty in retaliatory self-defense and self-help cases, a penalty that fulfills the same function as punitive damages. What, then, are the common markers of 'dangerousness'?

2.1 *Neutralization Strategies*

In a famous 1957 paper, Sykes and Matza argue that juvenile delinquents often justify their deviant behavior using "techniques of neutralization." For delinquents, "social controls that serve to check or inhibit deviant motivational patterns are rendered inoperative, and the individual is freed to engage in delinquency without serious damage to his self-image."[19] Sykes and Matza described five neutralization strategies, the first of which was 'denial of responsibility.' Using this strategy, delinquents attribute their behavior "to forces outside of the individual and beyond his control such as unloving parents, bad companions or a slum neighborhood."[20] Whether the delinquent's assertions about his life circumstances, and the manner in which those circumstances deprived him of effective agency, are reasonable means little to either the delinquent or behaviorists. What matters is that the denial of responsibility strategy provides the delinquent with extraordinary freedom to engage in actions that he actually perceives to be illegal or impermissible. Another neutralization strategy employed by delinquents is 'denial of injury,' in which "wrongfulness ... [turns] on the question of whether or not anyone has clearly been hurt by his deviance."[21] The denial of injury strategy is roughly comparable to *malum prohibitum* (illegal but not immoral),[22] 'harmless error' (a procedural mistake that did not cause injury or harm) or obsolete laws that

[19] Sykes and Matza (1957, 667).

[20] Id.

[21] Id.

[22] Id. See also Boyce and Perkins (1989, 536), Cyanamid (1973, 1205) (finding that scienter is not required as an element of a malum prohibitum crime), and Emory (1994, 1254) (stating that intent is a formal element in every crime except malum prohibitum crimes).

remain on the books even though they are never enforced.[23] The denial of injury strategy is also crudely aggregative suggesting that there was an injury to the victim, but that the injury was insignificant compared to a larger increase in social welfare.

The third neutrality strategy is 'denial of victim,' in which the delinquent conceptualizes his behavior as "a form of rightful retaliation or punishment."[24] The delinquent transforms himself into a hero and the victim into a wrongdoer, mimicking the rough justice of mythical heroes like Robin Hood.[25] Role reversal strategies carry the additional benefit of allowing the delinquent to perceive himself as an enforcer, rather than a critic, of the dominant social structure. It also shifts the burden to victims who must prove that they are 'real' victims, in spite of any apparent injuries or losses from the attack. For the fourth neutralization strategy, 'condemning the condemnors,' the "delinquent shifts the focus of attention from his own deviant acts to the motives and behavior of those who disapprove of his violations."[26] Authority figures like teachers or police officers are morally compromised, and material success comes from "pull or luck."[27] The implication is that if the condemnors hadn't deprived them of certain advantages, the delinquent would have been equipped to follow dominant social norms. Ironically, the delinquent wants us to believe that the absence of certain advantages forced him down an unsavory life-path. For the fifth and final neutralization strategy—'appeal to higher loyalties,' delinquents believe that they are "sacrificing the demands of larger society for the demands of the smaller social groups

[23] There is a difference between an obsolete law and a law that merely reflects the preferences of an extremely small or politically disfavored minority. A clear example of the former, for example, would be a law that prohibits drinking or smoking on Sunday, restrictions that would have been completely acceptable in the early nineteenth century, but hardly command respect today. Another example would be a law that prohibits individuals from using more than two gallons of water on the weekends. Such a law would also be considered obsolete, not because governments never require citizens to ration precious resources like water during emergencies, but because the three-gallon limit would be almost impossible to follow, and therefore equally impossible to enforce. Only inattention, inadvertent neglect or indifference can explain the presence of truly obsolete laws based on unrealistic or outdated assumptions.

[24] Sykes and Matza (1957, 668).

[25] Id.

[26] Id.

[27] Id.

to which the delinquent belongs."[28] Contrary to the popular maxim that 'there is no honor among thieves,' delinquents try to convince experts that they are only 'partial rule-breakers.'

Neutralization theory has expanded into many areas, including the development of new or revised neutralization techniques,[29] versions for different kinds of rule-breakers,[30] the rationalization and coping strategies of victims,[31] research suggesting a connection between low self-control, rule-breaking and neutralization,[32] neutralization to discourage stereotyping rather than alleviating guilt,[33] whether neutralization is crime-specific,[34] less likely to occur for "offenses against property"[35] or dependent on the subject's belief that deviant behavior is actually wrong.[36] But regardless of the context, neutralization strategies share three characteristics. The first is that neutralization, in some form, can and has been used by individuals, organizations or nation-states. It doesn't matter if it's a juvenile delinquent, chronically mismanaged, toxic companies that violate regulations and social norms, or large illiberal states like Iran, China or Russia. Second, neutralization has strategic value even if the user does not believe that their deviant behavior is wrong. And third, neutralization provides rule-breakers with more freedom (or autonomy) to engage in deviant conduct. Versions of neutralization have also been used by attackers in open world zones, particularly as a rationale for stealing intellectual property and digital files across borders.[37]

To be fair, neutralization is not an attack. It is just one of many possible rationalizations for attacks. It could also, at least in the minds of some culpable aggressors, serve as a partial justification for imposing the costs of retaliation upon the victim rather than the attacker. VRN reverses this logic because it would require attackers who use neutralization strategies to pay the costs of retaliation, especially if they recognized that their attacks were wrong. While it's true that some attackers who employ

[28] Id at 669.

[29] Moore and McMullan (2009, 443).

[30] Id. at 444.

[31] Id.

[32] Hinduja (2007, 189).

[33] Copes (2003, 104).

[34] Id. at 106.

[35] Sykes and Matza (1957, 670).

[36] Moore and McMullan (2009, 444).

[37] Id. and Hinduja (2007, 189).

neutralization strategies are more 'free' than others, and that some people who don't use neutralization tactics are actually more free to attack than attackers who do employ neutralization tactics, an appreciable percentage of 'dangerous' attackers use neutralization strategies.

2.2 *Repeat Offenders*

For hundreds of years, societies have punished repeat offenders more than one-time (or first-time) offenders. Many states, for example, authorize judges to send repeat juvenile offenders to adult courts,[38] while the federal government increases the penalty for repeat offenders convicted of a crime of violence or drug trafficking crime who use, carry or possess a firearm during the commission of that crime.[39] Societies usually justify punishing repeat offenders more severely than first-time offenders on two grounds. The first rationale is deterrence. While every society has its share of hardened criminals who are more likely to be deterred by the likelihood of getting caught than the amount of time they are likely to remain in jail, most people dread the prospect of imprisonment (and even home arrest), and are likely to reconsider committing a crime more than once if they are likely to spend much more time in jail for the second offense than the first. In addition to the stigma of imprisonment and the loss of income both during and after incarceration, most people also recognize that imprisonment itself can transform someone into a hardened criminal, even if the prisoner is genuinely remorseful.

Studies have shown that the severity of sanctions reflects common beliefs about the kinds of penalties likely to discourage targeted individuals from committing specific offenses. The prospect of going to prison for six months might be enough to discourage the average person from robbing a bank, but it may not be sufficient to discourage a seasoned bank robber who has already served time in prison. Only a more severe sanction, such as minimum sentences of twenty years for bank robbery, will discourage many seasoned criminals from robbing banks.

People also tend to comply with legal rules because they want to avoid the sanctions or penalties that come from non-compliance.[40] Imagine, for

[38] Waid-Lindberg et al. (2011, 45).

[39] Enhanced (2000).

[40] Posner and Rasmussen (2000).

example, if a society decides that every high school student with a valid driver's license will be expelled from school for a year if they are caught speeding. For the average student, the fear of losing a year of school alone is enough to discourage speeding. For other students, the expulsion rule will be effective only if it is combined with punishments from one's parents. For the most incorrigible students, the expulsion rule may only be effective if it prevents rule-breakers from going to college for five years and prohibits them from re-instating their driver's licenses for five years. But regardless of whether the expulsion rule works as is or requires the added burden of additional punishments, the rule itself is a form of sanction capable of inducing the compliance of a specific segment of the population. As compliance enforcement mechanisms, sanctions may work regardless of whether the sanctions are actually applied. For some people, the fear that a sanction will be applied is enough to produce compliance.[41] There are studies suggesting, however, that criminals who have broken rules and escaped punishment in the past are much less likely to comply

[41] The counter-narrative to sanctions-based compliance is legitimacy-based compliance ('internalization'). Consequently, a normally law-abiding citizen confronted with a warrant to search his home may be less willing to comply with governmental search efforts if he believes that the government has unfairly used warrants to raid the homes of certain groups more than others regardless of the facts in the case. Since most police departments recognize that they cannot prevent crime without the cooperation of otherwise law-abiding individuals, they will always try to balance legitimate probable cause suspicions held by the police in a particular case against generalized suspicions of police misconduct held by a entire community, typically using language and enforcement tactics that preserve some degree of trust between the people and its government. Legitimacy-based models make two crucial assumptions. The first is that adherence to rules entails a degree of reflection or personal scrutiny. When someone decides to pay taxes, in other words, the assumption is that the act itself is a form of ratification or approval, rather than a reflexive reaction to the mere fact of legal compulsion, the social force of legal obedience or the fear of sanction. Consequently, legitimacy-based models require a fairly robust model of human agency, a model in which individuals conceptualize their direct consent as part of any compliance regime. Legitimacy itself may also be based on the institutional pedigree of rules (i.e. rules must emanate from legitimate organizations or entities), the process by which rules are generated (i.e. through a pluralistic, transparent, democratic decision-making process), substance of rules (based on moral or core ideological criteria) or even the degree to which rules reflect the cultural preferences of the majority. But regardless of the criteria used, legitimacy-based models assume that individuals actually select a basis of legitimacy, and are prepared to resist or non-comply with legal rules that fail to satisfy the legitimacy criteria selected.

with a rule in future if the process by which sanctions or punishments are applied remains the same.[42]

The second rationale for punishing repeat offenders more severely than first-time offenders is that an enhanced sentence sends an unmistakable signal[43] of moral wrongfulness. Societies routinely impose enhanced penalties for particularly egregious conduct, including hate crimes, gun possession or violence toward children. Enhanced penalties are part of the expressive function of law,[44] and punishing repeat offenders more severely than first-time offenders conveys the message that society not only considers certain conduct particularly offensive and harmful to the social compact, but that it also wants to discourage people from engaging such conduct more than once. Every society grades rules violations along a scale, and the most egregious violations typically lead to the most severe sanctions. In the criminal arena, for example, murder may lead to the death penalty or life imprisonment, while civil infractions may lead to extremely large fines and or lifetime exclusion from desirable professional or political activities. The severity of sanction for norms violations reflects the degree of moral outrage experienced by a particular community, even to the point of calling for punishments proportionate to the violation.

[42] Criminals who escape punishment repeatedly are unlikely to fear sanctions in the future. See Paternoster (1987) and MacCoun (1993).

[43] Informal signaling is an important social norm. Social norms theory started as a response to pervasive collective action problems in small groups, and shortcomings in traditional 'rational actor' models. Collective action problems occur when the self-interest of an individual conflicts with the welfare of the group to which the individual belongs. This disconnect causes pervasive free-rider problems that only grow worse as group size increases. See e.g. Olson (1965) and Elster (1989, 126). Under the traditional collective action model, only external coercion can prevent individuals from pursuing their own interests at the expense of the welfare of the entire group. Social norms theory, however, proposes an alternative source of social cooperation: 'norms,' otherwise understood as social conventions, established modes of cultural practice, or symbolic behavior. Those who cooperate with social norms receive benefits such as esteem, or the avoidance of legal sanctions, ridicule, ostracism or punishment. McAdams (1997) As the benefits of social cooperation increase, and benefits are transmitted to a broad segment of society via 'signaling mechanisms,' norms 'cascades' or the active encouragement of 'norm entrepreneurs,' the cost of social cooperation decreases. See e.g. Posner (1998, 768), Sunstein (1996, 912), Eisenberg (1999). For a discussion of norms cascades on human rights discourse, see e.g. Finnemore and Sikkink (1998).

[44] See e.g. Cooter (1988, 587), Chong (2000, 19), Posner (2000, 3–4, and Ellickson (1992).

2.3 *Rule of Law Violations Caused by Political Failure*

Along with repeat offenders and attackers of critical infrastructure, open world communities should also consider attacks from a third entity particularly dangerous and deserving of 'enhanced' punishment: rogue, illiberal regimes plagued by widespread political failures. The concept of political failure can be defined in at least three ways. A narrow definition occurs when formal institutions charged with political responsibilities fail to reflect the popular will. Such a definition assumes that the popular will can be determined with a fair degree of precision, and that the most important purpose of a political institution is to represent the demonstrated preferences of the majority.[45] An intermediate definition of political failure is that failure occurs when political institutions fail to govern in a 'political' way given the historical expectations of the people. In modern democratic societies, for example, governing in a political way would have to include accountability, transparency and some form of procedural, administrative or bureaucratic rationality, all markers of political legitimacy regardless of the substantive content of laws.

A weaker version of the intermediate view is that political failure occurs when institutions that are obligated to work together in the public interest fail to do so, undermining the long-term efficacy, reputational capital and responsiveness of governmental operations. The broadest definition of political failure is that failure occurs when majority interests or groups completely tyrannize minorities. This definition does not require the mediating influence of any formal institutions to transform either ordinary or extraordinary socio-cultural or economic preferences into political sentiments, but it does assume that majoritarian and minoritarian preferences are political, and that some majoritarian checks on minority views may be politically and/morally inappropriate. In addition, the broadest definition of political failure does not equate failure with anything other than the failure to protect minoritarian interests rather than the public's best interest as expressed through demonstrated public opinion or a larger, objective determination of public welfare.

[45] In a non-democratic society, or a democratic society in which political institutions pursue objectives consistent with a particular understanding of the public interest rather than the stated preferences of the people, the third assumption would be almost entirely irrelevant.

Neither the narrow nor intermediate definitions of political failure limit political failure to any particular branch of government. Since all three branches contain a democratic component, either at the appointment, indirect ratification or decision-making phase all three can fail to reflect the popular will, fail to govern in a political way given the expectations of the people or even fail to work together effectively as they conduct the people's business. In addition to a quasi-democratic component, all three definitions suggest that political failures occur when an important threshold of unaccountability, operational dysfunctionality, abandonment of public welfare concerns or excessive majority control over minority interests has been reached. Depending on the circumstances, this threshold may define the boundary between well-functioning and dysfunctional governance, or it may define the boundary between acceptable and extreme dysfunctionality. All three forms of political failure demand either remedial steps to correct the behavior of an offending institutional political entity, or steps to shift responsibility and quite possibly law-making authority itself, to a third party authorized to make political decisions.

The specter of political failure raises two important issues. The first concerns the boundary between well-functioning and dysfunctional governance. And the second is more sobering: does political failure provide dysfunctional societies with more administrative or strategic freedom than better governed societies? As to the first issue, it is impossible to construct a perfect political system, either in the sense of a system that perfectly represents the will of the people (proportionally and in intensity level), or in the sense of a system that completely insulates public officials from private interference or rent-seeking. Every defense of state power, therefore, must consider threshold questions, identifying tipping points beyond which ordinary political dysfunctionality cannot be countenanced, ignored or even repaired absent a wholesale re-examination of institutional and/or administrative processes.

Once that tipping point is reached, however, theorists and decision-makers must determine if there are any advantages to dysfunctional, authoritarian rule. The disadvantages are breathtakingly obvious, especially in the open world: loss of political legitimacy and reputational capital, reduced access to developmental assistance from international institutions or cultural bodies, and reduced access to valuable goods, services and information across international borders. All of these restrictions affect the distribution of goods and services internally, exacerbating environments in which civil rights abuses and corruption are already likely. But theorists have also identified various 'authoritarian advantages' during crises.[46] Relative immunity from internal political dissent and international pressure dramatically increases the likelihood that rogue, illiberal regimes will engage in recklessly provocative, symbolic demonstrations such as North Korea's strategically timed, ballistic missile testing program, or dangerously escalatory military campaigns like Russia's invasion of Ukraine. In spite of their diminished status on their world stage, therefore, authoritarian regimes do enjoy certain administrative advantages over democratically elected, non-authoritarian regimes, and those advantages make their attacks on open world zones particularly dangerous.

2.4 *Critical Infrastructure or Public Safety*

Many countries have passed laws that prohibit someone from launching transnational attacks on cyber-control system (CCS)[47] that cripple or destroy physical systems like dams, nuclear plants and missile systems. Congress has also passed laws that prohibit control system attacks, either as an unintended byproduct of cyber attacks designed to only affect software, or as a potential threat to national security. These laws attempt to not only prevent the government from violating the rights of incidental users of computer systems who might get swept up into a CCS by accident, but they also send a clear message that it is the ends (the compromise of control systems), rather than simply the means (access to computer networks), of transnational CCS attacks that pose the greatest risk to national security. They try to cleanly separate, in other words, the 'signal from the noise,' drawing a sharp distinction between merely undesirable computer hacks that expose commercial or governmental secrets, and completely impermissible CCS attacks that could, and perhaps should, lead to war.

Experts have identified many cyber-attack varieties, from so-called soft attacks to hard attacks like malicious remote encryption of data,

[46] See e.g. Wrage (1999)

[47] For more on the programmable logic devices that drive cyber-control systems, see Moon (1994).

widespread 'denial of service' outrage,[48] the deliberate destruction of confidential information on private or governmental servers, scrambling navigations systems on aircraft in flight, and most famously in the Stuxnet case, manipulating code that controls vital hardware systems like nuclear reactors, traffic systems, electrical grids and flood control networks.[49] By definition, soft attacks seem to produce less harm (usually defined systemically, economically or militarily), and in many cases, the harm created by soft attacks falls within the tolerable 'margin for error' that people have come to expect in the modern world. Perched precariously between soft and hard attacks are so-called semantic attacks that modify code to produce incorrect data but do not damage to existing data,[50] and more ambitious hacks that modify code and data, produce desired results with the modifications, store the new information on a distant server and restore both the original code and data before exiting the system.

Like repeat offenders and reckless authoritarian regimes with distinct administrative advantages, 'super cyber attacks' that cripple or destroy vital military systems or computer hardware are extremely worrisome in the open society. Sophisticated hacking networks orchestrate these attacks, and some of these groups receive clandestine support from nation-states.[51] Super cyber attacks orchestrated by hacking networks are extremely dangerous because they could inflict major damage on American assets, and if perpetrated on a massive scale, nudge the United States into the use of military force.[52] Relying on preventive measures to detect and prevent transnational cyber hacks from affecting vital US assets may not be enough to deter super cyber hacks in future.

[48] Hathaway et al. (2012, 837–38).

[49] Vida (2005).

[50] Hathaway et al. (2012, 824).

[51] Bracknell (2007) ("1. China or other foreign powers have the desire to acquire foreign intelligence, both security-related and economic intelligence, from the United States; 2. China or other foreign powers have the technological capacity to create or manufacture technical espionage devices, apparatus or computer code of the type required to capture and transmit information to the receiving foreign entity clandestinely; 3. China or other foreign powers have the analytical capacity to turn massive volumes of raw data captured from US systems into useful intelligence, beneficial to the foreign power's security or economy, to the detriment of US security and economy."), Gross (2011) (describing successful Chinese hacks of public and private systems).

[52] Lin (2010, 70–82) and Nakashima and Warrick (2012).

3 The Logic of Deterrence and Non-Escalation

It is not at all uncommon for the victims of an attack to retaliate against the suspected attacker in the belief that retaliation is the best way to deter future attacks. In an ideal situation, the victim can identity the attacker, retaliate in a manner that is proportional to the attack and retaliate based on the reasonable belief that other alternatives, such as law enforcement, 'hardening' or even simply walking away, are less likely to deter future attacks than retaliation. Retaliation, in other words, is part of the logic of deterrence for many victims. It is also not uncommon for an attacker, fully aware of the possibility of retaliation after an attack, or a third part6 observer with his own ideological agenda, to argue that the logic of deterrence demands non-retaliation.

The idea that retaliation undermines deterrence typically takes four forms. The first variant is that retaliation necessarily escalates, or increases the intensity of an ongoing conflict between two parties. One can imagine, for example, a confined, one-time exchange between X and Y rapidly turning into a prolonged battle that not only blurs the initial distinction between victim and attacker, but also encourages, due to the length of the conflict, third parties to intervene on behalf of X or Y. Non-retaliation might appear to deter escalation even during an ongoing conflict. Such an argument has been made by Ukranian defense officials who refuse to retaliate against Russian attacks in Ukraine by attacking targets inside Russia because of the fear that retaliation could provoke Russia into an even more intense ground and air assault in Ukraine and precipitate Russia's allies to intervene on Russia's behalf.

Another variant of the idea that retaliation undermines deterrence is that the attacker is actually righting a wrong by attacking the victim. In spite of appearances, the true 'victim' in this scenario is actually the attacker, not the other way around. By attacking, the aggressor is 'closing the circle' between himself and the victim, righting an injustice or erasing a benefit, already enjoyed by the victim. Since the attacker has already, in his own mind, already retaliated, any further retaliation is pointless. Non-retaliation by the victim deters future conflict by signaling, for the attacker, the victim's willingness to absorb damage or harm as payment in compensation for a prior attack. The third variant is that for some attacks, precise attribution is impossible. Without adequate precision, victims of attacks run the risk of creating massive spillover effects on innocent third parties, which will significantly undermine public confidence in security and law enforcement. A fourth variant, closely related to the third, is that retaliation is always less effective as a deterrent of future attacks than law enforcement, litigation or hardening.

In the wrong hands, all four variants of the idea that retaliation undermines deterrence could create as many problems as they appear to solve. The 'wrong hands,' in the current geo-political environment, are people on opposite ends of the open world scale: zealots and opportunists. Zealots, if you recall, are open world maximalists who are absolutely determined to prohibit almost any restriction or limitation—regardless of how commonsensical, zone-specific or time-delimited it might be—on the transfer of persons, information or goods across borders. Opportunists, by contrast, are open world minimalists, not because they want to ban all cross-jurisdictional transfers, but because they are only interested in exploiting gaps, conflicts and ambiguities in any formal regulation of cross-jurisdictional transfers or informal community agreements impacting the transfer of goods, persons or ideas across borders. The zealots' perspective on open world discourse is over-permissive; the opportunists' perspective on open world discourse is parasitic. In their hands, equating non-retaliation with effective deterrence is, at best, a recipe for abuse, and at worst, a ticking time bomb.

Both zealots and opportunists, for example, might take advantage of inconclusive data on the relative success of retaliation in different zones, or with regard to different parties, to discourage retaliation for attacks in the open world. (fourth variant). Their motivations differ but their policy recommendations would be the same—discourage any retaliation that could compromise the free flow of goods, services, persons and ideas across borders, even if those transfers deprive victims of a legitimate, controllable alternative to spotty law enforcement and expensive, slow litigation. Zealots and opportunists might also (third variant) try to characterize victims who want to retaliate against attackers when precise attribution is difficult, or potential retaliators who derive significant benefits from open world policies, as either 'super-sensitive plaintiffs' over-reacting to violations that could have come from a variety of different sources, or as 'free-riders' who want to take advantage of open world principles without paying the 'modest' cost of a temporary loss in privacy or security. These arguments are the equivalent of accepting the likelihood that the number of criminals and terrorists entering your country will probably increase if you open your borders for humanitarian reasons. Anticipating the decreased effectiveness of screening procedures at the border during a humanitarian crisis (*ex ante*) should translate into a stronger national commitment to eliminating criminals and terrorists from the general population, not a weaker commitment. (*ex post*). Along the same lines, permitting retaliation for open world attacks when attribution is reasonable (as opposed to absolutely precise), proportionality constraints are followed and immediate deterrence is likely, is preferable to either driving victims toward an unregulated retaliatory underground or driving them toward authoritarian regimes with law enforcement rules that quickly punish attackers, not because they want to protect open world expectations, but because they want to harden their own national boundaries.

Zealots and opportunists will also argue that retaliation is inadvisable because the open world, like the world that came before it, is a 'sea of past grievances' that blurs the line between victims and attackers (second variant). The problem with this argument is that it ignores persistent, chronic capability gaps between the victim and attacker that might have not only discouraged the victim from relying on law enforcement, litigation or hardening, but also might have lead the attacker to believe that he could attack the victim without consequence. VRN identifies capability-deficient victims in the open world, and asserts that the victim's claim to retaliation should be considered presumptively valid if it was done out of necessity, and satisfies reasonable proportionality and attribution requirements. It also avoids problematic aggregative critiques of retaliation that mirror equally problematic aggregative welfarist arguments suggesting that retaliation undermines total utility. The same logic applies to arguments that retaliation necessarily leads to escalation and undermines the general welfare (first variant).

References

Andropopolis, H. 1994. *The Laws of War: Constraints in Warfare in the Western World*. New Haven: Yale University Press.

Barnett, R. 2003. The Original Meaning of the Necessary and Proper Clause. *University of Pennsylvania Journal of Constitutional Law* 6: 183–227.

Bentham, J. 1843. Principles of Death Law. In *J. Bentham's Works*, Part II, Bk. 1, ed. J. Bowring. Edinburgh.

Bothe, M. 2003. Terrorism and the Legality of Pre-Emptive Force. *European Journal of International Affairs* 14: 227–231.

Boyce, R., and R. Perkins. 1989. *Criminal Law and Procedure*, 7th ed. New York: Foundation Press.
Bracknell, R. 2007. Trust Not Their Presents, Nor Admit the Horse: Countering the Technically-Based Espionage Threat. *Roger Williams University Law Review* 12 (3): 832–851.
Bradley, C., and M. Flaherty. 2004. Executive Power Essentialism and Foreign Affairs. *Michigan Law Review* 102: 545–580.
Carter, S. 1983. The Political Aspects of Judicial Power: Some Notes on the Presidential Immunity Decision. *University of Pennsylvania Law Review* 131: 1341–1384.
Chong, D. 2000. *Rational Lives*. Cambridge: Harvard University Press.
CIC. 1776. U.S. CONST. art. II, § 2, cl. 1.
Cooter, R. 1988. Expressive Law and Economics. *Journal of Legal Studies* 27: 585–604.
Copes, H. 2003. Societal Attachment, Offending Frequency, and Techniques of Neutralization. *Deviant Behavior: An Interdisciplinary Journal* 24: 101–110.
Crimm, N. 2005. Post-September 11 Fortified Anti-Terrorism Measures Compel Heightened Due Diligence. *Pace Law Review* 25: 203–224.
Cyanamid. 1973. United States v. American Cyanamid Co., 354 F. Supp. 1202 (S.D.N.Y. 1973).
Eisenberg, M. 1999. Corporate Law and Social Norms. *Columbia Law Review* 99: 1253–1286.
Ellickson, R. 1992. *Order Without Law* (Oxford University Press,1992).
Ellickson, R. 1992. *Order Without Law*. New York: Oxford University Press.
Emory. 1994. Emory v. State, 647 A.2d 1243, 1254 (Md. Ct. Spec. App. 1994).
Enhanced. 2000. 18 U.S.C. $924©(1)(2000). Act of November 13, 1998, Pub. L. No. 105-386, $1, 112 Stat. 3469, 3469-70.
Finnemore, M., and K. Sikkink. 1998. International Norm Dynamics and Political Change. *International Organization* 52: 887–898.
Gross, M. 2011. Enter the Cyber-Dragon. *VanityFair.com*, 9.
Harmelin. 1991. Harmelin v. Michigan, 501 US 957.
Hathaway, O., Rebecca Crootof, Philip Levitz, Haley Nix, Aileen Nowlan, William Perdue, and Julia Spiegel. 2012. The Law of Cyber-Attack. *California Law Review* 100: 817–856.
Hinduja, S. 2007. Neutralization Theory and Online Software Piracy: An Empirical Analysis. *Ethics and Information Technology* 9: 187–197.
Lin, H. 2010. Offensive Cyber Operations and the Use of Force. *Journal of National Security Law and Policy* 4: 63–86.
MacCoun, R.J. 1993. Drugs and the Law: A Psychological Analysis of Drug Prohibition. *Psychological Bulletin* 113 (3): 497–512.
Material. 2006. 18 U.S.C. § 2339A (2006) (terrorists) and § 2339B.

McAdams, R. 1997. The Origin, Development, and Regulation of Norms. *Michigan Law Review* 96 (2): 338–433.
McCulloch. 1819. McCulloch v. Maryland, 17 U.S. (4 Wheat.) 316, 413–414, 420–421 (1819).
Moon, I. 1994. Modeling Programmable Logic Controllers for Logic Verification. *IEEE Control Systems Magazine* 14 (2): 53–59.
Moore, R., and E. McMullan. 2009. Neutralizations and Rationalizatoins of Digital Piracy: A Quantitative Analysis of University Students. *International Journal of Cyber Criminology* 3 (1): 441–451.
Nakashima , E., and J. Warrick. 2012. *Stuxnet Was Work of U.S. and Israeli Experts*, Officials Say, WASHPOST.com. 6/3/2012.
OLH. 2001. *Int'l & Operational Law Department*. U.S. Army, Operational Law Handbook. http://www.jagcnet.army.mil/TJA65A.
Olson, M. 1965. *The Logic of Collective Action*. Cambridge: Harvard University Press.
Paternoster, R. 1987. The Deterrent Effect of the Perceived Certainty and Severity of Punishment: A Review of the Evidence and Issues. *Justice Quarterly* 4 (2): 173–217.
Posner, E. 1998. Symbols, Signals and Social Norms in Politics and the Law. *The Journal of Legal Studies* 27: 765–798.
Posner, E. 2000. *Law and Social Norms*. Cambridge: Harvard University Press.
Posner, R., and E. Rasmussen. 2000. Creating and Enforcing Norms, with Special Reference to Sanctions. Working Paper No. 96, March.
Prakash, S., and M. Ramsey. 2001. The Executive Power over Foreign Affairs. *Yale Law Journal* 111: 231–356.
Protocol_1a. 1977. *Protocol Additional to the Geneva Conventions of 12 August 1949, and Relating to the Protection of Victims of International Armed Conflicts (Protocol I)*, adopted June 8, 1977, art. 51(1), 1125 U.N.T.S. 3, 26, (lawful combatants); art. 51(3), 1125 U.N.T.S. at 26 (unlawful combatants).
Protocol_1. 1977b. *Protocol Additional to the Geneva Conventions of 12 August 1949, and Relating to the Protection of Victims of International Armed Conflicts (Protocol I)*, adopted June 8, 1977, art. 51(1), 1125 U.N.T.S. 3, 26, (lawful combatants); art. 51(3), 1125 U.N.T.S. at 26 (unlawful combatants).
San, S., M. Bastug, and H. Basil. 2021. Crisis Management in Authoritarian Regimes: A Comparative Study of Covid-19 Responses in Turkey and Iran. *Global Public Health* 16 (4): 485–501.
Schmitt, M. 2014. Below the Threshold Cyber Operations: The Counter 1233 Measures Response Option in International Law. *Virginia Journal of International Law* 54 (3): 697–732.
Sunstein, C. 1996. Social Norms and Social Roles. *Columbia Law Review* 96 (4): 903–968.

Sykes, G., and D. Matza. 1957. Techniques of Neutralization: A Theory of Delinquency. *American Sociological Review* 22 (6): 664–670.

Vest. 1776. U.S. CONST. art. II, § 1, cl. 1.

Vida, A. 2005. Defining the Parameters of Cyberwar Operations: Looking for Law in All the Wrong Places. *Naval Law Review* 51: 132–174.

Waid-Lindberg, C., R. Dobbs, and T. Shelly. 2011. Blame The Media? The Influence of Primary New Source, Frequency of Usage, and Perceived Media Credibility on Punitive Attitudes. *Western Criminology Review* 12 (3): 41–59.

WarCrimesA. 1907. Article 75 of the Annex by the Hague Convention IV. *Respecting the Laws and Customs of War on Land*, signed 18 October 1907, Article 23(e).

WarCrimesB. 1977. 18 USC $2441.

Wrage, S. 1999. Examining the 'Authoritarian Advantage' in South East Asian Development in the Wake of Asian Economic Failures. *Studies in Conflict and Terrorism* 22 (1): 21–31.

Yoo, J. 2002. War and the Constitutional Text. *University of Chicago Law Review* 69 (4): 1639–1674.

PART II

Domestic and International Hack-Backs

CHAPTER 6

Surveillance and Disruption

1 Global Justice in Cyberspace

1.1 Familiar Guideposts

Cyberspace is arguably the most fluid and contested open world zone, and yet in spite of its dynamism and fluidity, cyberspace discourse is still dominated by the familiar logic of primary goods/rights and wealth maximization/utility. Internet access, for example, has been characterized as a fundamental human and political right, an implied primary social good of modern existence. Controlling data across jurisdictional borders, guaranteeing the operational security of computer systems, and protecting the accidental or intentional disclosure of personal information have also been characterized as rights. Conceptualizing communicative expectations in cyberspace as rights or primary social goods has its advantages, the most important of which is the recognition that 'digital rights' may trump majoritarian sentiment. We also assume that most digital rights have been codified into formal laws; that some digital rights are secured by sovereign powers while other are secured by international institutions or human rights norms; and that digital rights generally adhere to all persons, regardless of their political or economic status. Finally, we assume that the central focus of digital rights is information and that most of the people who want to ensure the free flow of information across borders are operating in good faith, secure in the belief that relatively unfettered information will ultimately benefit humanity.

A. J. Thomas, *The Open World, Hackbacks and Global Justice*, https://doi.org/10.1007/978-981-19-8132-6_6

What we quickly discover, however, is that mere possession of digital rights or digital primary goods is not enough to ensure equality or equal access in cyberspace; that the entities and individuals best positioned to give meaning to digital rights are often the biggest violators of rights-based expectations; and that information creation and access are not separable from pre-existing cultural, social and economic biases and disparities. Information may travel freely, but entrenched cultural restrictions do not. And interpersonal comparisons, one of the traditional shortcomings of Rawls primary goods framework and utilitarianism,[1] are still difficult to make in cyberspace, both because the pseudonymization of personal identity is relatively easy in digital domains, and because the 24-hour news cycle, fed by massive amounts of digital information, makes it difficult to measure the preferences of communities that only exist online.

The familiar counter-narrative to digital rights is 'digital utility.' Utilitarians prioritize the maximization of general welfare, but they differ on whether welfare should be measured by increases in wealth or efficiency. Emphasis on the former in the digital realm raises the same questions that appear in the kinetic or physical realm: whether wealth accumulation necessarily leads to happiness to all persons regardless of context, whether significant wealth accumulation is even available to most people, and whether it is possible to determine the minimum amount of wealth (or income) sufficient to adequately protect individuals, organization or even nation-states that choose to conduct their affairs online. If the metric is efficiency rather than wealth maximization, on the other hand, the question is whether a particular digital regulation is Pareto-optimal or consistent with Kaldor-Hicks.

But regardless of whether utilitarians measure general welfare using efficiency or wealth maximization, utilitarian justifications for digital regulations, transactions and even simple expectations suffer from three shortcomings. The first is that wealth maximization and efficiency rest on assumptions about voluntariness, state intervention and the superiority of unfettered market relations that many digital communities find morally objectional, even if they are well-positioned to take advantage of them. Second, attempts to measure the efficiency of data-related regulations that encourage wealth maximization frequently overlook pockets of persistent data under-utilization, pockets that persist because of entrenched cultural

[1] Sen (1999b, 325) ("the primary goods approach seems to take little notice of the diversity of human beings").

that transcend wealth or the possession of rights/primary goods. And third, the manner in which wealth maximization and efficiency proposals gather empirical digital information about parties—'subjective preferences'—often misinterpret adaptive preferences that victims use to avoid constant threats of harassment, violence or ostracism.

Capability theory is a useful alternative to rights/primary goods and utility/wealth maximization in cyberspace because it identifies actual constraints that individuals, entities and nation-states encounter everyday, constraints that persist in spite of wealth and the possession of rights/primary goods. Capability-based rationales for 'digital justice,' like their familiar counterparts, now face an old challenge with a new face: retaliatory hackbacks. Retaliatory self-defense and self-help in the physical or kinetic world have existed since the dawn of human civilization. In a thoroughly digitized world, however, the retaliatory impulse has never been more powerful because it combines successive waves of political, cultural and economic decentralization with the technological capacity to execute one's version of retributive or corrective justice with the simple push of a button, from virtually any location. At the same time, the retaliatory impulse in digital space confronts a world that has never been more interconnected legally, economically and technologically, dramatically increasing the possibility that a localized retaliatory digital operation escalates into a full-scale military conflict between nation-states. Most jurisdictions have enacted laws designed to punish or deter cyber hacks. Far fewer laws on retaliatory hackbacks have been introduced at the state and federal level. Almost all of these laws reflect the logic of primary goods/rights or wealth maximization/utility.

1.2 Digital Capabilities

In order to assess alternative understandings of justice and freedom in cyberspace, scholars weary of the pervasive influence of rights, primary goods, utility and wealth maximization have turned to capability theory. Narrow inquiries focus on the minimum requirements of minority participation in cyberspace, and whether those requirements correlate with Sen's understanding of capability. Earlier, I discussed three points on the capability spectrum: disability or cognition-based capability deficiencies, situational capability deficiencies and existential capability deficiencies. Many studies suggest that people with significant disability or cognitive difficulties find it very difficult to take advantage of, or participate in cyberspace, even if they have wealth, primary goods and legal rights to technology and digital information under statutes like the American With Disabilities Act. Cognitive—disability-based capability deficiencies can prevent individuals from using a computer, fully understanding the digital information stored on the computer. It may be impossible to fully equalize the distributive outcomes of cognitive/disability-based capability-deficient people, but even if we could equalize outcomes in cyberspace, cognitive—disability-based capability-deficient people will still suffer from social stigma, the kind of stigma that they could not outrun even if they had wealth, primary goods or rights.

Broad inquiries, however, acknowledge that "technology is an input that enables capability, used alongside other resources like food to exploit its particular characteristics."[2] conceptualizing capability in the digital space also requires an inquiry into the relationship between "technology inputs" and "conversion factors."[3] From this perspective, digital technology like cell phones or the Internet are instrumental tools, comparable to bicycles and cars. Not everyone is capable of using them, but those who can find it easier to realize their capabilities and take advantage of opportunities afforded by luck, economic fortune, natural ability or geographic advantage. Digital technology is also central to the choices we make, to agency and freedom, and perhaps most importantly to the formation of adaptive preferences when intractable capability deficiencies become apparent. Familiarity with, and access to, digital technology may not be as important[4] to capability realization as food and bodily insecurity, but it is extremely important nonetheless, as important to the modern age as literacy. All of these objectives extend individual capability.[5]

The digital equivalent of someone with existential capability disabilities, comparable to someone who has food shelter or bodily insecurity issues, is more difficult and envision because so many societies, regardless of their affluence or international status, have prioritized citizen access to computer and digital information. In many first world and second world societies, it is also relatively difficult to find accidental data deserts that cannot be reached by receivers, cell towers or GPS. You are more

[2] Haenssgen and Ariana (2018, 100).

[3] Id. at 100.

[4] Id. at 101.

[5] Id. at 103.

likely to find deliberately created semi or total data deserts, such as non-signal zones near military installations or areas in which the government has severely restricted Internet access. Data overload, rather than data starvation, is the common scourge of modern societies. It is also not uncommon, for people to look for places where they can live off the grid, relatively free from utility companies, digital signals and even human contact.

Situationally constrained, capability-deficient people confront the limits of the new normalcy in digital spaces: widespread anonymity dramatically increases the likelihood of online harassment, abuse and deception; ubiquitous computer use and always on Internet devices make it easy to monitor confidential communications and ordinary activities; the volatility of cryptocurrency slowly erodes public confidence in market transactions; AI and machine learning enable companies to mimic human communications, making it more difficult to know who or what you are actually communicating with online. Open world zealots and opportunists constantly tell us that all of the aforementioned advantages have made us more free, but so many people feel less free in their everyday lives, trapped in the digital equivalent of Foucault's carceral.[6] Situational constraints that people faced in the pre-digital era, constraints that prevented them from taking advantage of opportunities afforded by wealth, access to primary goods and formal rights guarantees, might be amplified, rather than eliminated, in digital spaces. Situationally constrained capability-deficient people also confront burden shifting strategies that place most of the blame for digital dysfunctionality on victims rather than the system or aggressors. Retaliation, as I mentioned before, is only a challenge in capability discourse because of the common perception that people who retaliate are more capable than people who do not, but digital retaliation practices are an even greater challenge because effective trace back and attribution strategies appear to require a level of technical expertise, access

[6] For more on modern carcerality, see Gottschalk (2006). See also Foucault (1977, 196–217) (disciplinary discourses), Foucault (1991), O'Malley (1996), and Rheingold (2002, 188–190 (Panopticon metaphor and the development of surveillance technologies). In prisons and mental health facilities, administrators organize populations based on the possibility of violence. The spatial organization of violent inmates or patients in prisons and hospitals reflects rational sensibilities of administrative communities, which tend to place responsibility for the management of risk in the hands of institutions rather than individuals.

to wealth or right-based authorization that simply exceeds the capability of most people, regardless of where they are on the capabilities spectrum.

Even though situationally constrained capability-deficient people are probably more 'capable,' loosely speaking, of arranging for or engineering a retaliatory hackback, the retaliatory impulse is probably the same across the entire capability spectrum. VRN's presumptive principle legitimizes this shared retaliatory impulse, while the desert principle factors good faith, responsibility, spillover effects, and the cost of damages caused by retaliation into the entitlement equation.[7] VRN also solves the capability spectrum problem by targeting four capability trigger-points in retaliatory hackback disputes:

1. Whether the victim experienced any capabilities deficiencies before the aggressors initial attack.
2. Whether the victim experienced any capability deficiencies affecting his choice to retaliate.
3. The aggressor's knowledge of the victim's capability deficiencies before the initial attack.
4. Whether the victim experienced any capability deficiencies affecting their ability to exercise due care during the retaliation.

1.3 The Political Tilt of Capability Theory and VRN

Every cyber security dilemma provides decision-makers with a perfect opportunity to re-examine the tension between morality and politics. Hackback authorizations, like all cyber security questions, balance deterrence against competing moral agendas. But it is also not uncommon for decision-makers with different moral or political sensibilities to endorse the same cyber security policy. The same logic applies to a wide variety of public policies, ranging from aggressive judicial review and anticipatory self-defense, to drone strikes and free trade. Any discussion of the 'political tilt' of retaliatory hackbacks, therefore, must account for two distinct possibilities. On the one hand, people may accept or reject the general idea of hackbacks based on their moral or political sensibilities, but on other hand, people with opposing moral sensibilities may find specific hackback strategies politically attractive for the same reasons, the reasons

[7] Mansell (2002, 426).

being effective deterrence given the weaknesses of law enforcement, and retributive justice.

Proponents of capability-based, welfarist and fairness-based defenses of active defense policies may find hackbacks politically palatable once the third-party spillover, attribution and negative externalities problems are resolved, even though they don't share the same set of moral beliefs. Proponents of fairness, for example, may find retaliatory hackbacks politically appealing because they advance progressive priorities. Welfarists may find retaliatory hackbacks politically appealing because they reflect neo liberal or neo conservative priorities. Capability theorists, however, may find VRN's capability-based defense of retaliatory hackbacks politically appealing because capability theory combines cosmopolitanism, left-populism and global economic justice.

VRN reflects a skeptical view of the open world, particularly with regard to the way that all people with capability deficiencies, regardless of kind (cognitive/disability, existential or situational), consistently find it difficult to either protect themselves or take advantage of opportunities in cyberspace. For some many of these people, and by extension, organizations and even nation-states, a largely unregulated, poorly policed, internet is both a blessing and a curse. The idea that capability-deficient entities that are targeted in cyberspace deserve protection, and should receive limited immunity if they choose to hackback, regardless of where they or their attacker resides, seems to combine cosmopolitanism and human rights discourse. The latter, as I mentioned earlier in this chapter, is just another instance of the overlap problem that haunts all capability theories: in the same way that some capability theorists seem to merely expand on, rather than completely reject, well-being or welfare, other capability theorists seem to conceptualize Sen and Nussbaums' formulations of functioning, flourishing and need satisfaction as basic human rights.[8] The cosmopolitan character of capabilities theory, however, is more complicated, both generally and as a basis for any active defense of national security policy.

[8] For the best discussion of the relationship between capability theory and human rights, see Sen (2004) and Sen (2005).

The central theme of cosmopolitanism[9] is statelessness, whether the state is defined in traditional terms, using borders, boundaries[10] and locally defined juridical systems, or in more exotic terms, involving the intentions or consciousness of certain individuals, seemingly alienated from their immediate environments. Conceptually, cosmopolitanism can take three distinct forms: peripatetic (acquiring rights in one jurisdiction that can be exercised in another jurisdiction), instrumental (temporary denizens, brought into an environment or jurisdiction in order to perform a specific function) and preemptive ('universal' human rights that protect someone from hunger, torture or persecution). Politically robust versions of cosmopolitanism are also sensitive to restrictive norms of conduct shared by all persons regardless of background, and restrictive contexts in which equality norms are formed against the backdrop of larger institutional, economic or administrative expectations. Under these conditions, cosmopolitan identity is derived not from the symmetrical categorical requirements of the liberal neutral state, or from cultural identities filtered through specifically designed accommodation policies, but instead from institutional and discursive factors that require participation in deliberative practices and fluency in the dynamics of pressing, short-term social problems.[11]

Sensitivity to the dependence of cosmopolitan expectations on global institutions, as well as restrictive social conditions 'on the ground,' can explain the historical evolution of cosmopolitanism and the link between certain kinds of cosmopolitan thought and populism. In terms of the former, I.R. Rivera[12] has identified three distinct 'ages' in cosmopolitan thought. The first, 'Stoic Cosmopolitanism' (SC), is attributable to

[9] Brock and Bighouse (2005).

[10] Miller (1999, 36) and Pogge (1992, 48-9)) ("Three elements are shared by all cosmopolitan positions. First, individualism : the ultimate units are human beings .. second, universality : the status of the ultimate unit of concern attaches to every living being equally.. [and third], generality : this special status has global force.")

[11] For more on transnational citizenship and multicultural discourse, see Kymlicka (1995) and Habermas (1992).

[12] Rivera (2012).

Seneca, and emphasizes personal integrity, virtue, empathy and the political autonomy, all guaranteed and reinforced by the Roman Empire.[13] The second age, 'Kantian Cosmopolitanism' (KC), attributable to Kant, emphasized citizenship, hospitality, political autonomy and war-avoidance norms held by rational persons and guaranteed by a community of sovereign nation-states.[14] The third age, 'global age cosmopolitanism' (GAC), embraced by Amartya Sen, Martha Nussbaum and Peter Singer, links cosmopolitan expectations such as economic justice and socio-economic responsibility across borders, to global institutions, but GAC also places primary responsibility for the development and actualization of cosmopolitan expectations in local institutions and cultural practices.[15] The universal legitimacy of capability claims may be reinforced by global institutions, but capability theorists are more than prepared to criticize global bodies and local institutions that fail to alleviate the suffering of capability-deficient people in their everyday lives, both online and offline.[16]

In addition to global age cosmopolitanism, capability theorists are also, either implicitly or explicitly, sympathetic to populist formulations that attempt to channel more of the wealth and opportunities generated by capitalist societies into the hands of ordinary citizens. In The Six Faces of Globalization,[17] Anthea Roberts and Nicolas Lamp identified two versions of populist thought: 1) the left-wing populist narrative (wealth for ordinary persons increases modestly in developed nations, wealth for elites increases significantly in developing and developed nations, wealth tends to move from ordinary people to elites in both developed and developing countries), and 2) the right-wing populist narrative (wealth for ordinary persons increases modestly in developed nations, wealth for

[13] The Stoics "accepted the idea of a single humanity united by a model community, and while this by no means required a world government, it did require cultural and moral empathy of the part of individuals." Id. at 92.

[14] Id. See also Kant (1784).

[15] Rivera (2012, 92).

[16] Id. at 98 ('Economic justice is ... a primary aim for GAC. To bring about economic justice, one of the most important duties of the global cosmopolitan and other critically educated thinkers is to find information available through global communication technology and engage in practices of observation and denunciation that are even now being developed.")

[17] Roberts and Lamp (2022).

elites increases significantly in developing and developed nations, wealth tends to move from ordinary people in developed countries to ordinary people in developing countries). Both versions of populism reject the establishment narrative that 'everyone wins from globalization' and the global threats narrative that 'everyone loses from unsustainable economic growth.' They also focus their attention on the actual conditions and constraints that ordinary people face every day.

Capability theorists tend to subscribe to the left-wing populist narrative that protects all capability-deficient people, regardless of whether they live in the developed or developing world. They also tend to reject the moral scapegoating and demonizing rhetoric that play such a prominent role in right-wing populist thought.

2 Empirical and Legal Overview

2.1 *The Mechanics of Cyber Attacks and Hackbacks*

Programmers have always found ways to break into, or 'hack,' computer systems without authorization. Among the most common is a DDoS (denial of service attack) that floods systems with enormous amounts of traffic, eventually preventing authorized persons from using the system at all.[18] Hackers have also installed keystroke software to capture passwords, initiated spear-phishing, ransomware and malware schemes, and even planted code to monitor and shutdown grid systems and nuclear plants.[19] In addition to stealing information and possibly undermining computer systems, hackers employ elaborate means to hide their tracks, using fake IP addresses, virtual private networks and rootkits.[20] The long-terms costs of data breaches, loss of intellectual property, consumer confidence in both the private sector and law enforcement, and even national security vulnerabilities, are almost incalculable,[21] and that cost may not account for the expense of cyber defense software and mitigation. As private firms implement policies to protect themselves from future attacks and

[18] Jayawal et al. (2002, 380).

[19] Valeriano and Jensen (2019, 8).

[20] Shackleford (2019).

[21] Valeriano and Jensen (2019, 15) (national cyber security estimates), Shackleford, Danuvasin, Waite and Zhang (2019, 8) (Equifax data breach costs $125 million; NBER estimates that firms lose 1.16% or market value and 3.2% of sales growth).

mitigate the damage caused by previous attacks, they often complain that they are better equipped to protect themselves and retaliate against hackers than law enforcement because they already know their vulnerabilities and capabilities.[22] But private hackbacks violate Section 1030 of the Computer Fraud and Abuse Act (CFAA),[23] a federal law passed in 1986 and amended in 2008 that prohibits unauthorized manipulation of, or entry to, computer systems.[24] The CFAA defines damage caused by unauthorized access as "any impairment to the integrity or availability of data, a program, a system, or information",[25] distinguishes between harm caused by authorized and unauthorized users, and establishes whether mere access (as opposed to manipulation) to data necessarily constitutes harm.[26]

Cyber attacks typically fall into three categories: espionage ("[alters] the balance of information in a way that enables coercion"), disruption (often takes the form of "DDoS attacks and website defacements that harass a target to signal resolve and gain temporary position of advantage"), and degradations that "degrade or destroy some aspect of the target cyberspace networks, operations and functions."[27] Passive[28] measures like monitoring transactions for suspicious activity, regularly scheduled password changes, and routine scans of downloaded applications are non-invasive (i.e., they don't require gaining access to the system

[22] Shackleford (2019, 3, 18 and 20) (private firms may know their vulnerabilities, but they still experience attribution problems); and Eichensehr 2018, 5) (private/state cyber defense initiatives, ranging from active collaboration to state forbearance of private action and state responsibility for private actors under international law); ACDC (2017, Section 2(8) (CDC acknowledges need for government to be more responsive to private sector complaints about hacking).

[23] Shackleford (2019, 6) (private sector hackback violates the CFAA).

[24] Shackleford (2019, 11).

[25] 18 USC 1030(e)(8).

[26] 18 USC 1030 (a)(5)(a) (prohibits damage caused by authorized persons), 1030(a)5)(B) and (C) only prohibit damage by unauthorized persons). See also Thomas (2017) (disgruntled employee is convicted for violating CFAA after sabotaging his employers computer system by deleting files, disrupting backup systems, and preventing log-in world before submitting his resignation. Defendant had access but not permission to damage system).

[27] Valeriano and Jensen (2019, 8) (cyber attacks lie along a spectrum from most to least intrusive).

[28] Shackleford (2019, 10) (some responses are both passive and active).

of a suspected hacker or a third party).[29] In fact, 'target hardening'[30] is almost mandatory for private firms, part of a broader cyber-hygiene deterrence regime that firms are expected to implement regardless of whether they have already been hacked or expect to be hacked in the future.

Gathering empirical information about the impact of hackbacks can be difficult because very few hackback incidents are actually reported to the police or described in the press. In September 1998, for example, the Pentagon shutdown a browser-based DDoS attack engineered by a hacktivist group called Email Disruption Theater (EMT) by placing offensive applets on the EMT's system to disrupt their browsers.[31] In January 2000, a group called Electrohippies launched a DDoS attack on a server owned by the WTO and operated by a company called Conxion.[32] After tracing the attack back to Electrohippies using an IP trace, and discovering that Electrohippies had sent out email instructing people to mail bomb the WTO servers, Conxion redirected Electrohippies email traffic back to the group, shutting down their system from a few hours.[33] More recently, the US government shutdown servers belonging to Russia's Internet Research Group during the 2018 midterm election in response to widespread Russian interference with the 2016 presidential campaign.[34] The United States has also placed destructive code in software systems that run Russian physical infrastructure installations in response to Russia planting similar code in American installations.

It is not difficult to understand why so little empirical information is available about hackbacks because the practice could reveal embarrassing institutional and technological vulnerabilities.[35] But that has not

[29] Temple (2011).

[30] Valeriano and Jensen (2019, 16, 18) (target hardening).

[31] Jayawal et al. (2002, 381).

[32] Id.

[33] Himma (2007, 112) (Conxion response was retaliatory / punitive, not defensive because "it was not needed to stop the attack").

[34] Sanger and Perlroth (2019) (US installs malware on russian power grid and shuts down Russian troll farm before 2018 election).

[35] Prince (2012) (companies avoid disclosure of past hackback strategies, so statistics on HB are unreliable), Elgin et al. (2012). Victimized firms can also share zero-day information safely and hopefully avoiding future intrusions by joining a vulnerability equities process. See VEP (2017) ("The Vulnerability Equities Process balances whether to disseminate information to tne vendor/supplier in the expectation that it will be patched, or

prevented nation-states and private firms[36] from either engaging in hackbacks far away from the public gaze, or tacitly condoning hackbacks by endorsing the use of active cyber defense. After tentatively embracing active defense for its own purposes while saying little to discourage private firms from following suit,[37] the United Kingdom formally established an Active Cyber Defense Program.[38] At a Blackhat conference in 2012, 36% of the respondents admitted that they had engaged in hackbacks (or 'retaliatory hacking').[39] As early as 2013, the United States implicitly condoned firms tracking down and disabling servers containing stolen intellectual property.

Targeted studies of hackbacks have yielded interesting results. Researchers at the CATO Institute, for example, discovered that even though hacks are increasingly common and expensive, most responses to cyber attacks are proportional,[40] and only 33% of cyber attacks generated any response at all. The same study also demonstrated that most hackbacks focused on disruption and espionage, not degradation. Espionage generated the fewest number of responses, and all attacks tend to generate either a less intrusive response or a non-kinetic response like sanctions.[41] In another study, researchers found marked differences between public responses to cyber attacks and similar responses to conventional or nuclear

o temporarily restrict knowledge of the vulnerability to the US government, and potentially other partners, so that it can be used for national security and law enforcement purposes, such as intelligence collection, military operations and/or counterintelligence."), Ambastha (2019) and Pell and Finocchiaro (2017) (proposing version 3 of the CVSS as a baseline metric for evaluating vulnerabilities). Hacking victims can circumvent CFAA now by enlisting the aid of governmental agencies, who have the authority to act under Section 1030. The agency can then deputize private firms to identify hackers, shutdown on-going hacks and even retrieve stolen data. Such an approach does not provide victims with guidance about what they can do on their own beyond standard passive defense techniques. See Gandhi (2017).

[36] See West (2012) (government quietly deputizing private firms to hackback), and Dipshaw (2018).

[37] NCSS (2016) and Carera (2019) (government pledges to develop "lawful and carefully governed offensive cyber capability to combat and deter the most aggressive threats.").

[38] ACDC (2019).

[39] Boose (2012).

[40] Valeriano and Jensen (2019, 9).

[41] Id. at 10.

attacks. Using an effects-based, as opposed to a means-based approach to cyber conflict, the authors discovered thresholding dynamics that are more likely to lead to either a proportional response or no response at all, rather than an unpredictable, uncontrollable escalation of conflict.[42]

2.2 *Domestic Statutory Response*

Proposed authorizations of hackback operations have been introduced at the state and federal level. State efforts offer a mixed bag of successes and failures. Most states have already passed laws prohibiting hacking, including unauthorized access, ransomware, viruses, spear-phishing, spyware, DDoS and malware.[43] Many states also impose a duty to report data breaches,[44] maintain data security practices[45] and dispose of data as early as possible.[46] The state of Georgia also passed a bill that authorized active-defense measures "designed to prevent or detect unauthorized computer access",[47] but the bill was vetoed by Governor Neal at the urging of technology firms.[48]

At the federal level, the most controversial and comprehensive proposed legislation is the Active Cyber Defense Certainty Act

[42] Kreps and Schneider (2019, 3).

[43] Shackleford (2019, 14) (summarizes state anti-hacking legislation).

[44] Skelrov (2009, 14).The GDPR imposes similar duties for data breaches. Article 33(1) requires controllers to notify supervisory authorities of data breaches, if feasible, unless the breach is "unlikely to result in a high risk to the rights and freedoms of the natural persons." Under Article 33(3)(d), such notification should include steps taken to "mitigate possible adverse effects." Article 36(1) also requires controllers to notify supervisory authorities prior to processing if their data impact assessments indicate a high probability of risk to personal data that cannot be adequately mitigated by the controller.

[45] Shackleford (2019, 12).

[46] Id. at 14.

[47] Georgia (2016).

[48] Google (2018): "Network operators should indeed have the right and permission to defend themselves from attack, but before Georgia endorses 'hackback' authority in 'defense' or even anticipatory potential attacks with no statutory criteria, it should have a much more thorough understanding of the ramifications of such a policy. Provisions such as these could easily lead to abuse and be deployed for anticompetitive, not protective purposes"). Google (2018).

(ACDC).[49] The language of the ACDC amends the CFAA to authorize three kinds of hackbacks. The first consists of hackbacks designed to determine the identity and/or location of suspected hackers.[50] 'Attributional' hackbacks are only authorized if 'defenders' share any information about potential criminal activity that they discover about 'attackers' with law enforcement, and they can include software 'beacons' that "return locational or attributional data."[51] Attributional code cannot damage the attackers system in any way, including destroying data, impairing core system functions or installing backdoor world for subsequent unauthorized world.[52] The ACDC Act also authorizes 'disruptive' hackbacks that terminate on-going unauthorized access to another computer system,[53] as well as hackbacks that monitor foreign systems in order to "assist in developing future intrusion prevention or cyber defense techniques."[54] All of these hackback strategies are subject to a number of constraints, including destroying or irreparably undermining the attacker's system, recklessly causing physical injury or financial loss, creating a threat to public health or safety, persistently disrupt internet connectivity or undermine the operation of an intermediary's system.[55] The ACDC Act also establishes an affirmative defense for active-defense users and excludes from protection those who user beacons in a reckless or destructive manner. Only the latter sunsets after two years.

[49] ACDC (2017).

[50] *Id*. at Section 4(1)(3)(B)(i)(aa).

[51] Attributional data refers to "any digital information such as log files, text strings, time stamps, malware samples, identifiers such as user names and internet protocol addresss and metadata or other digital artifacts gathered through forensic analysis." Id at Section 3(K)(2).

[52] Id. at Section 3(K)(1)(A-B).

[53] Id. at Section 4(1)(3)(B)(i)(bb).

[54] Id. at Section 4(1)(3)(B)(i)(cc).

[55] Id. at Section 4(1)(3)(B)(ii)(1–6). In June 2019, Reps. Graves and Josh Gottsheimer (D-NJ) reintroduced the ACDC Act. A similar act was reintroduced in 2022. See SCAROA (2022).

2.3 International Standards

Cyber attacks implicate the laws of war at two crucial points. The first point is whether a cyber attack produces enough harm to be considered an attack of war, triggering be inherent right to self-defense (*jus ad bellum*). The second point of contact consists of restraints on the use of force once a war caused by a cyber attack has formally commenced (*jus in bello*). Articles 2(4)[56] and 51[57] of the UN Charter define the inner parameters of when a nation is justified to use force in retaliation for a cyber attack. Article 2(4) rests upon a prior determination that a specific cyber attack constitutes an attack, as it is usually defined by international law. Most cyber attacks fail to inflict as much physical damage (either directly or indirectly) as kinetic attacks, but the inflicted harm component of governmental cyber-attack guidelines are granular enough to encompass a wide variety of harms that could be serious enough to trigger war, especially if a cyber attack affected a vital military installation. Cyber infittrations that inflict harm during (or as part of) on-going armed conflict,[58] of both the international and non-international varieties[59] may also be considered armed attacks under Article 2(4). The difficulty of ascertaining whether a cyber-attack produces harm comparable to a kinetic attack on a case basis, however, has prompted experts to develop guidelines, including severity, immediacy, directness, invasiveness, measurability and presumptive legitimacy.[60]

Once a determination is made that a cyber attack rises to the level to the level of an armed attack under Article 2(4), and that a retaliatory measure based on self-defense is justified, international law requires states to observe two constraints when exercising force. The first is that self-defense must be based on necessity. In the context of cyber attacks,

[56] The provision states "All Members shall refrain in their international relations from the threat or use of force against the territorial integrity or political independence of any state, or in any other manner inconsistent with the Purposes of the United Nations." UN Charter, Art. 2, para 4.

[57] The provision states "Nothing in the present Charter shall impair the inherent right of individual or collective self-defense if an armed attack occurs against a Memher of the United Nations, until the Security Council has taken the measures necessary to maintain international peace and security"). UN Charter, Art. 51.

[58] ICRC (2008, 5).

[59] Solis (2010, 58).

[60] Schmitt (1999, 914).

jus ad bellum necessity rules may be difficult to satisfy. One can easily imagine, for example, a scenario in which a cyber attack accidentally destroys classified information on a Defense Department server, but the DOD has a backup copy of the data stored offsite. On the one hand, the data has been stolen and damaged, thereby destroying its value as exclusive and confidential content owned by the government, but on the other hand, a perfectly usable copy is available for immediate use, preserving the government's ability to function with adequate, if not optimal, operational security. Under these circumstances, it is not clear if the government could claim necessity if it launched a retaliatory, kinetic strike against the perpetrators of the attack. Launching an anticipatory attack against a state planning a major cyber attack might also be difficult to justify under the principle of necessity, especially if the attacking state possesses limited kinetic capability. The latter point is extremely important because weak states and non-state actors prefer cyber attacks because of their fundamentally asymmetric character.[61]

The second constraint on the use of force is proportionality.[62] Once again, the nature of cyber attacks will make it difficult to satisfy proportionality requirements. If, for example, there are significant attribution problems after a cyber attack (i.e., the perpetrator is a weak state that successfully eludes detection or a private, non-state actor operating either on its own or with the clandestine support of a national state), determining how much force to use in self-defense, and against who will be extremely difficult. In some situations, proof of overall control over the apparatus used in a cyber attack may be sufficient to prove attribution. It might be even more difficult to fashion a proportional response in self-defense if a cyber attack takes the form of a widespread denial of service hack that exploits multiple computers at different locations. Given these difficulties, it is not surprising that retaliatory responses to cyber attacks can combine kinetic and non-kinetic actions, as long as the proportionality principle is satisfied.

As of this writing, no country (or non-state actor) has formally declared war and launched a retaliatory strike under Article 51 based solely on a hostile cyber attack,[63] although a number of nations have reserved

[61] Jensen (2013, 201).

[62] Wedgwood (2002, 227).

[63] Hathaway et.al. (2012, 840).

their right to do so under international law. If an international or non-international armed conflict is launched based on a cyber attack, *jus in bello* rules that that regulate the actual conduct of war itself will apply. *Jus in bello* rules (alternatively called the law of armed conflict or international humanitarian law) (IHL) focuses on targeting and treatment of prisoners, and it is based primarily on the principles of necessity, distinction and proportionality. Article 49 of the Additional Protocol 1 provides a very basic definition of 'attacks,' to include "acts of violence against the adversary, whether in offence or defence." Such a definition is broad enough to encompass cyber attacks, but the crucial question is how do we envision an armed conflict conducted primarily, if not exclusively, in cyberspace? A full-fledged cyber-war might be difficult to imagine, but such a conflict would certainly require both parties to target specific cyber assets, either acting alone or attached to physical objects like planes, dams or nuclear facilities. Under the principle of distinction, any targeting of cyber assets must clearly distinguish between civilian and military cites.[64] The proliferation of 'dual use' computer facilities, however, will make it difficult to guarantee the protection of civilian cites or individuals in a cyber-war, as required by separate provisions of Additional Protocol I and II.

There is also the question of identifying combatants during a cyber-war. After the attacks of 9/11, the United States initiated a global kinetic war against Al Qaeda, resulting in the capture of many individuals with vastly different combatant designations. Some qualified for full protection under the Geneva Convention, while others only qualified for partial

[64] Article 51 of Additional Protocol I states: (1) The civilian population and individual civilians shall enjoy general protection against dangers arising from military operations. To give effect to this protection, the following rules, which are additional to other applicable rules of international law, shall be observed in all circumstances. (2) The civilian population as such, as well as individual civilians, shall not be the object of attack. Acts or threats of violence the primary purpose of which is to spread terror among the civilian population are prohibited. (3) Civilians shall enjoy the protection afforded by this Section, unless and for such time as they take a direct part in hostilities. (4) Indiscriminate attacks are prohibited. Indiscriminate attacks are: (a) those which are not directed at a specific military objective; (b) those which employ a method or means of combat which cannot be directed at a specific military objective; or (c) those which employ a method or means of combat the effects of which cannot be limited as required by this Protocol; and consequently, in each such case, are of a nature to strike military objectives and civilians or civilian objects without distinction.

protection under Additional Protocol I and II.[65] One can only imagine the definitional chaos of a 'DDOS' cyber attack conducted during a cyber-war that exploits the computers of individuals residing in multiple jurisdictions, some of whom gave their passive consent to be part of the cyber attack. If captured, such persons certainly would not be members of the armed forces, but it would also be difficult to define them as generic POW's who 'took up arms to resist the invading forces' as defined by Article 4(6) of the Third Geneva Convention.[66] Perhaps their passive consent to allow their computer to be used in a widespread DDOS attack would transform them into civilians who can be targeted under IHL because they took "a direct part in hostilities," as defined by Article 51(3) of Additional Protocol I.[67]

Under international law, nation-states can respond to the "undesirable" actions of other states, without committing an act of war, in two ways: retorsion and countermeasures. Retorsion, the legal option, flows from the sovereign power of any state. The formal definition of retorsion is "unfriendly conduct, (1) which is not inconsistent with any international obligation of the state engaging in it, and (2) which is done in response to an International wrongful act."[68] Sovereign states can engage in retorsive, retaliatory action even if another state has only committed unfriendly, but not wrongful, acts.[69]

[65] Prisoners captured during traditional international armed conflicts between states receive protection under Common Article 2. Prisoners captured during non-international conflicts only receive protections from Additional Protocol II and Common Article 3 of the Geneva Conventions.

[66] The provision states that "Inhabitants of a non-occupied territory who, on the approach of the enemy, spontaneously take up arms to resist the invading forces, without having had time to form themselves into regular armed units, provided they carry arms openly and respect the laws and customs of war").

[67] Additional Protocol I, Article 51(3) ("Civilians shall enjoy the protection afforded by this Section, unless and for such time as they take a direct part in hostilities").

[68] DODWar (2016, 1110).

[69] Damrosch (1997, 54).

Given its flexibility in terms of the triggering acts (unfriendly or wrongful), timelines (potentially unlimited in duration),[70] and mechanisms (sanctions,[71] expelling diplomats), retorsion is considered to be more political than militaristic. It is a soft power alternative to hard power exertions. It also reflects the complicated political relationship between two nations because a retorsive act (such as a diplomatic protest after the violation of a treaty between two nations) might give rise to counter—retorsive acts in one context and a formal declaration of war in another context.[72] Attribution requirements are also flexible: State C is not required to specify if their retorsive acts are a response to the conduct of State A or State B, and proxies may be the target of retorsive acts even if State C cannot establish, with absolute precision, that the proxy was operating on the behalf of State B.[73]

Unlike retorsive acts, countermeasures can only be undertaken in response to acts that violate international law. Countermeasures also require "reasonable attribution",[74] must be conducted proportionally,[75] and can only be conducted for a limited time.[76] Some international law provisions prohibit countermeasures from violating "fundamental human rights norms, obligations under the international human rights prohibiting reprisals, or other jus cogent norms."[77] Even though countermeasures must satisfy reasonable attribution requirements, states are under no obligation to target the source of a triggering violation. They may choose, instead, to target a proxy operating with the authority of, or in cooperation with, a sovereign state.[78]

[70] This is the modern view. For an argument that the retorsive measures should only last as long as the triggering act, whether unlawful or merely unfriendly, see Oppenheim (1912).

[71] Law and Martin (2014) (measures "taken against a state to compel it to obey international law or to punish it for breach of international law").

[72] Ruys (2016, 5).

[73] Johnson and Schmitt (2021, 27).

[74] Schmitt (2014, 727).

[75] ILCDraft (2000a).

[76] ILCDraft (2000b).

[77] Ruys (2016, 13).

[78] Johnson and Schmitt (2021, 25).

There is an enormous amount of disagreement in the international law community about the legal, as opposed to moral, status of retaliatory hackbacks. The moral/legal distinction is important because retaliatory hackbacks, standing alone (i.e., without any consideration of spillover effects or recognition under international law), are usually based on a morally recognizable claim to self-defense made by sovereign nations. Almost no one believes that retaliatory hackbacks undertaken by states necessarily constitute an act of war, but it is not clear if they should be defined as 'below the threshold' retorsions or 'below the threshold' countermeasures.

Both retorsion and countermeasures, for example, must be 'non-forcible,' but hackback measures like DDoS or ransomware attacks are more 'forcible' than hackback measures like honey pots or passive beacons. States may also use retorsion in response to cyber attacks that it does not consider to be a critical threat to national interests (such as Obama's response to the 2016 Russian election attack, or Bidens response to the 2020 Solarwinds attack), while reserving targeted countermeasures, including retaliatory hackbacks, for attacks against critical infrastructure or military facilities. Not unlike domestic state and federal laws on of hackbacks, international law treatments of hackbacks are unsettled, but three crucial 'grey' areas are particularly concerning: (1) the wisdom of launching hackback countermeasures against states in response to cyber attacks on critical, but privately held, infrastructure facilities, (2) designating repeated cyber attacks on domestic political processes as per se violations of the principle of 'non-intervention,' thereby justifying the use of countermeasures and (3) as the number of cyber attacks increase, states may opt to use aggressive countermeasures, as opposed to retorsion, in order to signal their willingness to use military force.

3 Cyber-Espionage, Squatting and Surveillance

Zone 3
permissible attacks
impermissible retaliations

3.1 Cyber-Enabled Trade Secret Theft

In a hypothetical situation, Company A has been spying on Company B, a major competitor, for years, gathering as much information as it can about Company B's trade secrets. Company A seizes upon a perfect opportunity to gather even more information about Company B's trade secrets—participation in a cyber vulnerabilities program. Congress is considering two information sharing cyber vulnerability statutes—the Cyber-Security Information Sharing Act (CISA) and the Cyber intelligence Sharing and Protection Act (CISPA)—that contain provisions designed to prevent companies participating in the program from sharing any trade or vulnerability secrets they learn about other firms with third parties. But the legislation has flaws that Company A plans to exploit if the legislation passes.

Section 1104 (b)(3)(B) of CISPA, for example, states that shared information about cyber vulnerabilities "may not be used by an entity to gain an unfair competitive advantage to the detriment of the protected entity or the self-protected entity authorizing the sharing of information." The statute does not define unfair practices explicitly, but the authors of a report provided a practical example of an unfair response to a reported cyber vulnerability:

> Company A shares information about a cyber vulnerability in one of its products with Company B, a competitor in the same marketplace; Company B the next day puts out an advertisement saying, "Don't buy Company A's product because it has the following vulnerability ...instead, buy our product which doesn't have the same vulnerabilities.

In this scenario, unfair competitive advantage refers to using vulnerability information to steer potential customers away from competitors, but the report emphasizes that it would not be an unfair competitive advantage for one company "to build a better, more secure product that can be marketed without reference to the vulnerability shared by the particular entity." This language suggests that the federal government is prepared to countenance companies joining the CISPA network primarily, if not exclusively, to engage in cyber-espionage on competitors, as long as any internal improvements resulting from discovering the weaknesses of competitors do not result in products that specifically reference those weaknesses.

Company A concludes that the state-sanctioned cyber-espionage permitted by Section 1104 (b)(3)(B) is more efficient and cost effective than old-fashioned, cumbersome, commercial cyber-espionage tactics that require a company to physically break into a competitors office facility to find valuable trade secrets. In fact, CISPA seems to provide companies with a way to engage in practically unlimited and unchecked cyber-espionage on competitors without requiring them to contribute any vulnerability information themselves, either about their own operations or other companies cyber operations. CISPA also doesn't require participating companies to provide any vulnerability information about themselves in exchange for vulnerability information they might receive from the government about other companies.

Most experts think that the government would have to provide companies with a banquet of incentives to join a cyber vulnerability information sharing regime because companies wouldn't want to reveal information about their operations to competitors. But aggressive companies like Company A signal their willingness to join CISPA's information sharing regime because it allows them to engage in low-level cyber-espionage with few adverse consequences and no substantial expectation of reciprocity, either from the government or other firms. The absence of any real consequence is reinforced by the House Committee, which states that information sharing is not going to be

> accompanied by any private or public cause of action and thus may only be enforced by a private party choosing to limit or completely curtail sharing of cyber threat information with a bad actor (as well as encouraging others to do so also).[79]

CISPA does not provide vulnerable companies with any recourse (i.e., public or private cause of action) except to either stop sharing information with a bad actor, or withdraw from the entire CISPA network.

Company A also considers the possibility of liabilities from selling cyber threat information acquired through CISPA to non-participating entities. Section 1104 (b)(4) grants immunity to participating companies that 'act in good faith,' but that language only applies to "a protected entity, self-protected entity, cyber-security provider, or an officer, employee, or agent

[79] Farivar (2012).

of a protected entity, self-protected entity, or cyber-security provider."[80] In addition, 'lack of good faith' is defined as "any act or omission taken with intent to injure, defraud, or otherwise endanger any individual, government entity, private entity, or utility."[81] Under CISPA, therefore, only participating private entities that do not intentionally operate in bad faith enjoy complete immunity for their cyber-sharing activities. However, CISPA does not preclude a suit brought by the vulnerable company against a third-party purchaser that engages in unfair practices using cyber threat information from the CISPA network acquired by a participating entity. Consequently, passive cyber-squatting by private firms that chose to participate in the CISPA information sharing system primarily (if not solely) to engage in espionage on rival firms was going to be an on-going concern if CISPA passes, even when the language of the bill places considerable limits on what participating firms can do with any cyber threat information in the economic marketplace.

After a considerable amount of debate, CISPA passes both Houses of Congress and is signed into law by the President. Company A and B join the CISPA information sharing network. Company B dutifully and responsibly reports its own cyber vulnerabilities, but Company A does not. It also concludes, correctly, that Company B stores vital trade secrets on a specific server. Using this information, Company A devises a more aggressive, but still 'traditional,' espionage operation, and conducts the operation in-house, without the assistance of third party. It does not sell what it learns about Company B to another party.

Eventually, Company B discovers what Company A has done, but CISPA does not provide it with a remedy to sue Company A. Company B wants to remain in the CISPA information sharing network for strategic, reputational and technological reasons. It also concludes that if it informed law enforcement, any resulting litigation could take years, providing Company A with enough time to exploit what trade secrets it found on Company B's servers. And even if Company B eventually prevailed in court, it may never retrieve its stolen secrets. A private suit using existing trademark law would also take years. Given these choices, Company B decides to launch a retaliatory hackback against Company A's computer system. The hackback successfully and carefully retrieves

[80] Id. at 3.

[81] Id.

stolen trade secrets from Company A's server. Company A encourages law enforcement to file charges against Company B for illegal access to a computer system under state law and requests further authorization from the government to retaliate against Company B. If there is evidence that Company B was capability-deficient, then the dilemma faced by Company B falls squarely into Zone 3 of VRN. Company B's hack of Company A's computer system is morally permissible if the attack was executed with 'due diligence,' but Company A's request for legal authorization to hackback against Company B's system is morally impermissible because Company A provoked Company B's cyber attack.

3.2 Provocation, Capability and the Presumptive Rule

The saga of Company A (aggressor) and Company B (victim) is, unfortunately, all-too-familiar in intellectual property discourse. In our hypothetical, the alleged theft of intellectual property by Company A is, at least initially, accomplished in a relatively novel setting—a software vulnerability disclosure program—instead of using old-fashioned espionage techniques like dumpster diving, private surveillance or paying off former employees. Our hypothetical also involves the use of retaliatory tactics by both parties: Company B in response to Company A's use of trade secrets learned from the vulnerability program, and Company A in response to a cyber attack engineered by Company B to retrieve trademark secrets acquired by Company A.

Besides capability concerns, the most important characteristic of zone three scenarios is provocation. Repeated IP theft by China, Russia and Iran is deliberately provocative, prompting retaliatory actions ranging from sanctions, and criminal cases against third-party proxies operating with the support of the attacking nation,[82] to occasional threats of military action.[83] In fact, provocation is an inextricable part of IP theft discourse, either as a response by an IP theft victim against an IP thief, or as a potential trigger for an escalating theft/disruption cycle between

[82] FBI reports the indictment of nine Iranian nationals for, among other things, "[stealing] proprietary data... and [selling] that stolen data to Iranian customers, including the Iranian government and Iranian universities" FBI (2018).

[83] Li and Alon (2020) Gundert et al. (2018).

IP competitors in cyberspace.[84] The classic definition of provocation in criminal law reduces murder to manslaughter if there's proof that the defendant acted in the heat of passion.[85] By contrast, provocation deprives the plaintiff of protection under VRN's presumptive rule because it undermines the assumption that a capability-deficient plaintiff's reaction to the defendants' attack was based on good faith. The shift in presumption is strongest for situationally challenged capability-deficient plaintiffs because developmental and existential capability-deficient plaintiffs may not recognize that their actions were considered provocative by the defendant. The severity of a capability deficiency, in other words, can determine the reasonable boundaries of the plaintiff's situational awareness in the context of provocation.

Similar logic applies to the defendant. In evaluating the defendants provocation defense, decision-makers have to distinguish between the reasonableness of the defendants fears about the plaintiffs provocation, and the reasonableness of defendants reaction to the plaintiffs provocation.[86] Provocation by someone with severe capability deficiencies is probably going to be considered less of a threat then provocation by someone with more capability.

3.3 *Opportunities Available to Victims of IP Theft*

In terms of assessing whether Company B's hack of Company A fits into zone 3, there are two preliminary issues: legal permissibility and moral permissibility. Even though Company B was arguably provoked into attacking Company A after discovering that Company A intended to take advantage of Company B's trade secrets, the legal permissibility of Company B's hack is questionable under federal law. The Defend Trade Secrets Act (DTSA),[87] passed in 2016, establishes a civil cause of action if the plaintiff can prove that the idea or concept under scrutiny is indeed a

[84] Rodrigues (2020, 11) (rejects the notion that permitting private actors to hackback is more likely to provoke an international cyber-war than the NSA's well-documented surveillance activities.

[85] Under a volition theory, provoked defendants lack the self-control required for cold-blooded murder. Under an evaluation theory, the provocation defense "embodies appraisals of moral quality." Kahan and Nussbaum (1996, 713).

[86] Robinson et al. (2009).

[87] DTSA (2016).

'trade secret,' that the plaintiff owns the trade secret, that the trade secret has economic value, was involved in domestic or foreign commerce, and was misappropriated by the defendant. The DTSA also provides a variety of remedies, ranging from a temporary restraining order and a preliminary injunction to *ex parte* seizure orders. Plaintiffs must make reasonable efforts to protect their trade secrets.[88]

Given the range of remedial options, why would Company B, faced with a clear case of IP theft, hack into Company A's computer? The chances of getting a temporary restraining order, for example, would be quite good if Company B could establish irreparable harm, a relatively easy task giving what they already knew about Company A's plans. Securing an *ex parte* seizure order would be more difficult, but not outside the realm of possibility. Company B's decision is also perplexing because the DTSA does not permit victims to break into the computer systems of suspected IP thieves, and in fact doing clearly violates the Computer Fraud and Abuse Act.

Establishing the moral permissibility of Company B's initial hack of Company A's server is less complicated than establishing the legal permissibility of Company B's actions. VRN's presumptive rule protects the entitlements of defendants if the initial attack was not designed to exploit the victims known capability deficiencies, or the defendant was provoked by a victim with sufficient capabilities. In the hypothetical, the suspected IP thief, Company A, clearly provoked Company B. If Company B's initial attack of Company A was not specifically designed to target the capability deficiencies of Company A, the presumptive rule will protect Company B's entitlement to cyber hack Company A's computers in order to retrieve stolen trade secrets. Company B's initial attack of Company A, in other words, is permissible under zone three of VRN. By extension, Company A's retaliatory entitlement is presumptively impermissible, even if Company A agrees to conduct a hackback using careful attribution and proportionality.

88 Rowe (2016, 409–10).

3.4 *The Continuing Appeal of Retaliatory Hackbacks in Trademark Secret Law*

Given the availability of so many legal remedies for trade secret theft under federal and state law, what accounts for the continuing appeal of retaliatory backpacks in trade secret law? In spite of the increasing frequency of intellectual property infiltrations like the Solarwinds Orion attack, and the growing awareness that domestic intellectual property victims should not always have to rely on the government,[89] efforts to add hackback language to intellectual property legislation like the EEA and the DTSA have failed. Even authorizations for hackbacks as passive as a monitoring beacon have not made it into intellectual property legislation.

Proponents of retaliatory hackbacks in trade secret law have also watched courts try, somewhat unsuccessfully, to conceptualize the theft of other kinds of digital data (emails and web scraping) as 'cyber trespasses.' The concept of cyber-trespass is derived not from 'trespass to land' (real property), but from 'trespass to chattels (goods),' defined in the Restatement (Second) of Torts[90] as an action that either dispossesses[91] or 'intermeddles' objects owned by another person.[92] Section 218 of the Restatement only imposes liability if the possessor actually loses control of the chattel, intermeddling deprives the possessor of a material interest, the possessor has been deprived of chattel for a substantial period or dispossession/intermeddling has caused bodily harm to possessor or another person/thing in which the possessor has an interest.[93] Numerous courts have already concluded that trespass to chattels is appropriate for cyber cases involving the unauthorized use of long-distance telephone services (phreaking),[94] unauthorized transmission of spam,[95] unauthorized aggregation of consumer data[96] and unauthorized use of domain

[89] Rowe (2016, 409).

[90] Restatement Torts (1965).

[91] Id. at Section 217.

[92] Id. at Section 217 (requires physical contact).

[93] Id. at Section 218, comment i. (1965) ("A mere momentary or theoretical deprivation of use is not sufficient unless there is a dispossession").

[94] Thrifty (1996).

[95] Compuserve (1997).

[96] eBay (2000).

name services.[97] In Intel v Gandi, which involved an unauthorized screen-scraping operation, the California Supreme Court concluded that trespass of chattels did "not encompass, and should not be extended to encompass, an electronic communication that neither damages the recipient computer system or impairs its functioning".[98] The majority focused on harm and intent, but the decision also considered social costs, echoing an approach that many experts find preferable to rules-based 'trespass to chattel' models.[99]

There is also the question of whether an unauthorized hack creates an automatic right of retaliation by the victim under either a 'trespass of chattels' or 'cyber-nuisance' theory, whether that right should give rise to an affirmative defense, and whether the extent of harm caused by the hackback should expose the victim to further liability. Hackbacks are, first and foremost, a form of self-help, but self-help often takes the form of mitigation or negotiation in cyber-trespass cases. Intel, for example, tried to block the defendants emails in Hamidi; both Compuserve and Ebay traced the defendants' address and tried to block their behavior. None of the plaintiffs in these cases faced additional liability. But there is much less support for the idea that cyber-trespass creates a right of direct retaliation for victims, especially when retaliation consists of recovering stolen information from a suspected hacker or even shutting down the hackers' system. Under the Restatement (Torts) definition of trespass to chattel, however, possessors have a limited right to use force to re-establish exclusive possession after a trespass occurs, even if no harm occurs to the chattel, the possessor or any object in which the possessor holds an interest.[100]

Capability theory, which rarely appears in intellectual property theft analysis, may also indirectly explain the continuing appeal of retaliatory hackbacks in trade secret law. The general requirement that intellectual property theft victims must use "reasonable precautions" to protect their trade secrets largely ignores different degrees of capability deficiency. People with situationally constrained capability deficiencies, for example,

[97] Register (2000).

[98] Intel v Hamidi, at 1347, g, g: Intel Corp. v. Hamidi, 30 Cal 4th 1342 (2003). Intel (2003).

[99] Burk (2000, 53-4) (distinguishes between trespass to property and cyber-nuisance models for cyberspace) and Kam (2004, 427).

[100] Restatement (?) at Comment (a), Section 217.

are likely to have higher reasonable precaution thresholds than people with developmental or existential capability deficiencies, assuming that all three are capable of generating trade secrets with economic value. Courts that interpret the reasonable precaution requirement to mean that intellectual property victims have a positive, affirmative and continuing responsibility to protect their trade secrets are also perpetuating the 'blame the victim' ethos that open world zealots and opportunists, in particular, aggressively defend. Such an affirmative duty is not required by VRN to establish the presumptive legitimacy of either a permissible violation or a permissible retaliation. Affirmative, aggressive defensive tactics might be the most efficient way to administer a trade secret system, the best way to incentivize intellectual property owners to accumulate wealth in order to protect their assets, or the best way to enshrine idea protection as a primary good, but it understates the complexity of capability deficiencies in the trade secret arena.

In terms of general capabilities, small business owners are always at a disadvantage because they may lack the resources to aggressively protect their trade secrets[101] in an era of rampant, state-sponsored cyber-espionage and hacker 'arms races.'[102] Even though the consensus view is that trade secret law favors small businesses because of "initial lower cost and [the] lack of government formality",[103] it still operates on the assumption that intellectual property victims have to be more "capable" than most business owners in order to prevail against a suspected intellectual property thief in court.

Is there any possibility that capability-deficient victims, forced to deal with intellectual property theft on a regular basis, might turn to retaliatory hackbacks, even though they have wealth, primary goods and access to formal legal rights? The answer is yes if IP theft victims believe that the litigation process itself may reveal more valuable secrets. This is a capability deficiency that many IP theft victims face every day, and it is strong enough to pull them toward retaliatory hackback solutions, and away from cooperating with law enforcement, private suits or hardening. Even though they may lack the capability to engineer a hackback themselves, their capability deficiencies may make hackback solutions more attractive.

[101] Rowe (2016, 413).

[102] Sharkey (2013, 702).

[103] Rowe (2016, 422).

3.5 *The Efficiency of IP Theft Hackbacks*

Capability-deficient IP theft victims may opt for a retaliatory hackback against their attackers even though they possess wealth and a generous endowment of primary goods. Under VRN, that hackback is permissible if the victim acts out of reasonable necessity, and executes the hackback with due diligence (i.e., proportionally and sufficient attribution after traceback). However, is a VRN-compatible retaliatory hackback in IP theft situations efficient? The weak answer to this question is that under a first reading of VRN, efficiency appears to be irrelevant. VRN only protects capability-deficient parties who hackback against their attackers in a responsible manner if they did not provoke the initial attack themselves. That protection is more important than efficiency. Conceding this question to the welfarists seems to be the prudent choice because granting legal and moral protection to capability-deficient people who hackback responsibly under VRN is very inefficient from the standpoint of traditional economics.

The only *ex ante* restraint on capability-deficient victims of IP theft who choose to hackback responsibly, for example, is that there must be proof that they did not provoke the initial attack. Economists, however, might conclude that capability-deficient IP theft victims can only hackback if they were the 'cheapest cost avoider.'[104] Placing a burden on victims to 'help themselves' before they pursue assistance from the state is not only consistent with the rule, applicable in "pure economic loss" cases, that "the common law does not assist a plaintiff who could have helped himself"[105] but it also, in some cases, provides the plaintiff with an incentive to pursue the most 'efficient' outcome: entering into a contract with the violator that does not require state intervention.[106] If it is difficult to determine who the cheapest cost avoider is, the most efficient outcome apportions responsibility between the plaintiff and defendant under one of the liability rules.[107]

The same logic applies to the allocation of costs and burdens in the cybersecurity arena. Assessing whether it is more efficient for the government or private industry to assume the primary burden of protecting

[104] Sharkey (2013, 683).

[105] Id. at 679, n. 17, citing Stapleton (1995).

[106] Id.

[107] Calabresi and Melamed (1972, 1120-21).

cybersecurity and/or internet access depends on which party can protect network systems more effectively at the lowest cost.[108] If it is not clear who the cheapest cost avoider is in terms of protecting cybersecurity, which is the case for many critical infrastructure operations that are owned by private industry but impact vital national security interests, the burden of security should be shared between stakeholders.[109]

Consequently, if an IP theft victim's self-help, in the form of a retaliatory hackback, is the cheapest way to either "mitigate or avoid"[110] further injury, economists will consider the hackback 'efficient.' The same efficiency considerations may lead to the requirement that victims establish control or possession of a contested object or entitlement before they can retaliate against someone for violating their possessory interests. Imposing a burden to positively establish control and/or possession of a contested item before engaging in self-help (or other authorized forms of retaliation) is efficient because it lowers the avoidance costs of third parties.[111] Avoiding these costs is the central advantage of property rule protection, especially when "the potential users of [a] resource are unidentified and large in number."[112] The same dynamic occurs when societies rapidly create new forms of property, which is a constant problem in the intellectual property arena. Fuzzy notice requirements may suppress the pace of innovation and increase the likelihood of litigation.[113]

VRN does not require capability-deficient IP theft victims who hackback responsibly to establish that their hackback was the cheapest way to avoid additional costs or damages. Nor does it require capability-deficient IP theft victims to aggressively establish their possession over

[108] Sales (2013, 1517) ("Economists often argue that risk should be allocated to the low cost avoider. If the government can reduce a vulnerability more efficiently than a firm, it should pay; if the firm can reduce the vulnerability more efficiently, it should pay.")

[109] Id. See also Katyal (2001, 1099) (considers whether it is efficient to impose the cost of deterring cyber crime on ISPs (internet service providers) as the 'cheapest cost avoiders').

[110] Sharkey (2013, 683).

[111] Id. at 699.

[112] Sterk (2008, 1293).

[113] Elkin-Koren (2005, 410) ("Each new property form may subject third parties to novel duties, thereby dramatically increasing their avoidance costs. The more diversity of terms we allow, the higher the costs of avoidance born by third parties. Avoiders must determine whether they invaded any rights of right holders.")

disputed intellectual property before employing a retaliatory hackback. These two policies may be more efficient in terms of lowering the costs of IP protection using the traditional criteria for efficiency, and they may even encourage innovation in the digital self-help arena.[114] But VRN's protection may achieve a level of efficiency by increasing the opportunity-freedom of capability-deficient IP theft victims.

In terms of expanding the range of reasonable opportunities available to capability-deficient victims, VRN permits the beneficiaries of VRN protection to fashion a retaliatory hackback response that the beneficiary, as opposes to law enforcement officials or the court system (if hackback or mitigation is permitted), prefers. The range of possible traceback and hackback strategies is quite extensive for IP theft, and as long as the beneficiary hackbacks in good faith, using proportional strategies and reasonable attribution, the victim's entitlement to hackback will enjoy protection. Critics could argue that the only credible repositories of expertise about proportionality and attribution reside in the government, but private security firms appear to be perfectly capable of accurately identifying suspected IP cyber thieves, and some IP theft victims may possess a comparable level of expertise themselves. Given data suggesting that responses to cyber attacks tend to be proportional, there is no reason to assume that all, or even most, private hackback undertaken without the explicit authorization of law enforcement officials will necessarily lead to a full-scale, uncontrollable cyber escalation.

In addition to increasing both the quality and quality of reasonable opportunities available to capability-deficient IP theft victims, limited immunity for hackbacks undertaken in good faith and with due diligence may also signal the victim's desire to not only improve his own well-being, but also the well-being of others, a concern that is distinguishable from any concern about externalities or third-party harms caused by hackbacks. One of the persistent criticisms of Sen's capability model is that it prioritizes independence over interdependence, even in his treatment of sympathy. Nussbaum's work on care-giving generally, and care-giver networks in feminist spaces more specifically, establishes a stronger link between capability theory and interdependence than Sen.[115] But regardless of whether the capabilities approach is based on a narrow

[114] Sharkey (2013, 703).

[115] Gasper (2007, 71-74).

or broad understanding of personality and sociality, retaliatory hackbacks for IP theft victims may increase the quantity of reasonable opportunities available to capability-deficient victims.

Evaluating the possible efficiency of a retaliatory hackback for capability-deficient IP cyber-theft victims can proceed in two directions: 1) the efficiency of increasing opportunities for unilateral self-help to victims when the possibility of a cyber escalation is quite low, and 2) the efficiency of encouraging cooperation between capability-deficient IP theft victims and perpetrators when the possibility of a significant cyber escalation is relatively high. When the chances of a significant cyber escalation are quite low, providing unilateral opportunities for retaliatory hackbacks can achieve a limited form of efficiency. Sen's opportunity-freedom model, for example, manages to achieve 'weak efficiency' by minimizing the regret caused by extremely limited opportunities. An intellectual property system filled with content creators who envy opportunities available only to a 'privileged' few who have enough resources or political clout to ensure that their most valuable ideas are protected using 'normal' legal channels will not survive for very long. Preserving the opportunity to hackback for IP theft victims who are not only capability-deficient, but who have had their intellectual property stolen from them repeatedly precisely because they are capability-deficient, might be the most efficient way to not only minimize the accumulated regrets of these victims, but also preserve the integrity of the intellectual property system itself.

When the possibility of a significant cyber escalation following the theft of intellectual property from a capability-deficient entity is quite high, however, encouraging cooperation is the most efficient strategy. Imagine, for example, an IP theft incident involving repeated, massive cyber attacks by a large country against a small country with valuable assets. A similar incident occurred in 2007 when Russia allegedly engineered a massive DDoS attack on many websites in Estonia. Nadiya Kostyuk produced a simple Prisoner's Dilemma[116] game describing four possible outcomes[117]:

[116] For a thorough overview of the Prisoner's Dilemma game, see Section 4.5 of this Chapter.

[117] Kostyuk (2022, 3).

Estonia ->Russia |v	*Cooperates*	*Does Not Cooperate*
Cooperates	Unlikely Scenario: 1) individual hackers are punished, 2) future hacks are deterred	Highly Unlikely Scenario: 1) Russia denies responsibility, 2) Russo-Estonian relations worse
Does Not Cooperate	Likely Scenario: 1) Estonia seeks help from Russia, 2) Mutual assistance treaty is worthless	Highly Likely Scenario: 1) The attacks escalate, 2) Countries are incapable of policing its cyberspace—stepping stone nations nations for future attacks by third parties, 3) Estonia seeks help from NATO and the EU

According to Kostyuk, "Estonia was stuck in the classic prisoner's dilemma—a situation in which a country can be a victim, but cannot afford to identify its victimizer" [p. 2]. The most efficient strategy for immediately de-escalating tensions was cooperation between Russia and Estonia, but that scenario is unlikely to happen, and in fact that is exactly what happened during the Russia/Estonia DDoS incident—Russia refused to participate in Estonia's investigation of the attack, even though Russia and Estonia had an MLAT (Mutual Legal Assistance Treatment) regime in place. Estonia also chose not to launch a retaliatory hackback, even though it was reasonably certain that Russian operatives had engineered the attack. Two versions of the 'no-cooperation/no-cooperation' scenario, one involving a retaliatory hackback, could have led to a major escalation of cyber and kinetic hostilities. Instead, Estonia opted to cooperate, not with Russia, but with other powerful cyber-nations, including the United States. When the possibility of a significant cyber escalation is high, capability-deficient IP theft victims, like smaller nation-states, face the same dilemma in cyber-space: "[they] do not have the ability to accuse their victimizer; instead, they either have to cooperate with the offender or seek support from powerful allies."[118] Minimizing the regret of victims who chose not to cooperate is less 'efficient' than minimizing the threat of a major shutdown in cyberspace through cooperation.

[118] Id. at 4.

4 Ransomware, Botnets and Disruption

Zone 1
impermissible attacks
permissible retaliations

4.1 Preventing Disruptions

Preventing network disruptions is one of the highest priorities of cyber security experts because customers—individuals, organizations and nations—cannot complete transactions, communicate effectively or monitor on-going kinetic or non-kinetic operations without reliable access to information packets transferred by servers connected to the network. Some internet disruptions are accidental, in much the same way that a single car crash can slow down an entire highway, or a large ship stuck in the Panama Canal can dramatically disrupt global supply chains. Network disruptions can also be caused by acts of God rather than acts of people, such as communications delays caused by lightning strikes on electrical equipment, or an undersea earthquake that destroys deep sea phone cables. But the disruptions that pose the greatest risk to modern civil society, and most likely to provoke a retaliatory hackback, are deliberate, malicious and carefully designed to dramatically slow down, if not completely disable, targeted network operations or user communications with their own computer systems.

There are three major forms of network disruption. The first consists of Disruption Denial of Service or 'DDoS' attacks that slow down system operations by flooding the network with forged or malformed data packets. All DDoS attacks contain four components: (1) the actual attacker, (2) victims or hosts, (3) "handlers or masters, which are compromised hosts with a special program running on them, capable of running multiple agents"[119] and (4) "attack daemon agents or zombie hosts, who are compromised hosts that are running a special program and a are responsible for generating a stream of packets towards the intended victim."[120] Network-transport-based attacks (using TCP, UDP, ICMP

[119] Douligeris and Mitrokotsa (2004, 646).

[120] Id.

and DNS protocols) exhaust network bandwidth; exploit existing vulnerabilities to over-consume customer resources; send forged requests to reflectors that relay the requests to servers, eventually overwhelming network resources; or re-broadcasting messages they receive from the user hundreds of times to other users, which send replies back to the original users, falsely amplifying messages in order to shutdown the possibility of normal communications.[121] Reflection/amplification attacks, on the other hand, exhaust network bandwidth by dramatically increasing session connection requests rates, session high-workload requests disguised within a single packet or single http session; sending thousands of tiny, fragmented http packets slowly so that the system times out; or sending empty messages with filled headers and then sending the message contents one byte at a time, forcing the system to wait until all of the contents have been received.[122] DDoS attacks may also shutdown a system by deliberately sending an ICMP echo request for data that exceeds the system's maximum standard IP size[123]; recursively execute the 'finger' command on the users' system[124] or even scramble a system's IP address using a well-known DNS exploit.[125]

Botnets are another popular network disruption tactic.[126] Botnets are responsible for a many cyber attacks and have become a source of income for hackers involved in DDoS attacks, ransomware, spam and clickfraud. Regardless of whether botnet attacks follow the client–server (controlled by a central botmaster) or the peer to peer approach (no central control), 'botnet herders' can control a variety of wireless devices (IoT or 'internet of things') infected with botnet malware, and can even modify code dynamically to take advantage of opportunities or environments.

Botnet attacks could escalate into kinetic conflicts between nation-states if an attack cripples critical infrastructure like power grids, nuclear plants or military installations, or involve a 'dead man's trigger,' which is or malware code residing on an infected device that automatically deletes data on the device if the code detects any attempt to delete it from a third

[121] Zargar et al. (2013, 3).

[122] Id. at 4.

[123] Douligeris and Mitrokotsa (2004, 645).

[124] Id.

[125] Id. and Zargar et al. (2013, 7).

[126] DDoS attacks are often executed by botnets.

party. And yet, in spite of these concerns, governments and private firms regularly employ shady 'white hat' firms to hack botnet swarms, often in order to avoid potential violations of the law.

Developers have employed a variety of methods to preventing botnets from attacking other systems. In order to shutdown the Njrat and Njworm malware bots, for example, Microsoft obtained a temporary restraining order and took control of no-IP domains run by Vitalwerks Internet Solutions because the malware used those domains to establish a command/control server for its bots. At other times, prevention requires actual modifications of the code on infected devices. To shutdown the Mirai botnet, researchers located a stack buffer overflow vulnerability and modified the offending code on infected devices to end all processes, but before those modifications took place, Mirai used thousands of infected IOTs like DVRs and webcams to compromise Twitter and Reddit with DDoS attacks. IOT companies can also minimize bot attacks by eliminating vulnerabilities in their own software, such as hard-coded passwords. Once discovered by attackers, those backdoor passwords can be used to infect the code of thousands of IOT devices.[127]

While DDoS and botnet attacks significantly slow down network operations, ransomware attacks prevent victims from using their system at all. Asymmetric ransomware was first developed in 1989 and has rapidly become one of the most traumatizing forms of malware on the internet. The basic purpose of ransomware is extortion. After the code is delivered by a Trojan (attachment, vulnerability exploited by a backdoor, phishing email),[128] victims are informed, by an attacker that they may never be able to identify with any precision, that either some or all of the victim's files have been encrypted by a private–public key combination that is unique to each victim. Victims are told that the private key necessary to decrypt their files will only be revealed if they pay a fee. To add insult to injury, victims who choose to pay the ransom amount, using a cryptocurrency that is very difficult to track like Bitcoin or Ethereum,[129] may never receive the private key, leaving them without resources or access to valuable files. Ransomware typically falls into five categories. The first, commonly referred to as PC Locker, prevents the victim from accessing

[127] Stark (2017).

[128] Kumari et al. (2019).

[129] Id. at 19. See also Jun and Kostyuk (2021).

data files on their system using the resident user interface, without actually modifying, encrypting or deleting the files themselves. Since advanced users can easily circumvent UI-based restrictions, PC locker attackers typically put enormous pressure on victims, including suggesting that they are under criminal investigation.

Unlike PC locker malware, 'data locker' ransomware actually encrypts targeted data files, rendering them useless unless the victim pays. Encryption of targeted files often takes place before the victim is told that ransomware software was already installed. Ransomware may also copy files from the victim's system, and threaten to distribute them unless the victim pays a fee ('doxware'); flood the victim's system with an endless series of notices and popups about security vulnerabilities unless the victim pays a fee ('scareware'); or prevent the victim from using his files until third parties pay a fee ('popcorn time').[130] Ransomware has grown so popular as a disruptive platform that entire industries have emerged called RaaS (ransomware as a service). RaaS operations, typically based on Random32, a ransomware toolkit written in Javascript, have enabled novice hackers to locate and hire advanced hackers to install ransomware on desirable targets without the necessary expertise to write and install the code themselves.[131]

4.2 *The Victim's Revenge*

In September 2019, Tobias Fromel, the victim of a cyber attack based on Muhstik ransomware package that targets exposed QNAP NAS devices running PHPMyAdmin, paid roughly seven hundred dollars after his computer was infected, but he also exacted revenge by infiltrating the attackers' command/control server. The attacker created new passwords for hack victims using a PHP script and then stored the password temporarily on a database that was accessed later after the victim paid the ransom. Web shells on the attackers server allowed Fromel to gain access to the script, create a new script and generate thousands of decryption keys that other victims could use to descrypt their files without

[130] Kumari et al. (2019, 17).

[131] Tandon and Nayyar (2019).

paying a ransom. Fortunately, Fromel published the decryption keys, and a decryptor program that victims used successfully.[132]

Fromel could have extorted the attacker, constructed a ransomware scheme that did not contain a webshell/PHP script vulnerability, or even use his newfound knowledge of the Muhstik malware to develop a worm that moved beyond the attackers' server to other systems. He also could have executed a retaliatory hackback without exercising due care, eschewing careful attribution or any concept of proportionality. But he did exercise due care.

Fromel's hackback clearly satisfies the structural requirements of zone 1 under VRN, but what about the substantive requirements of VRN that demand an inquiry into whether a cyber-attack victim lacked the capability to protect him/herself from a violation in cyberspace, but still had enough capability to engineer, request or even pay for a hackback after the initial attack? In regard to the first question, ransomware victims fall into a variety of categories. Fromel, a sophisticated programmer in his own right, clearly belongs on one side of the ransomware victim spectrum. Some ransomware victims resemble Fromel, either because their constant presence online (engaged in legitimate or illegitimate activities) makes them a convenient, easily identifiable target, or because they are considered to be wealthy enough to pay a requested ransom. But the ubiquity of computer use by individuals, businesses and governments has increased the number of potential ransomware targets exponentially. In fact, any entity with sensitive data and sufficient resources could be targeted for a ransomware attack. The reverse logic is also true: entities that don't appear to have many resources or possess valuable, sensitive data are less likely to be targeted. Given the vast array of potential victims, it should come as no surprise that many ransomware victims are extraordinarily

[132] For two more examples of cyber victims who hacked back, see Andy Greenberg, "North Korea Hacked Him. So He Took Down Its Internet." Security (2/2/22) (an independent hacker named P4x was hacked by North Korean operatives, and staged his own hackback six months later, but he ignored proportionality norms and brought down North Korea's entire internet temporarily), and Preetipadma K, "Lockbit Faces a DDoS attack from its Victim Entrust (8/25/22) (A ransomware outfit called Lockbit was counter-hacked by the security firm Entrust, one of its victims)." If P4x had been capability deficient, and the North Korean operatives' hack took advantage of P4x's capability deficiencies, the entire incident would fall under VRN's zone 4. P4x would be required to pay for half of the damages caused by his retaliatory hackback because he refused to execute a proportional attack.

unsophisticated, both in terms of the capacity to protect themselves from cyber attacks of all kinds, and their capacity to do a trace back and retaliate against the attacker.

4.3 *Capability Challenges in Ransomware Cases*

Ransomware operations ranging from malicious loan wolves and cyber criminal gangs to full-fledged ransomware as a service outfits always look for a particular kind of target. In the public imagination, the ideal target is just someone with enough resources to pay the requested ransom immediately,[133] but the ideal target may often have a more complex profile. Like a virus that doesn't want to kill its host before it has a chance to spread and infect other hosts, ransomware operators want to engage the target long enough to extract the maximum amount of money, but not long enough for the target to track back or locate the digital wallet that the ransomware operate uses to store extorted funds. They also want someone who is understandably afraid that confidential information will be disclosed or important files will be permanently lost. Institutions with security information, like law enforcement, technology firms, critical infrastructure and public health are prime targets. The same operators were also looking for targets that will not reveal that they are ransomware victims until after the ransom has been paid, if they were to be able to pay it at all. Beyond the cost of the ransom, reputational costs can result in the loss of careers.

Most of the literature on ransomware focuses on a typical cluster of events: (1) entities targeted for ransomware, (2) why victims pay the ransom and (3) what happens after they pay the ransom. But experts have devoted less effort to explaining why victims choose not to pay ransoms. The classic example of a situationally constrained capability problem is the abused mother who remains in her home, despite having wealth, primary goods and rights protection, because she is afraid of what her husband will do to her children immediately after she leaves. Even though she is essentially being held hostage by her abusive husband, she remains in the home and quietly develops adaptive preferences to help her tolerate her situation until she can find a solution to her problem, a solution that might lead to retaliation, either by herself or with the assistance of a

[133] Atapour-Abarghouei et al. (2019, 3) ("significant financial gain continuously drives the creation and spread of Ransomware").

third party. Wealthy organizations that have been harassed and threatened repeatedly by terrorist groups face a comparable dilemma. Many organizations have been told that a terrorist group has taken some of their employees hostage, and unless the organization pays the ransom, their employees will be shot. Even though these organizations have wealth, they may choose not to pay the ransom for variety of reasons. They may also avoid law enforcement. Instead, they may try to locate and liberate the hostages and eventually retaliate against the terrorist group.

A significant percentage of ransomware victims refuse to pay the ransom even though they have wealth, sometimes because they're afraid that the hacker will attack again in the near future, and sometimes because they're convinced that the hacker will not release their systems even if they pay the ransom. The same victim may choose not to engage law enforcement even though there are many legal protections in place for victims. Instead, victims may find a way to regain control over vital files and the operating system if the ransomware software is simply 'scareware' or 'lockerware' because neither technique encrypts files. If the victim determines that the ransomware is crypto ware that does encrypt all files, they may still decide to re-initialize their entire system, absorb the loss, resort to back up files and install improved hardening software. But regardless of what they find when they inspect their system, they don't pay the ransom or engage law enforcement. They also don't try to locate the crypto wallet that they're hacker intended to store the ransom because no ransom was paid. This kind of retaliatory hack starts with a trace back of a hacker's IP address, which is difficult but not impossible to discover once system control is restored, and ends with the retrieval of information about the hackers past deeds and funds if they're available.

Like the affluent abused married mother who refuses to leave her home, or the harassed wealthy company that refuses to negotiate with terrorists holding employees hostage, ransomware victims who refuse to pay are often criticized for not taking advantage of opportunities available to them because of their wealth, access to primary goods and rights. But capability theory can explain the decision not to pay as well as the decision to use a retaliatory hackback. In their every day lives, some victims who refuse to pay may be either incapable of, or unwilling to, take advantage of available opportunities. It might be more difficult to sympathize with affluent victims of cyber-crime than cognitive/disability or existential capability-deficient victims,[134] but the same could be said about most situationally constrained capability-deficient people, entities or a nation-states. Capability theory is multidimensional; it works from the bottom up and the top down. Even after your account for the overlap problem described in Chapter 3, capability theory can still provide a compelling explanation for the capability constraints of starving people in the third world, gender oppressed people in the second world and data deprived people in the first world.

4.4 The Classic Zone 1 Scenario

Ransomware attacks and retaliations for those attacks almost always fall into zone one of VRN: impermissible attack and permissible retaliation. With regard to the impermissibility of the ransomware attack, most jurisdictions have enacted laws prohibiting criminal extortion through technological means. The Lockit ransomware, delivered by email attachments in July 2022, exploited web browser vulnerabilities, in violation of state law.[135] VRN, however, requires two additional conditions before an initial attack is considered impermissible under zone one: (1) the aggressor must choose a method of attack that exploits the capability deficiencies of the victim (the case for VRN protection is stronger if the perpetrator deliberately chose a ransomware method that exploited the victim's known capability deficiencies) and (2) the plaintiff must not provoke the initial attack.

Capability-deficient victims may refuse to pay the ransom, or even contact law enforcement, due to situational constraints in their immediate environment. But ransomware operators may exploit victims' capability deficiencies in other ways, such as targeting affluent individuals with cognitive/disability issues, or forcing affluent people to pay the large ransom because the attacker knows how much the victim is worth and can afford to pay. Others may pay but not disclose to law enforcement or the press that they paid a ransom for variety of reasons (such as loss

[134] The decision making process of the former seems to be based on choice, whereas the decision making process of the latter seems to be based on dire necessity.

[135] D'Agostino (2022).

of reputational capital, insurance/liability problems, possible disclosure of additional vulnerabilities), even when required to do so by law.[136]

The permissibility of a retaliatory hackback after a ransomware attack in zone one depends on four factors: 1) the technological feasibility of a trace back effort given the nature of the ransomware, 2) whether a hackback can be done with due diligence, 3) evidence that the plaintiff, due to his own capability deficiencies, was forced to retaliate rather than use his wealth or law enforcement to pursue the attacker and 4) whether retaliatory strategies that require a back door or unauthorized access to secure information can receive any legal protection.

Retaliatory hackbacks after ransomware attacks tend to fall into two broad categories. The first consists of retrieving ransomware money stored in a cryptocurrency wallet using the bitcoin address supplied by the attacker. Once the victim, a third party or law enforcement obtains the bitcoin address, they simply 'sit and wait' for someone to download the victims money into the address, and then trace back the depositors to their source. Anonymous or pseudonymous protocols make this process difficult, but not impossible judging from the success of law enforcement efforts. On July 19, 2022, for example, the DOJ announced that it had obtained a seizure warrant to forfeit files in a cryptocurrency account controlled by North Korean hackers that contained ransomware money paid by healthcare providers in Kansas. A Kansas hospital paid $100,000 in bitcoin after the hacker encrypted their files. Using the information provided by the hackers to the victims the FBI traced the cryptocurrency address, and waited for a deposit to arrive, which it did on April 2022. The money was eventually returned to the Kansas hospital.

DOJ touts the success of its anti-ransomware wallet forfeiture operations,[137] but it does not reveal full details about how it successfully located the hackers who control cryptocurrency wallets. The FBI might have actually broken into a server to obtain the private key necessary to enter the wallets. What we do know is that the FBI did not immediately reveal that it was in possession of private keys obtained from Russian servers controlled by the hacker REvil, even though that information could have spared victims the agony of ransomware payments

136 See eg CIRA (2021).

137 FBI (2022) (Yaroslav Vasinski, a Ukrainian national and hacker associate with the rebel run somewhere gang, was it rain in Texas for arranging multiple ransomware attacks).

worth millions. Consequently, even though the FBI has been transparent about the legitimacy of securing cryptocurrency accounts used to pay for ransomware, and returning ransomware paid back to the victims, it has not been very transparent about the legality of the trace back procedures required to identify hackers and secure access to wallets. The FBI's success record in ransomware cases, however, suggests that the tactics that they use are legally permissible.

Another popular retaliatory response to a ransomware attack involves breaking into a command-and-control server. In order to formally establish their presence in, and mastery of, computer systems, hackers often create a command-and-control (C2) infrastructure to send commands between the victim's system and another platform that they already control. Sophisticated C2 frameworks try to mask their messages, mimicking communications used by other protocols like HTTP or DNS. C2 frameworks may establish control over zombie machines to communicate with other services clandestinely, or establish an elaborate messaging relay across many machines to control botnet swarms that use beacon or "phone home" capabilities to coordinate attacks. In addition to controlling message flow between compromised computers, C2 frameworks have also gathered data ('exfiltration') about victims, piggybacking server information onto messages sent through the control and command system that can be used for further extortion purposes. C2 frameworks can be centralized, operate peer to peer, or even consist entirely of messages sent on public social media sites like Twitter or Facebook, but regardless of the form, C2 systems are highly illegal and very popular among ransomware groups because they are so decentralized. Decentralization, however, comes at a cost because breaking into a C2 infrastructure can dismantle an entire cyber-crime network in one blow. This is essentially what Mr. Fromel did. He broke into a command-and-control shell server created by the Reservoir group, located an encryption program used by the C2 to generate keys for ransomware victims, and used a decryption program he wrote to help the victims generate keys.

Since there is evidence that some retaliatory hackback responses to ransomware attacks are probably legal and can be done safely, three 'zone one' requirements are satisfied: technical feasibility, due diligence and arguable legitimacy. The last factor requires proof that an individual was pulled toward retaliation, rather than law enforcement, a private suit or increased hardening, because of his capability deficiencies. This issue raises a question that I discussed in Chapter 4, which is the common assumption that people who retaliate are more capable than people who don't. The amount of expertise required to trace back a bitcoin address, locate a cryptocurrency wallet on an obscure server, decrypt files and break into a C2 system server is considerable. At first blush, it is very difficult to characterize ransomware victims like Mr. Fromel as 'capability-deficient.' And yet, he did not contact law enforcement, rely exclusively on hardening or initiate a private suit, even though he probably possessed some wealth, had access to primary goods and possessed legal rights as a ransomware victim. Instead he chose to retaliate on his own, relying on his own skills to secure some form of immediate justice.

One possible explanation for his behavior is that Fromel, as a professional programmer, might have harbored deep distrust of law enforcement, either for ideological reasons or because he had actually engaged in questionable cyber activities in the past. And this situational constraint might have limited his options in spite of his wealth, access to primary goods or access to formal rights. Given his own set of capability deficiencies, retaliation without the assistance of law enforcement might have been the most attractive option because he could quickly shutdown the ransomware operator, while avoiding the unwanted attention of the police.

4.5 *The Efficiency of Anti-Ransomware Hackbacks*

For a wide variety of reasons, capability-deficient ransomware victims may choose to execute a retaliatory hackback against their suspected attackers, even when they possess primary goods, formal rights and wealth. That hackback might take the form of breaking into the command-and-control server of the attacker, or breaking into a computer to obtain the private key of the wallet used to hold ransom proceeds, but regardless of its form, it is undertaken on its own, or with the assistance of private parties, rather than law enforcement.

Hackback operations against suspected ransomware operators are now routinely authorized by courts if they are done by law enforcement. The details of those operations are never completely revealed to the public. VRN strengthens the moral argument for authorizing retaliatory hackbacks against suspected ransomware operators if the operators specifically targeted the victim's capability-deficiencies, and the victim conducts the hackback in good faith, and with 'due diligence.' In this section,

I consider the question of whether private retaliatory hackbacks against ransomware operators are 'efficient.'

Research on this question focuses on the most efficient way to either extract money from an intended target or avoid paying the ransom. Ransomware negotiations are a perfect laboratory for studying the dynamics of two-player, one-shot games. Each player is rational, self-maximizing and modifies his/her actions based on an expected payoff. In order to maximize that expected payoff, players have to guess the intentions of other players and then decide to either cooperate or defect.

Game theorists have analyzed the negotiations between ransomware operators and the intended victim using the Prisoner's Dilemma. Under the Prisoner Dilemma, a single-shot, two-player, zero-sum game,[138] two isolated prisoners have to determine the best way to minimize their respective sentences. Both prisoners have an incentive to confess (which will lead to an intermediate sentence), even though remaining silent will lead to a minimum sentence, because if one party confesses and the other doesn't, the confessor will receive no sentence but the non-confessor will receive the maximum sentence. Since neither prisoner knows what the other prisoner will do, the wisest course of action is to confess in order to avoid the harshest sentence.

An alternative Prisoner's Dilemma game illustrates the meaning of Pareto efficiency. Imagine two parties negotiating to build a home. If one party refuses to cooperate (free rider), his payoff will be 35, while the contributing parties payoff will be 28. If both parties contribute, the payoff will be 32 for each. If neither contributes, the payoff will be 30. Pareto efficiency refers to a scenario in which at least one party is better-off, while no party is worse-off. Out of the four scenarios in our house building game,[139] one scenario is Pareto inefficient (player 1 contributes and player 2 does not) because it is the only scenario that makes one player worse-off if you move to another scenario. The other three scenarios are Pareto efficient because you can't move to another scenario (from a Pareto efficient scenario) and make at least one player worse-off.

[138] Zero-sum games are scenarios in which "the loss to one side is considered a gain by the other.". Rid (2012, 2)

[139] The four scenarios are – player 1 contributes / player 2 contributes, player 1 contributes / player 2 does not contribute, player 1 does not contribute / player 2 does not contribute, player 2 contributes / player 1 does not contribute.

Applying game theory to ransomware negotiations is more complicated than the basic and modified Prisoner's Dilemma games because ransomware operators and victims usually have multiple agendas and options besides extracting or paying a ransom. The victim, for example, will certainly modify his strategy if he has a recent, offsite backup of encrypted data, suspects that he can successfully traceback and locate the ransomware operator, finds a way to get around crypto-locked software, or suspects that the ransomware operator will not release the frozen data after getting paid. Similarly, the ransomware operator may decide to freeze the victim's system for a short period because prolonged negotiations increase the likelihood that the ransomware operator will be caught, thereby incentivizing the ransomware operator ro accept a lower ransom offer quickly.

With all of these options, the decision-tree for a ransomware game will be very complex. But a simple ransomware game based on three assumptions—there is only one guaranteed outcome (releasing the victim's computer), the game requires the cooperation of both parties (the victim must agree to pay X and the operator must agree to accept X) and the victim's eventual gain (restored access) fully compensates for the victim's initial loss (payment)—will resemble the following chart:

Ransomware Victim ->Perpetrator |v	*Pay Amount*	*Do Not Pay*
Accept Payment	40, 40 (PE)	35, 45 (PE)
Do Not Accept Payment	45, 35 (PI)	30, 30 (PE)

Every scenario is Pareto efficient except for the scenario in which the victim agrees to pay X and the perpetrator refuses to accept X because, from the other three spaces, it is impossible to go to another spot that doesn't make one party worse. From the Pareto inefficient spot, moving to another spot will necessarily make another party worse.

A third variant of the Prisoner's Dilemma is a game in which the players not only have many options (as well as multiple-shot interactions), but also perceive themselves as something other than rational self-maximizers. One could imagine a ransomware game in which an ecologically conscious, environmental ransomware operator discovers exactly how much a logging company can afford to pay in order to release data that the operator intends to encrypt, but refuses to request that amount because it would prevent the target from carrying out some other ecologically beneficial purpose that the target is both uniquely positioned, and completely willing, to perform. Instead of immediately demanding a ransom that the target can afford to pay, the ransomware operator stalls and eventually demands a smaller amount.

One could also imagine a scenario in which efficiency is measured not in terms of securing a maximum economic payoff through mutual cooperation, but instead in terms of maximizing the number of reasonable opportunities based on the decisional priorities of each player. If a ransomware victim believes that the number of reasonable opportunities he will have to successfully track the perpetrator (including a retaliatory hackback) actually increases with the passage of time, and the perpetrator believes that he will have more opportunities to extract a high ransom from the target with the passage of time (assuming that the victim's appreciation of the value of encrypted data increases), then the efficient, optimal strategy for both the perpetrator and the victim is to slow-walk, but to continue, negotiations. At least in the short term, maximizing reasonable opportunities would mean more than maximizing economic gain.[140]

References

ACDC. 2017. Active Cyber Defense Certainty Act, H.R. 4036, 115th Congress.

ACDR. 2019. *Active Cyber Defense Report*, July. http://www.ncsc.gov.uk/report/active-cyber-defence-report-2019.

Ambastha, M. 2019. Taking a Hard Look at the Vulnerability Equities Process and Its National Security Implications. *Berkeley Technology Law Journal (BTLJ Blog)*, April 22.

Atapour-Abarghouei, A., Stephen Bonner, and Andrew Stephen McGough. 2019. Volenti Non Fit Injuria L Ransomware and Its Victims. arxiv:1911,08364v1 [cs.CR], November 10.

[140] This scenario is the functional equivalent of a Nash equilibrium, in which "no one can improve their situation by changing strategy : each person is doing as well as they possibly can, even if that does not mean the optimal outcome for society." See GTPB (2016), and Pathology (1994, 25) 25 ("A Nash equilibrium occurs if there is a potentially self-reinforcing agreement whereby each actor 'does what is best for her given what others [would] do." It can be understood intuitively as an agreement from which no party has an incentive to defect."), citing Przeworski (1991, 20).

Boose, S. 2012. Blackhat Sources: 36% of Internet Security Experts Have Engaged in Retaliatory Hacking. *BusinessWire.com*, July 26.

Brock, G. and Bighouse, H. 2005. *The Political Philosophy of Cosmopolitanism*. (New York, Cambridge University Press).

Burk, D. 2000. The Trouble with Trespass. *Journal of Small and Emerging Business* 4: 27–58.

Calabresi, G., and A. Melamed. 1972. Property Rules, Liability Rules and Inalienability: One View of the Cathedral. *Harvard Law Review* 85: 1089–1097.

CIRA. 2021. The Cyber Incident Reporting Act of 2021, S.3875, September 2021, 117th Congress, 1st Session.

Coase, R. 1960. The Problem of Social Cost. *Journal of Law and Economics*. 3: 1–44.

Compuserve. 1997. Compuserve, Inc. v Cyber Promotions, Inc., 962 F.Supp. 1015.

Cordes, J. 2011. An Overview of the Economics of Cybersecurity and Cybersecurity Policy. *Report GW-CSPRI-2011-6*. June 1, 2011.

D'Agostino, S. 2022. An Outrage, Silence and Finally Acknowledgment of a Cyber Attack. *Inside Higher Education.com*, August 18.

Damrosch, L. 1997. Enforcing International Law Through Non-Forcible Measures. Working Paper Series.

Dau-Schmidt, K. 1990. An Economic Analysis of the Criminal Law as a Preference-Shaping Policy. *Duke Law Journal*. 1–38.

Dharmapala, D. and Garoupa, N. 2002. Penalty Enhancement for Hate Crimes : An Economic Analysis. *Economics Working Papers*. 2002, 12.

Dipshaw, R. 2018. A Federal Policy Loophole Is Supporting the Hack-For-Hire Market. Can It Be Closed? *Slate.com*, June 20.

DODWar. 2016. *US Department of Defense, Law of War Manual* (June 2015, updated in 2016).

Douligeris, C., and A. Mitrokotsa. 2004. DDoS Attacks and Defense Mechanisms: Classification and State of the Art. *Computer Networks* 44: 643–667.

DTSA. 2016. Public Law 114–153 (114th Congress) 130 Stat. 376.

eBay. 2000. eBay, Inc, v, Bidders Edge, 100 F. Supp. 2d 1058 (ND Cal. 2000).

Eichensehr, K. 2018. Today's Revolution: Cybersecurity and the International Order—A Review of Lucas Kello's the Virtual Weapon and International Order. *LawfareBlog.com*, February 8.

Elgin, B., D. Lawrence, and M. Riley. 2012. Coke Gets Hacked and Doesn't Tell Anyone. *Bloomberg.com*, November 4.

Elkin-Koren, N. 2005. What Contracts Cannot Do : The Limits of Private Order Facilitating a Creative Commons. *Fordham Law Review*. 74(2): 375–422.

Farivar, C. 2012. CISPA Advances in House, as EFF Decries Bill's Revisions. *Ars Technica.com*, April 26. http://arstechnica.com/tech-policy/2012/04/cispa-advances-in-house-as-eff-decries-bills-revisions/.

FBI. 2018. *FBI Report*, March 23.

FBIReportB. 2022. DOJ/FBI Report, March 9.

Foucault, M. 1977. *Discipline and Punish*, trans. Alan Sheridan. New York: Pantheon Books.

Foucault, M. 1991. Governmentality. In *The Foucault Effect: Studies in Governmentality*. New York: Harvester Wheatshaf.

Gandhi, D. 2017. Legal Hack Back: Balancing Risks and Benefits. *Information and Privacy Law Group (Au Law Blog)*, November 7.

Gasper, D. 2007. Adding Links, Adding Persons, and Adding Structures : Using Sen's Frameworks – A Response to Qizilbash, *Feminist Economics*. 13 (1) : 67–85.

Georgia. 2016. Georgia Bill SB 315. Section (b.1)(2)(C).

Google. 2018. *Letter from Google and Microsoft to Governor Nathan Deal*, April 16.

Gottschalk, M. 2006. *The Prison and the Gallows: The Politics of Mass Incarceration in America*.

GTPB. 2016. 'Game Theory: Prison Breakthrough. *The Economist*. 5/12/2016.

Gundert, L., S. Chohan, and G. Lesnevich. 2018. Iran's Hacker Hierarchy Exposed: How the Islamic Republic of Iran Uses Contractors and Universities to Conduct Cyber Operations. *Recorded Future*. CTA_2018-0509.

Habermas, J. 1992. Citizenship and National Identity: Some Reflections on the Future of Europe. *Praxis International*. 12: 1–19.

Haenssgen, M., and P. Ariana. 2018. The Place of Technology in the Capability Approach. *Oxford Development Studies* 46 (1): 98–112.

Hathaway, O., Rebecca Crootof, Philip Levitz, Haley Nix, Aileen Nowlan, William Perdue, and Julia Spiegel. 2012. The Law of Cyber-Attack. *California Law Review* 100: 817–856.

Himma, K. 2007. The Ethics of Hacking Back: Active Response to Computer Intrusion. In *Internet Security: Hacking, Counterhacking and Society*, ed. Kenneth Eimar Himma. Sudbury: Jones and Bartlett Publisher.

ILCDraft. 2000a. ILC Draft Articles of Responsibility, Article 51.

ILCDraft. 2000b. ILC Draft Articles of Responsibility, Article 49.

Intel. 2003. Intel Corp. v. Hamidi, 30 Cal 4th 1342 (2003).

ICRC. 2008. How Is The Term "Armed Conflict" Defined In International Humanitarian Law? *International Committee Of The Red Cross*.

Jayawal, V., W. Yurcik, and D. Doss. 2002. Internet Hackback: Counter Attacks as Self Defense or Vigilantism? *Proceedings of the IEEE International Symposium on Technology and Society*, June.

Jensen, E. 2013. Cyber Attacks: Proportionality and Precautions in Attack. *International Law Studies*. 89 : 198–217.

Johnson, D., and M. Schmitt, 2021. Durand Johnson and Michael Schmitt, Responsibility for Proxy Cyber Operations Under International Law. *The Cyber Defense Review* (Fall): 15–30.

Jun, J., and N. Kostyuk. 2021. *The Pros and Cons of Mandating Reporting from Ransomware Victims*, November 1.

Kahan, D., and M. Nussbaum. 1996. Two Concepts of Emotion in Criminal Law. *Columbia Law Review* 96 (2): 269–375.

Kam, S. 2004. Intel Corp. v. Hamidi: Trespass to Chattels and a Doctrine of Cyber-Nuisance. *Berkeley Technology Law Journal* 19 (1): 427–453.

Kant, I. 1784. Idea for a Universal History With a Cosmopolitan Purpose, in *Kant : Political Writings* (Cambridge Texts in the History of Political Thought, 2d. ed) (Cambridge, Cambridge University Press, 1991).

Katyal, N. 2001. Criminal Law in Cyberspace. *University of Pennsylvania Law Review*. 149: 1003–1114.

Kostyuk, N. 2022. The Digital Prisoner's Dilemma : Challenges and Opportunities for Cooperation. *Worldwide Cybersecurity Initiative*.

Kreps, S., and J. Schneider. 2019. Escalation Firebreaks in the Cyber, Conventional and Nuclear Domains: Moving Beyond Effects-Based Logics. *Journal of Cybersecurity* 5 (1): 1–11.

Kumari, A., Md Zakirul Alam Bhuiyan, Jigyasa Namdeo, Shipra Kanaujia, Ruhul Amin, and Satyanarayana Vollala. 2019. Ransomware Attack Protection: A Cryptographic Approach. In *Security, Privacy, and Anonymity in Computation, Communication, and Storage*, ed. G. Wang et al., 15–25. Atlanta, GA: Springer.@

Kymlicka, W. 1995. *Multicultural Citizenship: A Liberal Theory of Minority Rights* (Oxford, Oxford University Press).

Law, J., and E. Martin. 2014. *A Dictionary of Law*. (London : Oxford University Press). (7th edition).

Li, S., and I. Alon. 2020. China's Intellectual Property Provocation: A Political Economy View. *Journal of International Business Policy* 3: 60–72.

Mansell, R. 2002. From Digital Divides to Digital Entitlements in Knowledge Societies. *Current Sociology* 50 (3): 407–426.

Miller, D. 1999. *Bounded Citizenship in Cosmopolitan Citizenship* (eds. Kimberly Hutchings & Roland). (London, Palgrave Macmillan).

NCSS. 2016. National Cyber Security Strategy, 2016–2021, UK Government, 12. See also Gordon Carera, *UK Moves to Active Cyber Defence*, *BBCNews.com*, September 13.

O'Malley, P. 1996. Risk and Responsibility. In *Foucault and Political Reason: Liberalism, Neo-Liberalism and Rationalities of Government*. London: University College of London Press.

Oppenheim, L. 1912. *International Law: A Treatis*. Oxford: Oxford University Press.
Pathology. 1994. Pathologies of Rational Choice Theory: A Critique of Applications in Political Science. Edited by Donald Green and Ian Shapiro. New Haven: Yale University Press.
Pell, S., and J. Finocchiaro. 2017. The Ethical Imperative for a Vulnerability Equities Process and How the Common Vulnerability Scoring System Can Aid the Process. *Connecticut Law Review* 49 (5): 1549–1572.
Pildes, R. and Sunstein, C. 1995. Reinventing the Regulatory State. Reinventing the Regulatory State. *University of Chicago Law Review*. 62 : 1–129.
Pogge, T. 1992. Cosmopolitanism and Sovereignty. *Ethics*. 103(1) : 48–75.
Polinsky, M. and Shavell, S. 1998. The Fairness of Sanctions : Some Implications for optimal Enforcement Policy. Olin Program in Law and Economics, *Working Paper* 167. September 1998.
Prince, B. 2012. BlackHat: Hacking Back: The Best Defense May Not Be the Best Offense. *Security Week*, July 27.
Przeworski, A. 1991. *Democracy and the Market*. Cambridge: Cambridge University Press.
Register. 2000. Register.com, Inc. v. Verio.com, Inc., 126 F. Supp. 2d 238 (SDNY 2000), aff'd 356 F.3d 393.
RestatementTorts. 1965. *Restatement (Second) of Torts*.
Rheingold, H. 2002. *Smart Mobs: The Next Social Revolution*. Cambridge: Basic Books.
Rid, S. 2012. Kashmir: The Prisoner's Dilemma for India and Pakistan. *Peace Prints: South Asian Journal of Peacebuilding*. 4 (2). http://www.wiscomp.org/peaceprints.htm
Rivera, I. 2012. Global Age Cosmopolitanism. *International Journal of Business and Globalization*. 9(1) : 90–105.
Roberts, A. and Lamp, N. 2022. *Six Faces of Globalization : Who Wins, Who Loses, and Why It Matters*. (Cambridge, Harvard University Press).
Robinson, P., S. Garvey, and T. Ferrari, 2009. Measurable Provocation and Self-Defense: Recognizing the Distinction Between Act Reasonableness and Emotion Reasonableness and Criminal Law. In *Conversations*, ed. Paul Robinson, Stephen Garvey, and Timberway Ferrari. Oxford: Oxford University Press.
Rodrigues, A. 2020. If the Law Can Allow Takebacks, Shouldn't It Also Allow Hackbacks? *Marquette Intellectual Property Law Review* 24(1) : 1–13.
Rowe, E. 2016. RATs, TRAPs, and Trade Secrets. *Boston College Law Review* 57: 381–426.
Ruys, T. 2016. Sanctions, Retorsion and Counter Measures : Conceptualizing and Interpreting A Legal Framework. ssrn.com/abstract = 2760853.

Sales, N. 2013. Regulating Cyber-Security. *Northwestern University Law Review*. 107(4) : 1503–1568.

Sanger, D., and N. Perlroth. 2019. US Escalates Online Attacks in Russia's Power Grid. *N.Y. Times*, June 15.

Schmitt, M. 1999. Computer Network and The Use of Force in International Law: Thoughts of a Normative Framework, *Columbia Journal of Transnational Law* 37 : 885–937.

Schmitt, M. 2014. Below the Threshold Cyber Operations: The Counter Measures Response Option in International Law. *Virginia Journal of International Law* 54 (3): 697–732.

Sen, A. 1970. The Impossibility of a Paretian Liberal. *Journal of Political Economy*. 78 : 152–57.

Sen, A. 1999a. *Development and Freedom*. Oxford: Oxford University Press.

Sen, A. 1999b. Equality of What? In *Equal Freedom*, ed. Stephen Darwall. Ann Arbor: University of Michigan Press.

Sen, A. 2004. Elements of a Theory on Human Rights. *Philosophy and Public Affairs*. 32(4) : 315–356.

Sen, A. 2005. Human Rights and Capabilities. *Journal of Human Development*. 6(2) : 151–166.

Shackleford, Scott J., Charoen Danuvasin, Tristen Waite, and Nancy Zhang. 2019. Rethinking Active Defense: A Comparative Analysis of Proactive Cybersecurity Policymaking. *University of Pennsylvania Journal of International Law* 41 (2): 377–427.

Sharkey, C. 2013. Trespass Torts and Self-Help for an Electronic Age. *Tulsa Law Review* 44 (4): 677–706.

Skelrov, M. 2009. Solving the Dilemma of State Responses to Cyberattacks: A Justification for the Use of Active Defenses Against States Who Neglect Their Duty to Prevent. *Military Law Review* 201: 1–85.

Solis, G. 2010. *The Law Of Armed Conflict: International Humanitarian Law In War*. London : Cambridge University Press.

Stapleton, J. 1995. Duty of Care: Peripheral Parties and Alternative Opportunities for Deterrence. *Law Quarterly Review*. 111(2) : 301–345.

Stark, H. 2017. A Bipartisan Bill to Strengthen Cyber Security for the Internet of Things. *Forbes.com*, August 20.

Sterk, S. 2008. Property Rules, Liability Rules, and Uncertainty About Property Rights. *Michigan Law Review*. 106(7) : 1285–1335.

Swygert, M. and Yanes, K. 1998. Focus on Fairness, Efficiency and the Law: The Integration of Fairness into Efficiency. *Washington Law Review*. 73(2) : 249–328.

Tandon, A., and A. Nayyar. 2019. A Comprehensive Survey on Ransomware Attack: A Growing Havoc Cyberthreat. In *Data Management, Analytics and Innovation*, ed. V.E. Balas et al. Singapore: Springer.

Temple, J. 2011. Hackers Getting Hacked by Security Firms. *SFChronicle.com*, November 30.

Thomas. 2017. US v Thomas, 877 F.3d. 591 (2017).

Thrifty. 1996. Thrifty-Tel v Benezek, 54 Cal Rptr. 2d 468 (Cal. Ct of Appeals, 1996).

Valeriano, B., and B. Jensen. 2019. The Myth of Cyber Offense: The Case for Restraint. Cato Institute Policy Analysis, No. 862, January 15.

Van Staveren, I. 2007. The Ethics of Efficiency. *SCEME Working Paper* No. 18.

VEP. 2017. *Vulnerability Equities Process and Policy for the United States Government*, November 15.

Wedgwood, R. 2002. Proportionality, Cyber-war, and the Law of War. *International Law Studies*. 76 : 219–232.

West, Z. 2012. Note, Young Fella: If You're Looking for Trouble, I'll Accommodate You, Deputing Private Companies for the Use of Hackback. *Syracuse Law Review* 63: 119–146.

Zargar, S., J. Joshi, and D. Tipper. 2013. A Survey of Defense Mechanism Against Distributed Denial of Service (DDoS) Flooding Attacks. *IEEE Communications and Tutorials* 15 (4): 2046–2089.

CHAPTER 7

Exceeding Authorized Access Under the CFAA

1 A Perplexing Asymmetry

The Computer Fraud and Abuse Act (CFAA) appears to prohibit retaliatory hackbacks, the notable exception being hackbacks that fall under the language of Section 1030(f), which states that the CFAA

> does not prohibit any lawfully authorized protective or intelligence activity of a law enforcement agency of the United States, a State or a political subdivision of a State, or of any intelligence agency of the United States.[1]

Section 1030(f) authorizes governmental bodies, at both the state and federal levels, as well as private organizations working with a governmental body, to engage in retaliatory hackbacks, but it does not provide a similar 'exception' to private individuals or groups acting without governmental authority.

No other language in the CFAA either directly or indirectly prohibits retaliatory hacks, and no additional language on the topic seems necessary because the CFAA appears to prohibit all non-governmental, unauthorized computer accesses, which include attacks executed in response to a previous attack. After a recent Supreme Court decision, however, the CFAA's purported ban on retaliatory hackbacks is less clear. In Van Buren,

[1] CFAA (1986, 1030[f]).

A. J. Thomas, *The Open World, Hackbacks and Global Justice*,
https://doi.org/10.1007/978-981-19-8132-6_7

the Supreme Court addressed the question of whether an individual who legally accessed one section of a local governmental database, and only retrieved information from that one section for an improper purpose, violated the CFAA's prohibition on 'accesses that exceed authority.' The Supreme Court concluded that 'exceeding access' violations that remain within the area so authorized do not violate the CFAA, even if they were undertaken for an improper purpose. Van Buren essentially permits non-exploratory 'insider' hacks of governmental databases.

But the Van Buren majority, either intentionally or accidentally, created a new, perplexing asymmetry. How can the CFAA permit authorized accesses undertaken for an improper purpose while prohibiting retaliatory hackbacks against the same aggressor, and under the same circumstances (i.e., 'authorized,' counter-hacks into the aggressors computer system)? This counter-intuitive asymmetry between illegitimate accesses undertaken by 'insiders' and illegitimate retaliatory hackbacks perpetrated by their victims rests on three basic assumptions: (1) Section 1030 of the CFAA does not authorize courts to inquire into the aggressors purpose or intent when the aggressor already received authorization to enter a computer system, (2) reading purpose into the language would impermissibly expand the reach of the CFAA, potentially criminalizing everyday uses of computers as well as forms of computer research that benefit the public, and (3) unauthorized retaliatory hackbacks undertaken for a morally legitimate, if unauthorized, purpose should be discouraged because they often lack attribution, increase the likelihood of an uncontrolled cyber escalation and erode public confidence in the rule of law. The implicit assumption shared by open world zealots is that the asymmetry also promotes a desirable narrative in cyberspace, a narrative in which privacy expectations are lowered, and decision-makers propagate the view that the best way to prevent cyber attacks is to blame the victim for inadequately 'hardening' (security) rather than programmers for exploiting backdoors and vulnerabilities. Either intentionally or accidentally, however, eliminating purpose inquiries in CFAA cases rewards bad faith 'winners' and punishes good faith 'losers.'

This chapter analyzes the CFAA's purpose inquiry framework and its effect on existing and proposed authorization for retaliatory hacks, through the prism of global justice. It explores the crucial link between purpose inquiries and justifications for retaliation (self-defense or self-help) as opposed to police intervention, private litigation or hardening. Under VRN, purpose inquiries of unauthorized or improper authorized access cases are necessary in order to find evidence that a perpetrator intentionally tried to exploit or harm capability-deficient individuals. This is the same decisional logic informs 'zone 2' of the VRN chart: scenarios in which the desert rule protects plaintiffs that retaliate after a particularly dangerous attack by the defendant even though the initial attack was arguably permissible. Without a purpose inquiry of an unauthorized or improperly authorized access, however, courts are unlikely to discover information that could trigger a 'permissible' retaliatory attack.

2 Competing Formulations of Access Under the CFAA

One of the most litigated section of the CFAA defines the meaning of unauthorized access. After years of litigation, courts developed four formulations of access: code-based, agency-based, contract-based and norms-based. Under the code model, access to information is not authorized if companies fail to develop software code limiting access, even if the company has policies in place that prohibit downloading and/or using targeted data.[2] Code-based cases not only elevate technical access over stated policies that explicitly prohibit downloads, but they also decrease the likelihood that casual employee usage of company computers for private purposes—like gaming or watching videos in YouTube—will become the basis for an unauthorized access claim by the company. In US v. Nosal,[3] for example, the Ninth Circuit found that an employee who downloaded confidential information and turned the information over to a former employee/colleague and rival did not violate the CFAA because the employer did not have software that restricted access. Under the agency model, authorization is constrained by the loyalty that an agent owes to a principal.[4] The most important agency-based case is IAC LLC v.

[2] WEC (2012) (court upholds employee who downloads proprietary information, resigns and then uses the stolen information to prepare a presentation for a competitor. The employee did not violate the CFAA because the former employee forbid downloading and use of data, but allowed technical access); LVRC (2009).

[3] Nosal1 (2012). For more clarification, see NosalII (2016).

[4] Restatement_Agency (1992): "[U]nless otherwise agreed, the authority of an agent terminates if, without knowledge of the principal, he acquires adverse interests of if he is obviously guilty of a serious breach of loyalty."

Citrin.[5] In Citrin, the Seventh Circuit concluded that an employee who deleted company information as well as software that could detect data tampering from a laptop computer given to him by a former employer before starting his own company violated the loyalty that he owed to his former employer, and therefore lost authorization to access company data. Only the Seventh Circuit adopted the agency approach, and a number of courts expressed reservations about treating casual computer misuse as grounds for automatic termination of the agency relationship.[6] Some courts embraced the agency model as an opportunity to consider informal access norms and expectations in determining the boundaries of authorization,[7] while others recognized that the agency approach makes it easier for employers to sue errant current or former employees suspected of stealing confidential information or trade secrets.

Contract-based formulations of unauthorized access under the CFAA prioritize explicit language specifying the limits of computer usage, ranging from employee contracts and terms of service contracts to confidentiality and non-disclosure agreements. In EF Cultural Travel BV v. Explorica, Inc.,[8] the First Circuit found that a former employee violated a confidentiality agreement entered into with his former employer by accessing confidential information and then using that information to develop software specifically designed to extract proprietary pricing data from the company's public website.[9] Using the same contract-based approach, the Eleventh Circuit found a violation of the CFAA when a SSA employee, in clear violation of SSA policy prohibiting access to confidential databases for personal reasons, located information about 17 personal acquaintances,[10] and the Fifth Circuit found a violation of the CFAA in the case of an employee who, in violation of bank-computer use restrictions, accessed the company's database and provided a relative

[5] Citrin (2006)

[6] Nosall (2012, 862) (9th circuit rejects), WEC (2012, 206) (4th circuit rejects) ("Although an employer might choose to rescind an employee's authorization for violating use policy, we do not think Congress intended an immediately end to that agency relationship and moreover, the imposition of criminal penalties for such a frolic.").

[7] VChip (2006).

[8] EF (2001)

[9] Id. at 579.

[10] Rodriguez (2010).

with confidential information. In Bittman v Fox,[11] however, the Seventh Circuit declined to find a violation of the CFAA when the defendant, a public library official, created a fake social media account on Facebook and posted pictures of the plaintiff as part of an ongoing feud over library policy.[12] The court found that the purpose of the CFAA was to "punish trespassers and hackers." The defendant, although in clear violation of Facebook policies, was neither a hacker nor a "virus or worm writer," and the defendant did not "access a computer to damage, steal or tamper [with]" the plaintiff's data.[13]

Courts also defined unauthorized access using norms, although the norms approach appears to straddle the other three approaches. Social norms, broadly defined, consist of informal or implied codes, or patterns of conduct, recognized and followed by members of specific communities. Failure to comply with social norms may carry no consequences or it may give rise to an elaborate set of punishments ranging from shaming or ostracism to financial penalties or exclusion. In 1991, the Second Circuit defined the outer boundaries of authorized access based on the violation of implicit norms in US v. Morris[14] Morris, a computer science major, wrote a worm program that exploited weaknesses in *sendmail* and *finger daemon* in order to demonstrate potential security shortcomings. The code was launched from MIT, but quickly spread, damaging the networks of many colleges. Morris was charged with violating Section 1030(A)(5), which prohibits anyone from accessing a computer system without authorization, and "damaging, affecting, or destroying information,"[15] Morris claimed that he had legal authority to access by virtue of his status as a code tester, but the Second Circuit concluded that Morris did not extend to exploiting security holes in system software. His worm, in effect, exceeded the intended function of the software that he exploited.[16]

[11] Bitman (2015).

[12] Kluben (2009, 8).

[13] Id.

[14] Morris (1991).

[15] CFAA (1988, Section 1030[a][5]), added in 1988, was amended in 2012 by CFAA (2012, Section 1030[a][5][A]), which bans the knowing transmission of code that damages systems when that transmission was done without authorization.

[16] Morris (1991, 510).

In US V. Phillips,[17] the Fifth Circuit refined the intended function test utilized in Morris. Philips, a student at UT-Austin, managed to steal bank account records, credit card numbers, passwords, student records and social security numbers by scanning computer networks, even though he had signed an acceptable use computer policy with the university that banned port scanning using computer accounts. The university detected Phillips' activities and warned him to stop on three occasions, but Phillips continued. He also hacked into a secure server for faculty and staff maintained by the university, exploited a login vulnerability and collected thousands of social security numbers, which he later used to access even more personal information about server users, donors and alumni. The hack dramatically increased load usage on the network, eventually leading to system-wide shutdowns. Phillips, who was charged for violating Sections 1030(a)(5)(a)(ii) and B(i), conceded that he wrote the hack but denied any intention to sell or use the collected information. After agreeing with precedent suggesting that courts should "[analyze] the scope of the users authorization to access a protected computer on the basis of the expected norms of intended use or the nature of the relationship established between the computer owner and the user,"[18] the Fifth Circuit found that Phillips program was not "an intended user of the UT network within the understanding of any reasonable computer user." His hack was "an unauthorized access to computerized data that he was not permitted to view or use."[19]

Phillips also provided further evidence of how difficult it is to define authorized access norms without looking to contractual or agency relationships. In response to Phillips' argument that his authority to generate capable of guessing passwords on a login page accessible to everyone, for example, the Fifth Circuit essentially stated that regardless of whatever norms, social expectations or 'generic authorization' he enjoyed as a member of the World Wide Web, the brute force attack program he wrote to harvest social security numbers by exploiting vulnerabilities on

[17] Phillips (2007).

[18] Phillips (2007, 219), *citing* Creative (2004) (harvesting of log in names and passwords by information administrator violates cfaa) and Explorica (2001, 582 n.10) (conduct "not in line with reasonable expectations of the website owners and its users" is unauthorized by default).

[19] Phillips (2007, 219).

the TX Class login page was not available to the average web user.[20] The primary set of norms and expectations governing his conduct came from the contract he signed with the university, and his conduct clearly violated university policy. Courts confronted with an apparently open-ended, norms-based authorization standard have also used contract as an alternative source of justification to find that the defendant violated the CFAA. In Explorica, for example, a district court in the First Circuit found that a data scraping operation was inconsistent with the "reasonable expectations" of the computer owner with regard to the use of his site, and therefore in violation of the CFAA. The Court of Appeals preferred to rely on a confidentiality agreement that indirectly prohibited data scraping, even though the defendant in Explorica did not sign the agreement. Nonetheless, the Court of Appeals found that the data scraper's conduct was not permitted. In a subsequent case, the same court expressed reservations about the reasonable expectations test because it appeared to place defendants at "the mercy of a highly imprecise litigation-spawning standard."[21] Other courts have expressed fewer reservations about norms-based intended use analysis.[22]

3 'Broadening' v. 'Narrowing' the CFAA

For many years, commentators and litigants debated the question of whether the courts should broaden or narrow the CFAA. Participants in this debate generally fell into three categories: (1) 'broad' approaches that respected contract-based constraints on authorized access, (2) 'narrow' approaches that leaned toward code-based constraints on authorized access and (3) intermediate approaches that limited any potential downstream risks of both contract and code-based authorizations using prosecutorial discretion, data protection principles or agency law.

[20] Phillips (2007, 220–21).

[21] Id.

[22] Jones (2001) and Merritt (2015)

3.1 *Insiders and Outsiders*

Proponents of the broad view believed that the CFAA insiders as well as outsiders. As evidence, they pointed to language in HR Rep 98-894, which states that in 1984 Congress intended to address "the activities of so-called 'hackers' who had been able to access (trespass into) both public and private computer systems."[23] The term 'hacking' usually refers to an unauthorized access by outsiders, but the language of the CFAA muddied common understandings of the hacker moniker. Section 1030(a) of the original CFAA, for example, prohibited individuals from "knowingly accessing a computer without authorization, or having accessed a computer with authorization, us(ing) the opportunity such access provides for purposes to which such authorization does not extend."[24] In 1986,[25] the insider focus of the CFAA was clarified by Section 1030(e)(6), which states that no one "can access a computer with authorization and to use such access to obtain or alter information in the computer that the accessor is not entitled so to obtain or alter."[26] Language in the Economic Espionage Act of 1996[27] and Section 1030(a)(2)(a), also amended in 1996,[28] provides evidence of Congress' continuing intention to prohibit unauthorized access by insiders. The legislative history of the CFAA since its initial passage in 1984 also suggests that Congress has always wanted to prohibit unauthorized accesses by insiders and outsiders. Staff reports and Congressional testimony from 1983 and 1984 provide evidence of Congress' ongoing concern about insider access to confidential computer files. House and Senate Reports on the 1986 update clarified the wording that described 'exceeding unauthorized access,' but did not change the underlying meaning, or suggest that Congress was more concerned about outsiders than insiders. When the CFAA was amended again by the Economic Espionage Act of 1996, Senators again expressed serious concern about the problem of unauthorized accesses by insiders.

[23] CFAAReport (1984, 10).

[24] CFAASuppI (1984).

[25] CFAASuppII (1986).

[26] VanBurenGov (2019, 4).

[27] EEA (1996, 201[1][B][ii]) ("exceeds authorized access").

[28] CFAASuppII (1986, 1030[a][2][a]).

Establishing Congress' ongoing interest in preventing unauthorized accesses by insiders and outsiders allowed proponents of the 'broad' view to reconcile traditional property crime principles with rapidly evolving computer crime principles. For decades, experts have argued that property in the cyber world is fundamentally different from real property.[29] Perpetrators can make 'copies' of computer data without depriving the original owner of access (as opposed to exclusive ownership/possession/knowledge), take advantage of the fact that computer users often leave their network connections 'on' for extended periods or rarely read the 'terms of service' agreements that define the right to control, access and monetize customer data.[30] Given the apparent uniqueness of cyber property, it seemed logical to conclude that the traditional concept of 'exceeding authorized access' should not apply in cyberspace as well, but there is considerable evidence that Congress intended to apply the concept of 'exceeding authorized access' to cyber property, particularly in Section 1030(a)(2)(c), which was added in 1996.[31]

Broad proponents also argued that once it is established that insider access falls within the scope of Section 1030(a)(2) (i.e., 'exceeds authorized access'), courts should also be able to discern the true meaning of 1030(e)(6), which refines the core language of (a)(2): "means to access a computer with authorization and to use such access to obtain or alter information in this computer that the accessor is not entitled so to obtain or alter." Section 1030(e)(6) had been interpreted by some courts to only prohibit insiders from accessing information that they absolutely have no right to access under <u>any</u> circumstances. But broad proponents interpreted 1030(e)(6) to prohibit insiders from accessing information under certain circumstances that they may have authority to access under <u>other</u> circumstances.[32] The latter construction more accurately reflected the complex permission environment of workplaces that depend on computers and continuously assign different levels of access

[29] Sandvig (2020, 10).

[30] Nosall (2012, 860).

[31] VanBurenGov (2019, 31), citing 1996 Senate Report 7 ("..the theft of intangible information by the unauthorized use of a computer is prohibited in the same way [as] theft of physical items.").

[32] VanBurenGov (2019, 13).

to specific employees. It also assumed that even if an insider can technically access a wide variety of information on his computer system (i.e., no technological barriers prevent access), he may still lack the authorization to view that information.

In terms of defining the meaning of insider access, broad and narrow proponents also differed on the meaning of three words in Section 1030(e)(6): "so to obtain." According to proponents of the narrow view, the "so to" language in Section 1030(e)(6) excludes people "who could identify a single circumstance in which [they] could legitimately obtain computer information no matter how remote or limited the circumstance might be."[33] Such a construction could lead to situations in which someone with legitimate authority to access one part of a database could access another part without violating the CFAA as long as the second access did not involve "stealing information," which is operationally similar to what outside hackers do when they break into computer systems. In contrast to the narrow CFAA view that Section 1030(e)(6) should only apply to individuals who have "no right at all to access the information",[34] broad proponents believe that the 'entitled so to obtain' language permits insiders to access a specific set of information in a specific manner, and that authorization does not confer blanket approval on any other access possibility regardless of how or when the latter access took place. The right to obtain <u>any</u> information only extends as far as the authorization to obtain <u>specific</u> information.

3.2 *Purpose and Motive*

Proponents of the broad view also believed that Congress wanted courts to consider the purpose of outsiders who hack into systems without authorization, as well as insiders who exceed the limits of their authorization to view and use specific information on their computer system. Purpose under the CFAA is relevant for three reasons. The first reason is that the scope of insider access depends on whether the insider, who already has some form of access, "uses the opportunity such access provid[ed] for purposes to which such authorization does not extend."[35]

[33] Id.

[34] VanBurenBrief (2020, 6).

[35] VanBurenGov (2019, 4).

The language itself seems to suggest that access authorizations have a purpose, and that purpose must align with the actual purpose of an individuals' access in order for the access to be authorized. The counter-argument is that if someone has a right to access information, the actual intent or purpose of the accessor at the time of access is legally irrelevant.

Improper purpose can lead to heightened penalties. If an individual violates 1030(a)(2) "for purposes of commercial advantage or private financial gain",[36] the maximum penalty for a first-time offense goes from one year to five years. Finally, purpose affects the meaning of 'misappropriation.' Insider access violations of the CFAA are often characterized as insider misappropriation[37] because disgruntled employees rarely access information stored on company computers in excess of their authorization simply for the sake of viewing it; they usually intend to modify, delete, share or sell the information.[38] Sharing and selling information would be considered misappropriation (theft, conversion, etc.), but should courts care about how insiders misappropriate the information they accessed in excess of the proper authorization?

Divining the purpose (or motive) of an individual accused of exceeding authorized access is not considered important by most proponents of the narrow view. Even though numerous circuits held that exceeding authorization prong of Section 1030(a)(2)(c) is operationally similar to saying that restrictions on computer use are designed to ensure that authorized individuals limit their use to specific purposes,[39] proponents of the narrow view believed that courts should not read a purpose or motive requirement into Section 1030 without specific language, particularly in a criminal statute, as demanded by the rule of lenity. For narrow proponents, Section 1030(e)(6) did not contain purpose or motive language and should not be read to include one.

[36] CFAA (1986, 1030[c][2][B]).

[37] VanBurenGov (2019, 40).

[38] Teague (2011) (DOT contractor accesses student loan records of President Obama) and John (2010) (Citigroup account manager accesses customer records to make fraudulent charges) and VanBurenGov (2019, 40–41).

[39] John (2010, 273) "exceeding the purpose for which access is authorized"; Rodriguez (2010) and Explorica (2001, 7–8) ("cannot use information 'for purposes prohibited by a confidentiality agreement'); Citrin (2006, 420–21) ("when a person accessing data on his work computer for a purpose his employer prohibits").

Narrow proponents also believed that any published information in a website or database connected to the internet is available to everyone in full (i.e., exceeding authorization is not possible), and that if someone with authority to access one part of a system hacks into another part containing public information, the hack itself should be considered exceeding authorized access; no further inquiry into purpose or motive is required. Perhaps most importantly, some narrow CFAA proponents believed that a searching inquiry into purpose or motive under could lead to a parade of horribles ranging from casual use of workplace computers for non-business purposes (like email or Facebook) leading to individual criminal liability, to group uses of workplace computers for non-business purposes (like group chats or markup sessions) leading to conspiratorial criminal liability.

Introducing purpose or motive into the definition of exceeding authorized access also, to narrow proponents, provided unscrupulous employers with an opportunity to harass employees. As an example, unscrupulous employers "that want to rid themselves of bothersome employees only need to identify a violation of the computer use policy and then 'threaten to report them to the FBI.'"[40] In addition to providing employers with a new mechanism for selectively harassing targeted employees, focusing on purpose or motive in CFAA cases also appeared to provide zealous prosecutors with a new mechanism for selectively prosecuting unpopular defendants.

US v. Drew[41] was a textbook example of the latter. In 2008, Drew was charged with conspiracy (felony) to violate MySpace terms of service agreement, separate aiding and abetting (*misdemeanor) charges for violating the terms of service agreement with others, and felony charges of intentional infliction of emotional distress for creating a fake profile to harass someone who eventually committed suicide. Drew was thoroughly vilified in the press, and even though many people believed that she should not be prosecuted in spite of her reprehensible character, California proceeded to indict her, apparently in response to intense political pressure. Drew was eventually indicted on the aiding and abetting charges. The government recommended the maximum sentence for each aiding and abetting conviction, but the District Court refused to issue

[40] Nosall (2012, 860).

[41] Drew (2009).

a sentence and granted the defendant's motion to dismiss because the government's interpretation of the terms of service agreement appeared to render CFAA 'void for vagueness.' The government eventually withdrew its appeal. Postmortems of the case revealed a messy trial record: the prosecutors sentencing memorandum focused on Drew's apparent lack of remorse in causing the events that led to a suicide rather than the defendant's violation of MySpace's terms of service agreement; the jury lacked sufficient evidence to convict on the felony IIED charge; the jury foreperson believed that Drew had committed wrong-doing in violating the terms of service agreement because the violations lead to death. Rather than clarifying the meaning of 'exceeding authorized use,' judicial scrutiny of purpose and motive in Drew, according to narrow proponents, did little more than provide ambitious prosecutors with an opportunity to selectively punish an unpopular defendant, hopelessly confuse jurors concerned about the long-term moral and institutional implications of violating terms of service agreements, and expose various shortcomings in the language of the CFAA.

3.3 *Vagueness, Notice and Lenity*

Broad proponents believed that the exceeding authorized access prong of Section 1030 should protect contractual expectations, whether the contract involves employers and employees, internet service providers and customers, security researchers and voluntary research targets, or participants in vulnerability disclosure programs. Hundreds of criminal cases had already been brought under the CFAA based on alleged violations of terms of service agreements, and the circuits were split on this issue. It also wasn't clear whether a violation of an employee agreement with computer use restrictions constituted a violation of Section 1030. In response, the Department of Justice issued a memorandum in 2014 stating that "federal prosecution may not be warranted" when defendants "[exceed] authorized access solely by violating an access restriction contained in a contractual agreement or term of service with an internet service provider."[42] But the litigation barrage continued, much to the chagrin of broad proponents who believed that equating the violation

[42] VanBurenGov (2019, 42) citing Intake (2014).

of computer use agreements with 'exceeding authorized access' was completely consistent with the language and history of the CFAA.

Vagueness challenges to contract-based CFAA violations tested the clarity of specific contract provisions or specific portions of Section 1030. Considerations of the latter are governed by the Due Process Clause, which requires that laws provide "ordinary people fair notice of the conduct it punishes."[43] Vague laws undermine public confidence in the rule of law[44] particularly when vague, overbroad laws begin to infringe on constitutionally protected areas like free speech.[45]

Proponents of the narrow view believed that the CFAA's vague language might chill vital First Amendment research and expression at some point in the future. Proponents of the broad view, however, did not believe that vagueness determinations should be based on hypothetical speculations about future, unlitigated applications of statutory language.[46]

3.4 *The 'Public-Ness' of Data*

Since the CFAA was passed in 1984, decision-makers have tried to reconcile the general concept of unauthorized access with four different understandings of 'public' data: (1) any information stored on a website, database or email server owned or controlled by a public entity (*institutional*), (2) any information, regardless of where it is stored or who controls it, deemed to be 'beneficial' to public well-being (*welfare-driven*), (3) any information that is not technologically restricted from public view, with the accessible threshold range from novice to expert (*technological*), and (4) any information that is not legally restricted from public view (classified information, doctor-client, attorney-client, non-disclosure agreement, plea agreement) (*categorical*).

Proponents of the broad view generally reject any definition of public information that ignores or de-prioritizes computer use restrictions in contracts voluntarily entered into by accessors of the information. Regardless of whether they are public or private, information providers have a

43 Johnson (2015, 2556).

44 Davis (2019, 2325).

45 Hoffman (1982, 499) and Amsterdam (1960, 75).

46 VanBurenGov (2019, 47–48).

basic right to enter into computer use agreements with potential users or accessors of information that the providers own or control. Entering into that use agreement signals to the world that the provider does not want the information in question to be considered 'public.' Enforcing that agreement in a court of law might be difficult if it is unclear or violates existing law, but the contract itself remains a presumptively valid way to transform public into private property. Computer use restrictions that require accessors to a password before viewing information effectively transform public into private property. Unlike information on the web that is "presumptively accessible to the general public," password-enabled mechanisms are precisely the kind of meaningful "affirmation enabling action" that triggers Section 1030's authorization.[47] Even if a password-based system is 'front-facing' (i.e., password user input screen can be seen by the general public), it will defeat any institutional, welfare-driven technical or categorical justification for defining information as 'public.'

Interpreting the CFAA narrowly also undermined corporate attempts to protect confidential trade secrets that the public was never meant to see as much as an overbroad definition of CFAA. The former, if adopted, would limit the CFAA to outsiders who 'hack' their way into another computer system to steal trade secrets, completely ignoring the ongoing problem of insiders who steal trade secrets having already acquired limited access to another part of the system. Any access to trade secrets by insiders would be authorized, effectively making company secrets public for purposes of Section 1030.

An overbroad interpretation of the CFAA, however, would apply the same 'exceeding authorized access' standard to completely public data available on corporate websites and confidential private trade secrets residing on internal company servers. In addition, any technological access to confidential trade secrets would effectively transform that information into public data, regardless of whatever measures were undertaken by the company to keep its information private. Many proponents of the broad CFAA reject any attempt to ignore "the understanding of the parties as to what actions are (and are not) permitted in relation to confidential data"[48]—essentially cultural understandings about the 'public-ness' of specific company data—especially when companies are

[47] VanBurenGov (2019, 37).

[48] VanBurenGov (2019, 17).

legally obligated to "proactively safeguard confidential and non-public data though contractual and policy controls."[49]

When the public-ness of data is defined solely in terms of technological access—the core assumption shared by many proponents of the 'narrow CFAA'—organizations face a difficult administrative dilemma: "[allow] legitimate users the greatest freedom to conduct their work efficiently" but risk insider abuse and forego legal recourse against insiders, or "locking down" systems completely in order to preserve the possibility of legal recourse against insiders, even though such an arrangement is inefficient, costly and discourages both cross-platform functionality and intelligence-sharing.[50] Defining the public-ness of data solely in terms of technological access deprives systems providers of the freedom to negotiate, though contract, computer permissions structures with potential and actual insiders, as well as maximize the efficiency of their computer systems. It also, perhaps most importantly, encourages a technological 'race to the bottom' for malicious or mischievous insiders and outsiders. By treating insufficiently protected data as private—the technological equivalent of abandonment—malicious actors will endeavor to transform valuable data that the company clearly intends to keep private into public property through increasingly intense, persistent and intrusive backdoor tactics.

All four mechanisms for defining the public-ness of data impose potential burdens on information providers. The institutional approach, for example, requires organizations to carefully consider the advantages and disadvantages of defining themselves as purely public institutions. By defining themselves as purely public, institutions run the risk of subjecting all of their data—including data created in collaboration with private parties—to accountability and disclosure rules defined by political majorities. The welfare approach taps into shared, intuitive or folk understandings about the kinds of data that informed citizens ought to know, even though political decision-making is often based on fear, bias or moral chauvinism rather than empirical data. The categorical approach imposes a burden on the legal system to develop formal permission rules for organized bodies or knowledge, even though these rules may lag far behind ordinary expectations about the same data.

[49] VanBurenGov (2019, 18).

[50] PoliceBuren (2020, 4–5).

The technological access approach is also a burden-shifting device. Instead of placing the burden on accessors to police themselves, honor contractual arrangements or respect traditional notions of non-public property, the technological access approach places the burden on information providers to anticipate every possible avenue of entry, surveillance and non-destructive copying available to cyber-savvy insiders and outsiders using the most recent, cutting-edge technology. The familiar self-policing ethos, operable even in the computer age, consists of the recognition that "anyone can stand on the street and observe that a building has windows that are open but crawling through one without authorization would constitute a crime."[51] Its alternative is a daunting task, even for extremely savvy, resource-rich information providers capable of monitoring insider and outsider efforts to capture public network traffic using "non-invasive scanning techniques."[52] Expecting information providers to prevent every invasive, penetrative effort to access private files stored on computers linked to the internet, without the expectation of self-policing on the part of computer-savvy insiders or outsiders, is not only an impossible task, but it also encourages bad behavior, especially for individuals with a strong financial, professional, emotional or reputational incentive to 'bend the rules.'[53]

Defining the public-ness of data through the prism of technological access also illustrates the peril of open world maximalism. Regardless of the context, open world maximalism imposes impossible burdens and incentivizes bad behaviors. In the immigration context, for example, we can imagine two versions of open world maximalism. In one universe, literal open world maximalism takes the form of completely open borders, with no restrictions on the kinds of people who can leave one country and no restrictions on the kinds of people who can enter another country. In another universe, constructive open world maximalism applies the technological access paradigm so coveted by proponents of the 'narrow CFAA' view to the acquisition of citizenship through immigration: anyone who

[51] Heart (2020, 17).

[52] Id at 16.

[53] Id at 16. Under the narrow CFAA approach, "anyone could penetrate the computer systems of businesses under the guise of security research. It takes little imagination to envision unscrupulous business people hiring so-called security experts to undermine the operations of their competitors."

managed to cross the border between two sovereign nations would automatically become a citizen of the receiving country, regardless of how they did it. One could easily imagine a parade of horribles ranging from killing someone to gain entry or committing numerous unpublished crimes in the sending country, to lying about one's identity, concealing the reason for one's migration, or agreeing to carry illegal contraband in exchange for sponsorship in the receiving country. In the same way that data automatically becomes 'public' the moment that enterprising, cyber-savvy insiders or outsiders manage to access it, all immigrants automatically become citizens the moment their feet touch the ground of the receiving country.

4 Intermediate Platforms and Bad Faith Determinations

4.1 *Prosecutorial Discretion*

For decision-makers who find merit in both the narrow and broad approaches, prosecutorial discretion is a promising, if risky, intermediate platform. Prosecutorial discretion provides the government with the authority to determine if a case should be brought after an alleged violation of the criminal law. That discretion extends both vertically and horizontally. In terms of the former, the government can issue guidance about the kinds of cases that prosecutors are likely to pursue (with varying degrees of specificity) and the kinds of actions that police should charge. In terms of the latter, prosecutors have the authority to decide which charges to pursue, and which charges to drop, in any criminal case. Both horizontal discretion and vertical discretion take a variety of factors into consideration, including the weight of evidence, the identity, availability and potential immunities of parties and witnesses, sentencing constraints, political pressure, the language of statutes and governing precedent.

Virtually all of these issues have made their way into criminal CFAA cases. In a number of these cases, prosecutors have been accused of pursuing CFAA charges, not because of the weight of evidence, but just to 'set an example.' Drew and Swartz are notable examples of largely performative CFAA prosecutions. In the latter, Aaron Swartz, a computer science student and self-appointed activist broke into the JStor archive and downloaded millions of files. Roughly six months later, he was indicted for wire fraud, computer fraud and reckless damage to computers.

He was also later indicted for violating Section 1030 of the CFAA. After Swartz committed suicide, his father said that "[Aaron] was killed by the government."[54] Others accused the government of prosecutorial overreach, either because the language of the CFAA failed to distinguish between 'high level' hacking violations and 'low level' casual violations, or because the government acted like a bully in choosing to prosecute a case that it had the legal authority, but not the moral authority, to pursue.[55] In fact, many critics of using prosecutorial discretion in CFAA cases believe that even if the government wanted to apply the DOJ's guidance conscientiously and fairly, it could not, for a wide variety of institutional and political reasons, actually do so.[56]

Before Van Buren, the centerpiece of the prosecutorial discretion platform, is DOJ guidance from 2014 that discourages prosecutions if the accused "exceeded authorized access solely by violating an access restriction contained in a contractual agreement or term of service with an internet service provider."[57] The DOJ guidance appears to support aspects of the narrow CFAA approach, but the text does not discourage prosecutors from factoring the language and intent of contractual agreements into 'exceeding authorized access' cases, which is consistent with the broad CFAA approach. In their Van Buren brief, the government did not provide evidence indicating that the pace of exceeding authorized access prosecutions had slowed significantly since 2014, but it does suggest that any momentum to increase the pace of prosecutions will not come from ambiguities in the language of Section 1030 itself. Insider cases like Van Buren, of which there are many, fall squarely within the language and meaning of Section 1030, and nothing in the 2014 DOJ guidance discourages prosecutors from facts suggesting that an individual accused of violating Section 1030 did not do so in a casual or accidental manner.

[54] Guy (2013).

[55] Lessig (2013).

[56] See Nosall (2012, 863–4) ("[We] shouldn't have to live at the mercy of our local prosecutor … and it is not clear that we can trust the government when a tempting target comes along."), and McDonnell (2016, 2372–73) ("we cannot construe a criminal statute on the assumption that the government will 'use it responsibly.'") citing Stevens (2010, 480).

[57] VanBurenGov (2019, 42).

For proponents of the narrow CFAA view, however, the checkered history of prosecutorial discretion provided little guidance in terms of preventing exceeding authorized access. Neither the CFAA nor the DOJ's guidance, for example, provides prosecutors with a mechanism for distinguishing between contractual/TOS cases that clearly violate Section 1030x and those that don't.[58] Without that guidance, the text of Section 1030 appears vague and imprecise—a perfect opportunity to over-criminalize questionable cyber activity by ambitious prosecutors determined to 'set an example.' Nor is there any indication that the DOJ has completely abandoned, or even committed to abandon, 'exceeding authorized use' prosecutions merely involving TOS violations (as opposed to prosecutions based solely on TOS violations, which is clearly discouraged by the 2014 guidance). The fear of such a prosecution was enough to convince a judge to grant standing for a pre-clearance challenge to Section 1030(a)(2).[59]

If we bracket, for a moment, the more controversial observation that prosecutorial discretion may never prevent governmental abuse or the over-criminalization of human activity in all cases, it is notable that so many proposals for reforming prosecutorial discretion prioritize the prosecution of particularly egregious or malicious crimes, and deprioritize casual or random violations in ordinary contexts that lack extreme malice or viciousness.

4.2 *Post-Citrin Agency Formulations*

Another promising intermediate platform that combines aspects of the 'narrow CFAA' and 'broad CFAA' views focuses on agency law. In Citrin, the Seventh Circuit conceptualized unauthorized access as a violation of a duty of loyalty owed to a principal. After receiving a laptop from his employer for business purposes, which included access to installed company files, as well as the authority to 'return and erase' the laptop upon termination of his employment, Citrin decided to start his own firm, but only after he deleted files from the laptop and installed security erasure software to obscure his steps. In a private civil action, Citrin was sued by IAC (former employer) for exceeding authorized access under

[58] Heart (2020, 21–22).

[59] Sandvig (2020, 4–5).

Section 1030(a)(2) and damaging a protected computer without authorized access under Section 1030(a)(5)(a). The Seventh Circuit Court of Appeals reversed a district court ruling that Citrin had not violated the CFAA because he still had authority to delete files from his laptop when he subsequently deleted files and installed security erasure software to mask his deletions. Citrin's actions, according to the Seventh Circuit, violated his duty of loyalty to his employer and therefore terminated his authorization to access/modify the laptop.

Citrin raised three important questions: (1) whether an agent/employee's actions, as opposed to the employer/principal's express (or implied) rescission of a specific authorization, can terminate the employee/agent's legal right to access a protected computer, (2) whether independent, self-serving action by an agent, which the Seventh Circuit interpreted as a breach of loyalty owed by an agent to his/her principal, involves 'bad faith, and (3) whether the perils of open world maximalism in the cyber realm can be avoided without inquiring into not only the accessor/agent's intent, but also the number of times he/she accessed the employer/principal's protected computer, and how much damage he/she caused while operating without authorization, or in excess of authorization—additional indicators of 'bad faith.'

In LVRC Holding, LLC v Brekka,[60] the Ninth Circuit considered the first and second questions. Brekka simultaneously ran internet operations for one company (LVRC) while running another company that consulted with rehabilitation centers. While working for LVRC, Brekka emailed, using a LVRC computer, LVRC's usage statistics to himself to aide his own company. One month later, Brekka left LVRC, but he was eventually sued by LVRC after they discovered that he had sent the emails under Sections 1030(a)(2) and 1030(a)(4). Brekka argued at trial that he was still a LVRC employee when he sent the email and was not required to keep the emails confidential in a separate agreement with LVRC. Both the District Court and Ninth Circuit Court of Appeals agreed, concluding that only the employer could terminate, through an affirmative act, authority to use computers in a particular fashion, even if the employee harmed the employer or violated state law.[61] In effect, how Brekka used the information (i.e., misappropriation to gain

[60] LVRC (2009, 1134).

[61] Id. at 1133.

a competitive advantage over his former employer) was less important than the fact that Brekka retained formal authority to access the stolen information, and was not formally required to keep the information to himself when the theft occurred. Unlike the Seventh Circuit, therefore, the Ninth Circuit answered the first question by shifting the burden onto employers/principals to either formally specify, with extraordinary precision and prescience, the exact boundaries or agent/employee authority to use protected computers, or automatically terminate a principal/agent relationship one evidence that the agent/employee damaged or deleted files from a company computer becomes clear. With regard to the second question, the Ninth Circuit substituted an inquiry into Brekka's motives as an agent—intent to misappropriate for competitive purposes that could damage the principal's business, clear indicators of bad faith—with an inquiry into whether the agent had formal authority to access and steal information owned and produced by the principal.

Nosal, extended the reasoning of Brekka.[62] In Nosal, the question was whether a former employee of an executive search firm could escape criminal liability under the CFAA after he encouraged three current employees of the firm (all colleagues) to access the forms database and provide him with confidential information about clients. Nosal argued that the exceeding authorized access language of Section 1030(a)(2) only applied to people who already had authority to view one kind of information, but exceeded that authority by viewing information on a different part of the computer system. The colleagues in Nosal had authority to view the information they passed on to Nosal. The mere act of passing that information along to someone else did not terminate their authority, even the purpose of stealing the information was clear misappropriation. By contrast, the government argued that when an employee with authorized access exceeds that authority, the authority automatically ends. Passing stolen, misappropriated information to a former colleague for the purpose of providing the recipient with a competitive advantage clearly exceeds whatever authority the employer originally granted to the employee.[63] The Ninth Circuit eventually sided with Nosal because the government's

[62] The District Court rejected NosalI (2012), reversed and dismissed after LVRC (2012). The Ninth Circuit Court of Appeals reversed after granting an interlocutory appeal. Eventually, the Ninth Circuit re-instated the District Court decision, reviewed the case de novo and eventually dismissed charges against Nosal.

[63] Nosal1 (2012, 856–57).

position, which opened the door to misappropriation and breach of confidential agreements, appeared to "broaden the CFAA beyond its original scope as an anti-hacking statute."[64] It also appears to encourage the criminalization of casual computer use, traditionally the province of tort and contract law.[65]

4.3 *Data Protection Imperatives*

Another promising alterative to the 'narrow CFAA/broad CFAA' dichotomy emphasizes the protection of data. According to this approach, the core purpose of Section 1030(a)(2) is "to protect personal information stored in record keeping systems."[66] Both the language and history of the CFAA indicate that Congress was not only determined to protect personal information stored on record keeping systems from both outsiders (anti-hacking language prohibiting unauthorized access) and insiders (exceeding authorized use language),[67] but it also intended to do so using both civil and criminal laws.

Data protection proponents point to evidence that Congress intended to protect sensitive data stored on computers from insiders and outsiders, as well as various data protection statutes passed either immediately prior to or around the same time, as the CFAA itself. With regard to the former, language in the 1986 and 1996 amendments indicated Congress' ongoing determination to protect "sensitive information stored on record keeping systems" using the language of Section 1030(a)(2). The 1996 amendments expanded Section 1030(a)(2) to fill an apparent 'gap' in the 'exceeding authorized access' language of Section 1030(e)(6).[68] Restricting the CFAA to purely technological/code-based incursion, therefore, undermines the purpose of the 1996 Amendments.

Data protection also avoids the 'low bar' risks of the narrow CFAA approach, as well as the 'high bar' risks of the broad CFAA approach. The former restricts Section 1030(a)(2) to outsiders, protects users who manage to circumvent technological barriers from liability even if they

[64] Nosall (2012, 859).

[65] Id.

[66] Epic (2020, 6).

[67] Id. at 8.

[68] Id. at 4–5.

have violated contractual rules and permits users authorized for one purpose to access all data stored on a protected computer regardless of the purpose of subsequent misappropriation of the data. The latter, on the other hand, applies Section 1030(a)(2) to outsiders as well as insiders, but runs the risk of over-criminalizing both casual/harmless unauthorized accesses and 'socially useful' data collection efforts. Framing the CFAA as a data protection statute successfully also avoids the low-risk/high-risk dynamic by prioritizing the identification of bad faith in the interpretation and enforcement of Section 1030(a)(2) cases.

5 The Van Buren Decision

After years of litigation and conflicts between the circuits on the meaning of unauthorized access and exceeding authorized access, the Supreme Court decided Van Buren v. US[69] While Nathan Van Buren was working as a Sergeant in the Georgia Police Department, he pursued a loan from a man named Alba, who subsequently informed the police department that one of their own was trying to "shake down" Alba for cash. In cooperation with the FBI, Alba asked Van Buren to search the police department's license plate database for a woman he purportedly met at a strip club. In exchange for the requested information, Alba agreed to pay Van Buren $5,000. Van Buren complied using valid police-car authentication credentials to access the database. Eventually, he was charged under Section 1030(a)(2) of the CFAA for 'exceeding authorized access' to a computer system as well as 'honest services fraud.' He was convicted in trial court, and the conviction was upheld by the Eleventh Circuit Court of Appeals, one of the circuits that endorsed a 'broad view' of the CFAA, as described in Section 1030. The Supreme Court reversed the Eleventh Circuit on June 3, 2021.

The majority opinion focused on five basic points: (1) the meaning of 'so to obtain,' (2) whether the interpretation of Section 1030(a)(2) should be 'circumstance-specific,' (3) whether specific provisions should reflect ordinary understandings of 'exceeding authorized access' or merely reflect the language of the CFAA itself, (4) spillover effects and the possibility of over-criminalization, and (5) whether Section 1030(a)(2)

[69] Van Buren (2021)

should reflect a property law-based understanding of exceeding authorized access or an alternative understanding that reflects the 'uniqueness' of cyber-crime. Discussion of the meaning of 'so to obtain' begins with Section 1030(e)(6), which states that "to access a computer with authorization and for use such access to obtain ... information in the computer that the accessor is not entitled so to obtain."

Federal law contains many 'so to obtain' clauses, but few have generated as much litigation as the 'so to obtain' clause in Section 1030(e)(6). The majority interpreted 'so to obtain' to mean that an insider who enters into a section of a computer system using authorization credentials specific to that section can do so regardless of their purpose. The court rejected every attempt to limit the breadth of one's 'entitlement' to use a section of the computer system for which one already has proper authorization credentials based on the accessor's intentions. The ordinary language meaning of exceeding authorized access, for example, was rejected both because the CFAA's language explicitly limits 'so to obtain' to "the manner previously stated" (i.e., "using a computer one is authorized to access"), rather than an ordinary language meaning, and because a 'computational' approach to access (physically separating files and folders) is also presumably well established in the computer community. The former argument is circular, while the latter does little to establish the dominance of the 'computational' approach to system access compartmentalization.

The majority also rejected the government's argument that the 1984 version of the CFAA explicitly mentioned purpose by suggesting that the 1986 version, which does not use the same language, was Congress' attempt to remove purpose, rather than a change in language designed to broaden, rather than narrow, the scope of 'exceeding authorized access' cases. Interpreting Section 1030(a)(2) consistent with a property law-based understanding of access in which 'entitlement' to use another person's property is 'circumstance-specific,' urged by the dissent, was rejected both because of the purported 'uniqueness of computer crime' and because relying on a circumstance-specific would require courts to anchor 'exceeding authorized use' in a test for which there were "no guardrails" (i.e., context for 'so to obtain' could come from "anywhere—in the US code, a state statute, a private agreement, or anywhere else"). Preserving a clear distinction between prohibited insider access violations and prohibited outsider access violations, or as the majority frequently characterized it, "a pure 'gates-up or down inquiry'", would also benefit from rejecting purpose, or circumstance-specific information during interpretations of Section 1030(a)(2). The majority concluded that imagining a third variant—exceeding authorized use for an improper purpose—as the government urged, not only blurred the 'clear' boundary between authorized insider access and unauthorized outside access for no compelling reason, but it could also lead to the slow over-criminalization of computer access issues. A broad reading of Section 1030(a)(2), for example, might lead courts to convict for 'innocent' uses such as someone who "[logs] on to the computer with the purpose of obtaining a file for personal reasons," or commonplace vices like surfing the web or viewing personal email accounts on a work computer in violation of company computer access policies.[70] The majority also did not believe that prosecutorial discretion was robust enough to avoid over-criminalization problems.[71]

Van Buren is a formalistic and intellectually unimaginative response to the inevitable complexification of computer crime. Its flaws are numerous and regrettable. Parts of the majority opinion are blinkered or circular, while other parts completely ignore the dangers of under-criminalizing certain kinds of 'exceeding authorized use' problems.[72] The majority opinion in Van Buren also ignored capability issues. Requiring computer system operators to compartmentalize differently authorized areas computationally ignores the fact that many firms and network operators lack the resources to physically re-engineer their systems. Compartmentalization through employee contracts and terms of service is often much less expensive than physical compartmentalization. Even firms with enough capital to physically re-engineer, as well as the legal right to physically re-engineer, may opt for the contractual TOS route because physical re-engineering signals profound distrust of their employees, which is a reputational stigma that some firms cannot survive.

[70] Van Buren (2021, 18).

[71] Id. at 19.

[72] As an aside, the majority opinion in Van Buren perfectly illustrates the semi-adage that formalism has no political tilt. Conservatives on the Supreme Court, for example, used formalism to restrict the free speech rights of anti-war activist protesting inside of shopping malls operating as virtual 'town squares.' Liberals used formalism to discourage executive over-reach during national security crises. Moderates and liberals on the Supreme Court now use formalism to protect free speech rights in cyberspace. Justice Barrett's majority opinion in Van Buren was joined by Breyer, Sotomayor, Kagan, Gorsuch and Kavanaugh.

The majority's preference for a formalistic 'gates up or down' approach also mirrors the capability-based idea that freedom and human flourishing don't necessarily flow from the possession of wealth, primary goods or rights. The situational complexity of 'third access' litigation eludes judicial formalism. Fully understanding the 'worse-case scenarios' that 'third access' cases can create requires an inquiry into purpose by courts and prosecutors. The Department of Justice took up this challenge a year after Van Buren was decided.

6 Department of Justice Guidelines After Van Buren

On May 19, 2022, the Department of Justice issued new guidelines for prosecutors on charging decisions related to the CFAA.[73] These guidelines clarified the DOJ's understanding of the meaning of Van Buren, particularly with regard to cases that the DOJ claimed it was prepared to charge for alleged violations of "unauthorized access" under 18 USC 1030(a)(2), a(2), a(3), a(4) and a(5), as well as "exceeding authorized access" under 18 USC 1030(a)(1), a(3) and a(4). Nathan Van Buren was charged with "exceeding authorized access."

Under the new guidelines, the DOJ will not charge for alleged violations of "exceeding authorized access" unless six conditions are met:

(1) "a protected computer is divided into areas, such as files, folders, user accounts, or databases;
(2) that division is established in a computational sense, that is, through computer code or configuration, rather than through contracts, terms of service agreements, or employee policies;
(3) a defendant is authorized to access some areas, but unconditionally prohibited from accessing other areas of the computer;
(4) the defendant accessed an area of the computer to which his authorized access did not extend;
(5) the defendant knew of the facts that made his access unauthorized; and

[73] DOJCFAA (2022).

(6) prosecution would serve the Department's goals for CFAA enforcement, as described below in B.3."[74]

Conditions (2), (5) and (6) raise a unique set of issues, some of which are relevant to VRN, primary goods/rights and wealth maximization/utility. Condition (2), for example, imposes a burden on computer systems to place the computational equivalent of a picket fence or iron gate between sections of a computer system with different authorization levels. Contracts, terms of service and employment agreements are not sufficient. At the outset, it is important to note that for many experts, terms of service and employee agreements are more efficient anti-hack deterrents than physical compartmentalization into separate sections with different authorization protocols. While there is considerable evidence that most people ignore terms of service agreements (especially for arguably free services like Twitter, Facebook and Google), far fewer people ignore employee contracts. Efficiency gains also flow from the fact that voluntary contracts may not require state intervention to enforce. Depending on the circumstances, contracts may be cheaper to implement for large-scale, complex computer operations. The physical boundaries' requirement for Condition (2) places an onerous and potentially insurmountable burden on victims that, in my view, reflects the 'blame the victim' ethos of open world zealotry and underestimates the complexities of capability deficits in cyberspace. One might also wonder if a victim controlling access to authorized space through computational rather than contractual or normative means is more likely to personally retaliate if attacked by an insider, bypassing law enforcement intervention, and therefore prosecutorial discretion entirely, even if they possess wealth, primary goods or rights.

Condition (3) states that the defendant must "[know] of the facts that make their access unauthorized." The chain of inference leading back to an implicit purpose inquiry in DOJ investigations post-Van Buren starts with Condition (5) because it will be impossible for a court to determine if a defendant knew that his access was unauthorized without inquirying into the defendant's purpose. Insiders familiar with authorization access protocols rarely wander into a part of a computer system to which they normally have no access by accident. Insider access to specific sections of

[74] Id. at 3.

a computer system is usually undertaken for a purpose, either because the accessor believes that his intended purpose falls within the stated purpose or well-established purpose of the section he's trying to enter, or because the accessor believes that he will be able to access the aforementioned system successfully and carry out his intended purpose, even though that purpose differs from the stated purpose of the section he has accessed, because he believes that he will not be caught. Van Buren undoubtably fell into the latter category: he knew that it was illegal to sell license plat information from the state database to a third party, and he didn't think that he would be caught. But in order to prove that the defendant knew that his access was unauthorized because his intended purpose was different from the stated purpose of the system to which he had access, courts need to determine what the defendant's intended purpose actually was.

The DOJ's intention to read purpose back into prosecutorial decisions or charges in exceeding authorized access cases in spite of what the majority said in Van Buren is even clearer for Condition (6). Section B.3 states all of the reasons why the DOJ would charge for exceeding authorized access "because a substantial federal interest would be served by prosecutors in a case in which the admissible evidence is expected to be sufficient to sustain a conviction."[75]

1. The sensitivity of the affected computer system or the information transmitted by or stored on it and the likelihood and extent of harm associated with damage or unauthorized access to the computer system or related disclosure and use of information;
2. The degree to which damage or access to the computer system or the information transmitted by or stored on it raises concerns pertaining to national security, critical infrastructure, public health and safety, market integrity, international relations, or other considerations having a broad or significant impact on national or economic interests;
3. The extent to which the activity was in furtherance of a larger criminal endeavor or posed a risk of bodily harm or a threat to national security;
4. The impact of the crime and prosecution on the victim or other third parties;

[75] Id. at 3.

5. The deterrent value of an investigation or prosecution, including whether the need for deterrence is increased because the activity involves a new or expanding area of criminal activity, a recidivist defendant, use of a novel or sophisticated technique, or abuse of a position of trust or otherwise sensitive level of access, or because the conduct is particularly egregious or malicious;
6. The nature of the impact that the criminal conduct has on a particular District or community;
7. Whether any other jurisdiction is likely to prosecute the criminal conduct effectively, if the matter is declined for federal prosecution; and
8. The attorney for the government should decline prosecution if available evidence shows the defendant's conduct consisted of, and the defendant intended, good faith security research. .... That is: "good faith security research" means accessing a computer solely for purposes of good faith testing, investigation, and/or correction of a security flaw or vulnerability, where such activity is carried out in a manner designed to avoid any harm to individuals or the public, and where the information derived from the activity is used primarily to promote the security or safety of the class of devices, machines, or online services to which the accessed computer belongs, or those who use such devices, machines, or online services. Security research not conducted in good faith—for example, for the purpose of discovering security holes in devices, machines, or services in order to extort the owners of such devices, machines, or services—might be called "research," but is not in good faith....[76]

The "substantial federal interest" list in Section B.3 is notable for several reasons. First, items 2, 3, 5 and 6 demand a searching inquiry into the defendant's intended purpose, rather than simply an assessment of the affects or impact of the access. Insider abuse of authorized access rarely affects national security, critical infrastructure, public health, international relations, in furtherance of a larger criminal endeavor, or is particularly egregious accidentally or randomly. These access types usually reflect the aggressor purpose or intentions. Second, items 1, 4 and 6 of the B.3 list suggest that the DOJ is particularly concerned about accesses or attacks

[76] Id. at 3.

on specific or uniquely positioned victims and communities, attacks indicating that the aggressor was extremely aware of the victim's weaknesses, vulnerabilities and even capability deficits.

Many items in the B.3 list also correspond to attacks that I defined as 'particularly dangerous' in Chapter 4, especially attacks on critical infrastructure (#2), repeat offenders (#5), persistent rule of law violations (#5, abuse of position of trust) and deliberate, high-damage, wide-impact attacks (possibly #2, #3 and #5). Attacks falling into the dangerous category are factored into 'Zone 2' of the VRN. But there is an important difference between the DOJ's guidelines and VRN: the former provides an additional justification (furthering a substantial federal interest) in charging insiders accused of exceeding authorized access whereas the latter provides an additional justification to not only permit capability-deficient people to retaliate after an attack, but also force the defendant to absorb the cost of all damages caused by the retaliation. For particularly dangerous cyber attacks, the motivation to charge seems to be as strong as the motivation to retaliate. One could argue, of course, that the DOJ's determination to charge for particularly dangerous attacks engineered by insiders who exceed authorized access should provide victims with a <u>disincentive</u> to retaliate. But many victims may not be prepared to put their trust, livelihood or even personal safety into the politically sensitive hands of a prosecutor, or wait for the results of a lengthy criminal or civil investigation. In fact, situational constraints and capability deficits such as severe time and disclosure pressures may push some cyber-victims to retaliate immediately, even if they have wealth, primary goods or rights.

Item #8 of the B.3 list removes all doubt about the DOJs' determination to incorporate purpose analysis back into exceeding authorized access charging decisions. In spite of what the majority said in <u>Van Buren</u>, the DOJ fully intends to bring charges against security researchers accused of exceeding authorized access in 'bad faith.' In order to distinguish between good faith and bad faith researchers, the DOJ must inquire into the researchers' purpose, not just the affects of the access. In many cases, the bad faith security researcher who abused limited access authorization granted by a computer owner for nefarious purposes could overlap with attacks based on "an abuse of a position of trust or obvious sensitivity level of access" under Item #5. This scenario is so potentially dangerous that the DOJ is prepared to charge, pushing the anti-purpose holding of <u>Van Buren</u>. Unfortunately, the DOJ did not consider the wisdom of granting hackback authority to victims of attackers engineered by bad faith security researchers. I address this issue in the next section.

7 Weak 'Zone 2' Hackbacks

The narrowest holding of Van Buren is that Section 1030(a)(2) permits authorized access undertaken for an improper purpose (i.e., access in excess of authorization) if the accessor remains in the computer section for which the accessor has authorization. 'Non-exploratory' insiders who abuse access expectations are permitted because they only accessed the computer section for which they have authorization. As the majority stated toward the end of their opinion:

> [A]n individual 'exceeds authorized access' when he accesses a computer with authority but then obtains information located in particular areas of the computer - such as files, folders, or directories – that are off limits to him.[77]

Most commentators came to the conclusion that Van Buren did not disturb the widespread view that the CFAA forbids retaliatory hackbacks unless they fall under Section 1030(f). But this initial reading of Van Buren is incorrect. If Van Buren now permits authorized accesses undertaken for an improper purpose that remain within the computational bounds of the authorization, then it should also permit an authorized access undertaken for an improper purpose by the victim of the first access in retaliation if the second access remains within the computational bounds of its authorization. A computationally-constrained, insider access with proper authorization credentials but an improper purpose that works in one direction, in other words, should also work in the opposite direction.

This claim makes intuitive sense given the language of Van Buren, but there are three important caveats to consider. The first is whether a retaliatory scenario consistent with the narrowest reading of Van Buren is even plausible. The easy answer to this question is that such a scenario is entirely plausible. It is not difficult, of example, to imagine an unscrupulous insider like Van Buren storing license plate information he downloaded from a government database temporarily on a work

[77] VanBuren (2021, 20).

computer owned, controlled and monitored by his employer. Such a practice might be common, and it would make particular sense for an unscrupulous employee who was convinced that no one suspected his true motives. It is also not uncommon for companies to retain access to work computers, access that allows them to periodically check all files, folders and directories on those computers (including web searches and emails). If the employer suspected that an employee was selling license plate information to third parties, they might try to locate those files on his work computer, and they would have both the right and ability to do so if they signed an agreement with employees granting them access to all files and folders stored on work computers.

The second caveat is that the DOJ might decide, on policy grounds, to prosecute all retaliatory hackbacks not authorized under Section 1030(f), even if they are simply a reciprocal hackback consistent with the narrow reading of Van Buren (i.e., non-exploratory, authorized insider access, purpose is ignored). Such a decision would not be surprising given the contentious status of retaliatory hackbacks under state, federal and international law, the absence of any Supreme Court decisions on the bounds of Section 1030(f) (i.e., what counts as a retaliatory hackback and what kind of retaliatory hackbacks might be acceptable without government authorization) and the absence of any hackback-specific legislation in Congress that made it to the President's desk.

The third caveat is directly related to VRN. It is very easy to imagine a reciprocal hackback scenario consistent with the facts of Van Buren that conforms to the parameters of zone 2: an arguably permissible initial attack and a permissible retaliation. But the desert rule under zone 2 only protects the entitlements of capability-deficient plaintiffs (i.e., retaliators) who exercise due care during retaliation. It also forces the defendant to absorb the total costs of all damages caused by the plaintiff's retaliation if the defendant's initial hack falls into the 'particularly dangerous' category. Since reciprocal hackbacks under Van Buren have to take place on a workplace computer that the victim is already authorized to use, attribution and traceback issues will be largely irrelevant. The challenge is characterizing the Georgia Police Department, or for that matter any governmental body or private organization that not only provides its employees with workplace computers but also software that authorizes employees to use company databases, as capability deficient. And that capability deficiency must apply in two scenarios: (1) the plaintiff's capability deficiency made it difficult for them to adequately defend themselves before the initial attack, and (2) the plaintiff's capability deficiency pushed them toward retaliation rather than law enforcement, a private suit or hardening after the initial attack.

In terms of the first scenario, entities like the Georgia Police Department, both public and private, might be considered capability deficient if political pressure, reputational concerns, legal naivete or lack of technological sophistication prevent them from adequately protecting themselves from authorized insider accesses undertaken for an improper purpose. These facts might have situationally constrained the Georgia Police Department, leaving them vulnerable to an attack by an insider, even if they had adequate funding, a considerable endowment of primary goods (at the institutional level) and a formidable arsenal of formal rights. The same set of situational constrains (or adaptive preferences) may have pulled them toward a low-visibility, immediately available, reciprocal hackback using their own legitimate authorization credentials on the workplace computer of an employee suspected of selling official, confidential information to a third party. It is not difficult to conceptualize the Georgia Police Department, or comparable government entities, as at least temporarily capability deficient if we accept the fact that entities we normally perceive as powerful might find themselves unable to take advantage of all opportunities in specific situations.

8 The Security Research Debate

Unlike reciprocal hackbacks, hackbacks against security researchers who not only exceed authorized access, but also engage in conduct that falls into the particularly dangerous category, are 'strong' zone 2 scenarios. Under zone 2, initial attacks are arguably permissible, but retaliatory hackbacks are also permissible. The desert rule protects the entitlements of capability-deficient people who retaliate if they use due diligence in their retaliate. The Van Buren majority did not address software researchers at all, but the danger posed by bad faith researchers given limited, temporary access was extensively briefed by *amici* in Van Buren.

8.1 *The Perils of Bad Faith Research*

In their amicus brief, Voatz draws a sharp distinction between authorized and unauthorized security researchers. Voatz's definition of authorized "[includes] private consulting firms and participants in organized bug bounty programs",[78] and Voatz actively takes advantage of these kinds of researchers when it investigates possible security vulnerabilities.[79] Most security researchers are explicitly authorized by contract or computer use agreements. From these two unassailable facts, Voatz makes two additional claims. The first is that regardless of how one interprets Section 1030(a)(2)(c), security vulnerability analysis undertaken by authorized researchers consistent with formal contracts or terms of usage agreements is always permitted. The second claim is that the first claim does not require a 'narrow reading' of the CFAA, which Voatz defines as protecting "everyone who attempts to attack applications and websites without authorization or exceeds authorized use (in order to provide a safe harbor for those like themselves who do so with good intentions)."[80] In other words, researchers engaged in vulnerability analysis that was explicitly authorized by a contract or computer use agreement already enjoy a safe harbor under Section 1030(a)(2)(c); they don't require additional safe harbor protection, provided by a narrow reading of the CFAA, to continue their vital work.

As further proof that authorized researchers don't require an additional safe harbor under the CFAA, Voatz documented the effectiveness of bug bounty programs that offer substantial awards to participants who identify vulnerabilities in the software of public and private entities.[81] Opposing amici seeking to narrow the CFAA disputed both the effectiveness of and legal protections provided by bug bounty programs. Proof of the effectiveness of bug bounty and private consulting efforts also lends credence to the argument that there is no need to extend additional safe harbor protection to independent researchers by narrowing the CFAA, even when they are operating in good faith. According to Voatz, the risk of unauthorized hacking always outweighs the benefits.

Voatz offered itself as an example of what happens when independent researchers, operating on their own, without authorization, 'hack' into a computer system looking for vulnerabilities. In one incident, an MIT researcher wrote code to access Voatz' voting system. Instead of

[78] Voatz (2020, 2).

[79] Id. at 3.

[80] Id. at 13.

[81] High-profile bug bounty programs like Hacker One, Hack the Pentagon, Hack the Air Force, the DC Board of Elections, Homeland Security/Def Con, and FDA/We Heart Hackers were very successful. Voatz (2020, 6–8).

analyzing actual vulnerabilities, the researcher constructed a hypothetical scenario describing how he would have accessed Voatz' system, and what shortcomings he likely would have found.[82] In another incident, two college students operating on their own attempted to access Voatz's mobile voting application, which was being used during a live election in West Virginia. The students did not receive authorization to break into Voatz' software, and they did not participate in any bug bounty program organized by Voatz. As a result, Voatz could not distinguish the students' attempted access from a hostile attack taking place during a live election and notified the FBI.[83] Both incidents appeared to demonstrate the costs, inefficiency and risk—particularly for attacks on critical infrastructure—of unconstrained, independent hacking of computer sites. Requiring security researchers to enter into formal agreements with intended hack targets also, according to Voatz, mirrors long-standing property expectations about what constitutes theft and illegal trespass.

In the last section of the brief, Voatz directly addressed the policy implications of creating a new safe harbor provision for independent security researchers by 'narrowing the CFAA.' Opposing amici asserted that the creation of such a safe harbor was appropriate given the importance of security vulnerability research, the ineffectiveness of bug bounty programs and the inadequacy of legal protections available to independent researchers. But they also argued that the courts should create such a safe harbor doctrinally by narrowing the CFAA so that contract breaches will not be considered 'unauthorized accesses' or 'exceeding accesses.' Voatz maintained, however, that if such a safe harbor provision is to be created, Congress should do it rather than the courts. Congress has already created a similar safe harbor provision in the Digital Millennium Copyright Act[84] that protects internet service providers "from liability if they meet certain specific requirements (e.g., establishing procedures for taking" down infringing materials on demand)."[85] Finally, Voatz addressed the central moral (and empirical) claim made by opposing *amici*, which is that the vast majority of independent security researchers operate in good faith,

[82] Id. at 9.

[83] Id. at 10.

[84] DMCA (2003).

[85] Voatz (2020, 13).

and that the likelihood of gray/black hat abuse by independent security researchers newly protected by a judicially created safe harbor for unauthorized hacking is relatively low.

8.2 The Prevalence of Good Faith Research

Unlike Voatz, the Computer Researcher Brief (CSR) argued that 'broadening the CFAA' "[chilled] essential computer security research by exposing computer security researchers to criminal and civil liability."[86] Two participants in the CSR brief characterized themselves as strong proponents of the 'open internet,'[87] while two other participants proclaimed their support for "open sharing of security vulnerability info."[88] Like Voatz, CSR draws a distinction between good faith and bad faith, or malicious, hackers but CSR makes two additional claims. The first is that the CFAA, from its inception, was only designed to prevent malicious hackers from breaking into computer systems. Over the years, many courts have 'broadened the CFAA' by not only conceptualizing unauthorized access through a variety of conceptual lenses (i.e., contract, code, norms), but also treating employer/employee disputes over computer use, as well as traditional hacker scenarios, as potential violations of Section 1030(a)(2)(c). Both developments have slowly obscured the initial purpose of the CFAA.

The second major point made by CSR at the beginning of the brief is that the software industry "relies heavily on the work of independent researchers ...to identify and fix flaws that malicious attackers could.. exploit."[89] This statement alone provides a rough estimate of how much security research ISRs perform relative to non-ISRs, and is based on the assumption that a significant proportion of independent security researchers don't themselves belong to the broader community

[86] CSR (2020, 1).

[87] Electronic Frontiers Foundation seeks to "protect innovation, free expression and civil liberties in the digital world" CSR (2020, 2); Center for Democracy and Technology "represents the public's interest in maintaining and open internet" CSR (2020, 2).

[88] Rapid 7 (Boston-based cybersecurity company) "openly sharing ... results ... help defenders reduce risks" CSR (2020, 3), and Scythe (Virginia-based cybersecurity company) believes that "the open sharing and reporting of research I a key element for organizations to analyze and achieve optimal security postures" CSR (2020, 3).

[89] CSR (2020, 4).

of malicious hackers ('black hats'). 'Broadening the CFAA' to prohibit "accessing publicly available data in a manner beneficial to the public yet prohibited by the owner of the data"[90] might discourage ISRs from performing vital security-related tasks. But some of these threshold claims are either overbroad, conclusory or self-contradictory. It still is not clear, for example, if CSR is asking the court to establish safe harbor protection (by narrowing the CFAA) for ISRs because they are responsible for most vulnerability testing (regardless of their intent), because ISRs already perform vital tasks in a 'non-malicious' manner, because ISRs liberate important 'public data' from the control of powerful entities, or because ISRs liberate important private data from the control of powerful private entities.

8.3 Research on a Spectrum

CSR's central task in the first part of its brief is to establish both the validity and vulnerability of security researchers operating in 'independent fashion,' whether that takes the form of "outsider researchers (working) in tandem or with the permission of ... in-house security teams" or independent software researchers operating "without any official permission".[91] Regardless of what form it takes, voluntary research could "occur in a manner unanticipated and even disapproved of by computer owners,"[92] but "when outside researchers discover a vulnerability, they often follow a disclosure process, exchanging information with vendors or others about the nature of the flaw so that it can be fixed or mitigated."[93] CSR then provides numerous examples of what happens when researchers choose, for whatever reason, to disclose vulnerabilities. In each case, the result was arguably beneficial,[94] but CSR failed to provide further empirical support for, or further socio-psychological analysis of, at least seven scenarios involving the potential disclosure, or non-disclosure, of security researchers operating in an "independent fashion":

[90] Id at 5.

[91] Id at 9.

[92] Id at 8.

[93] Id. at 9.

[94] Id. at 10–15 (medical devices, car safety, voting applications, military systems, etc.) CSR.

1. disclosure/non-disclosure after they themselves were hacked/affected;
2. disclosure/non-disclosure that might lead to a monetary award;
3. disclosure/non-disclosure for the purpose of extorting money from the target;
4. disclosure/non-disclosure for the purpose of repeatedly harassing target, followed by selective disclosure of found information to either the public or third parties;
5. non-disclosure in order to protect competitive advantages;
6. improvident disclosure because the researcher mistakenly over-reacts to the danger posed by a discovered vulnerability; or
7. improvident non-disclosure because the researcher mistakenly under-reacts to the danger posed by a discovered vulnerability.

All of these scenarios create risks that could be mitigated (or at least anticipated) by requiring researchers to enter into contracts, participate in bug buying bounty programs or conform to informal social norms that encourage vulnerability researchers to immediately (and anonymously) inform law enforcement if they discover exploits 'on their own.' The real challenge for decision-makers is whether courts should be empowered to consider the background specifics of the aforementioned scenarios because such a consideration would require judges to determine whether the facts of a hack satisfy the requirements of unauthorized access or access exceeding authorization under the CFAA, inquire into the actual motivations of ISRs and assess whether malicious disclosure/non-disclosure by ISRs in specific circumstances is more likely than non-malicious disclosure/non-disclosures.

Van Buren did not involve an ISR accused of hacking into someone else's computer system without authorization, or even 'hacking' in the traditional sense, but it does test the Eleventh Circuit rule that violations of Section 1030 (a)(2)(c), which prohibits unauthorized access as well as access in excess of authorization, can turn on whether the defendant violated a formal computer use agreement or contract. Contract-based constraints on computer use are not uncommon in the American computer marketplace, even when they are embedded in employment or terms of use agreements. In addition, numerous courts have held that violations of computer use agreements can be considered a violation of both tests in Section 1030(a)(2)(c)[95], even when the target/victim failed to implement complementary code-based restrictions.[96]

In spite of the relatively popularity of contract-based computer usage agreements for average users, and the bright-line advantages of equating authorized access with adherence to contract terms when those terms are clear, CSR is determined to prove that vulnerability researchers generally, and ISR more specifically, will languish if their work is on any way constrained by contractual obligations. In its zeal to create a 'roving right to hack' under the CFAA, or establish safe harbor protection for presumably good faith researchers who develop—and not all hackers do—an allergic reaction to contracts, CSR provides many examples of 'inadvertent' violations of computer use agreements that courts should not automatically consider unauthorized access, but these examples walk a fine line between three categories: (1) accidental violations likely to be committed by casual users, (2) inadvertent, deliberate/intended (but framed as accidental) or entirely foreseeable violations caused by serious researchers with a sophisticated grasp of both the software and any potential ambiguities in contractual terms, and (3) curious, semi-exhibitionistic tinkerers with anarchistic sensibilities.

9 Strong 'Zone 2' Hackbacks

It is not unusual for legitimate software researchers to receive limited, section-specific authorizations for access to computer systems. Researchers working for firms or the government do it regularly, and even outsiders or independents participating in bug bounty programs may obtain limited, section-specific authorization. There are also self-styled or self-labeled software researchers who break into systems without authorization or oversight. Authorized software researchers may also, with the proper incentive, decide to wander around the system without regard to security protocols or possible spillover effect. According to the Van Buren majority, inquiry into the intended purpose, as well as what the researchers ultimately did with whatever they collected, is irrelevant in terms of assessing a potential violation of Section 1030(a)(2).

[95] See, e.g., Nosall, Rodriguez, Drew (2009).

[96] CSR (2020, 12).

All that matters is that the researcher used proper authorization credentials to enter the system and remains within the section covered by the authorization.

In its guidelines, however, the DOJ stated that it would not hesitate to charge software researchers who obtain authorization access with a Section 1030 to violation if the research was "not conducted in good faith," which includes research "for that purpose of discovering security holes in devices, machines or service in order to extort the owners of such devices, machines or services.[Such activity] might be called "research", but it is not in good faith."[97] In terms of assessing the permissibility of an insider access, the DOJ's guidance on software researchers is important for three reasons. First, as I mentioned before, it clearly indicates the DOJ's intention to consider purpose in his prosecution of Section 1030 (a)(2) violations even though the Van Buren majority explicitly and repeatedly rejected purpose inquiries for textual and policy reasons. And DOJ's purpose inquiry will be determined by what researchers do and how they do it. Second, any serious inquiry into the purpose of a researcher has to consider the meaning of wandering around a computer system. Under all but the most extenuating circumstances, operating in good faith has to include remaining within the section covered by your authorization. Third, inferring bad faith from the decision to wander around a computer system, beyond the computational limits of the research authorization, is clearly important from the standpoint of capability because the researcher might know that his target is temporarily incapable of either recognizing a software vulnerability or fixing the vulnerability in areas not covered by the authorization. And that capability deficiency might exist even though the target has wealth, primary goods and formal legal rights.

The questionable permissibility of a software researcher who might be operating in bad faith, and willing to exploit the capability deficiencies of his target even though he has limited authorization to enter specific sections of the targets system, is only one part of the VRN calculus. The other part concerns the permissibility of any retaliation for the researcher's problematic access. If federal courts decide that the DOJ's decision to consider bad faith in prosecuting potential violations of Section 1030 is consistent with Van Buren, then the accesses of bad faith software

[97] DOJCFAA (2022, 4).

researchers who exceed authorization will be considered impermissible. If, however, there are some questions about the permissibility of the initial attack—and those questions will remain unanswered until the DOJ's guidelines are challenged in the federal courts—retaliation is still, under zone two, permissible if the initial attack falls into the 'particularly dangerous' category. The DOJ has already described what it considers to be particularly dangerous attacks, and many of those categories overlap with what I consider to be the most compelling justifications for retaliation: deliberately high risk or high impact attacks, attacks on critical infrastructure, egregious or malicious attacks, attacks on military installations, repeat offenders or attacks on public health.

The DOJ does not mention retaliatory hackbacks in its guidance, but it is entirely possible, if not probable, that the government, either operating on its own or with the assistance of a third party, has already engaged in authorized retaliatory backpacks after what it deems to be particularly dangerous attacks, secure in the knowledge that the full details of those hackbacks will never be revealed to the public. The strongest rationale for zone two of VRN, however, is not just that permissible retaliatory hackbacks undertaken in response to particularly dangerous attacks may have already taken place. Instead, the strongest rationale is that retaliatory hackbacks were undertaken to compensate for important capability deficiencies in American foreign policy, deficiencies exploited by strategic enemies and the illiberal nations.

10 The Efficiency of Weak and Strong Zone 2 Hackbacks

This chapter described two scenarios in which retaliatory hackbacks might be permitted post-Van Buren after an 'insider' abuses his authorized access to acquire information for an improper purpose: 1) a retaliatory hackback by an affected entity and/or victim with lawful access to the insider's computer system ('weak zone 2') and 2) the insider's attack falls into the 'particularly dangerous' category ('strong zone 2'). The former simply reverses the logic of the majority opinion in Van Buren. If an authorized, but improper, insider access is permissible under Section 1030(a)(2), then the same provision should also permit someone with authorized access to the insider's computer to enter that computer in retaliation. If purpose is irrelevant for permissible authorized accesses under Section 1030(a)(2), then it should be irrelevant for the insider and the retaliator. VRN strengthens the argument for scenario 1 if the victim was capability-deficient, the insider exploited the victim's capability-deficiencies, the victim did not provoke the insider, and the victim's retaliation is done with due diligence (i.e., reasonable necessity, attribution and proportionality). The second scenario reflects the DOJ's stated intention to prosecute accesses in excess of authorization if those accesses fall into a limited set of categories. VRN also strengthens this argument, not only permitting retaliation if the victim already had lawful access to the insider's computer, but also authorizing the imposition of all costs for the victim's retaliation given the particularly dangerous character of the insider's attack.

The desert rule of VRN protects the entitlements of capability-deficient individuals who retaliate after an attack by an insider operating with an improper purpose (zone 2). Capability theory can explain the appeal of a retaliatory hackback to individuals victimized by insiders, even when they have access to wealth and enjoy a generous endowment of primary goods and formal rights. But is it possible to characterize a symmetrical hackback against insiders under the CFAA as 'efficient'? This section analyzes the efficiency of symmetrical hackbacks from three angles: 1) whether symmetrical hackbacks are the most efficient way to deter insiders who abuse their authorization credentials, 2) whether the decision not to disclose vulnerabilities before a symmetrical hackback is a more efficient cyber-security strategy than the disclosure of vulnerabilities that usually accompanies cooperation with law enforcement, and 3) whether symmetrical hackbacks conceptualized as 'penalty enhancements' under VRN's zone 2 are efficient.

What is the most efficient way to deter insiders who exceed authorized access to a computer system?[98] Virtually, every organization has to be prepared for insider access abuse. One step removed from the

[98] A related inquiry—the most efficient way to enter a computer system to which one does not have full access—isn't really relevant because both the weak and strong zone 2 CFAA scenarios assume that the targeted victim of an insider attack has full, lawful access to the attacker's computer system. The only question is whether retaliation after the insider misappropriates company information, as opposed to some other justification for entering the insiders computer system (such as installing software updates or monitoring internet use), is morally and legally permissible. Examining the most efficient way to deter insider abuse, however, is relevant because most cyber retaliations (including retaliations permitted by VRN) are at least partially (if not entirely) designed to deter the suspected attacker from inflicting any additional harm on the victim or affected third parties.

broad question of whether companies should enforce security access using code, contracts, system architecture (computational separation) or informal norms is the narrower question of whether any of the aforementioned security strategies are 'efficient.' Conventional cost-benefit analysis weights the costs of instituting a particular access regime against its effectiveness.[99] But cost-benefit models also largely ignore 'expressive' considerations.

> "CBA deals with the material or quantitative dimensions, not the interpretive and expressive ones. CBA examines alternative end states; it compares, for example, how much it would cost to reach a state in which health was protected to a certain degree against a particular risk. It cannot take account of the meaning of the transition–the values the transition will be socially understood to express–from one end state to another."[100]

Discouraging insiders from abusing their authorization credentials using computational barriers or restrictions implemented by code without the consent or knowledge of employees, for example, may be efficient (balancing costs against effectiveness), but it sends a signal that the organization does not trust its employees. Such a policy may even discourage employees from trusting each other. Explicit contractual prohibitions on insider authorization abuse, on the other hand, might be more expensive and less efficient given their effectiveness as a deterrent mechanism, but they signal trust and a willingness to bargain about security procedures.

An extension of CBA is that law should discourage insider abuse of authorization credentials by increasing the costs of that abuse. If, as many have observed, computer technology decreases the costs of crime,[101] then a cost-benefit rationale for policies that deter insider abuse should

[99] Both attackers and defenders take advantage of cost-benefit analysis Cordes (2011, 2) ("Given some level of security, attackers balance the cost of additional attacks against the benefits from additional attacks. For analytical purposes, one can imagine an 'equilibrium' in which the number of attacks is the point at which the marginal benefit from the additional attack just equals the marginal cost. Similarly, given some level of attack volume, defenders balance the cost of attaining additional security against the benefits from added security. As in the market for attacks, one can imagine an equilibrium in which the chosen level of security S, is the point a which the marginal benefit from increments of security just equals the marginal cost.")

[100] Pildes and Sunstein (1995, 70).

[101] Katyal (2001).

channel, steer or nudge insiders toward more 'expensive' abuse strategies.[102] Examples of this approach might include steering insiders toward information that is easy to track offline, and away from information that is more difficult to track offline; requiring employees to consent to greater surveillance and monitoring of online and offline activities if they are granted authorized access to highly valuable or confidential information; or exposing insiders to criminal liability if they use their authorization credentials to exploit discovered vulnerabilities in their employer's computer system.

Would the possibility of a legally permissible, VRN-compatible symmetrical hackback (weak zone 2) nudge rogue insiders toward more 'expensive' insider abuse strategies (i.e., strategies more likely to result in convictions)? The answer is yes if the alternative, more expensive strategy involves not just retrieving data specifically from the sections covered by the insiders authorization credentials, but also using those credentials to 'wander around' the target's computer system gathering data as well as extensive information about the computational structure of that system. Fear of a hackback by one's employer if the former scenario occurs might discourage a certain percentage of rogue insiders from using their authorization credentials to retrieve any data at all for an improper purpose, but more sophisticated rogue insiders might decide that the risk of getting caught is worth it if they can gather more valuable information by not only 'wandering around' the employers system, but also taking steps to 'cover their tracks.' In terms of cost-benefit analysis, therefore, a policy stating that employers will routinely inspect the work computers of employees suspected of misappropriating data secured using their authorization credentials might be very efficient in terms of steering rogue insiders toward abusive insider access strategies that are clearly prohibited by Van Buren.

What about the apparent inefficiency of discouraging the disclosure of software vulnerability, which is likely to be the strategy of choice for victims who choose to hackback? For many game theorists, failure to disclose vulnerabilities is a classic collective action problem: mutual cooperation serves the best interests of affected parties, and yet those same parties choose to act solely in their own interest. The Prisoners Dilemma perfectly illustrates the perils of inefficient non-cooperation when the

[102] Id at 1006.

payoff is measured in terms of economic benefit, but as I discussed in the last chapter, one can imagine a Prisoners Dilemma scenario involving ransomware operators and ransomware victims in which non-cooperation with regard to economic gain actually increases the number of reasonable opportunities and options available to both parties.

Cyber experts often argue that the impulse to not disclose cyber attacks is as damaging, to both the victim and society, as the impulse not to cooperate under the Prisoner's Dilemma. In fact, the overall social value of immediately disclosing an attack seems to explain the proliferation of purely public, public/private, and purely private vulnerability regimes, as well as the appeal of good faith exceptions for security researchers who look for security blindspots and software backdoors that can be exploited by rogue programmers.

Given the weight of opinion and data on the wisdom of disclosing vulnerabilities, how is it possible to argue that a capability-deficient victim of an insider attack who chooses to retaliate and not disclose that attack has acted 'efficiently'? To rephrase the question, when, if ever, is non-disclosure after an insider cyber-attack efficient? At the outset, it is important to state that in spite of a growing consensus on the wisdom of disclosure, many private and public entities are very selective about the cyber attacks that they announce, and mandatory software vulnerability disclosure regimes remain controversial. One could also argue that the incentive to report an 'insider attack' is weaker than the incentive to report an outsider attack because of what an insider attack reveals about the level of trust and professionalism inside of an organization. But the most compelling counter-narrative to the consensus view that disclosure after a cyber attack is efficient (or at least well-founded) stems from what I described earlier as the 'most dangerous' cyber attacks—the factual backdrop for VRN's 'zone 2' and at least some portion of the DOJ's post-Van Buren prosecution guidelines. Public disclosure of the details of a cyber attack before any retaliatory hackback makes little logical sense because a suspected attacker could 'disappear into the woodpile' if he suspects that the victim has identified and located him. But public disclosure of the details of a cyber attack before retaliation makes absolutely no sense for 'particularly dangerous' cyber incursions like malware attacks on critical infrastructure or military nuclear installations. In fact, public disclosure of cyber attacks on critical infrastructure and military installations after any retaliatory measures have taken place should be limited. This is, in fact, the operating strategy of most nation-states faced with an attack on critical infrastructure or military installations, and for good reasons. In these situations, the benefits of detailed public disclosure (as opposed to limited disclosure to strategic partners) are clearly outweighed by the potential costs of an imminent national security risk.

Finally, there is the question of whether symmetrical hackbacks conceptualized as penalty enhancements are 'efficient.' In Chapter 4, I argued that the moral and arguably legal permissibility of hackbacks under zone 2 stems, at least in part, from the particularly dangerous character of specific cyber attacks. In the same way that jurisdictions often impose severe sentences for bias crimes, attacks on children, crimes involved the use of firearms and crimes linked to terrorism, VRN's zone 2 protects the entitlements of capability-deficient entities subjected to a 'particularly dangerous' attack by imposing all costs for any damages caused by the victim's retaliation on the perpetrator. From the standpoint of conventional law and economics, the efficiency of the protection provided by VRN's 'zone 2' would be measured in terms of deterrence and cost savings.[103] The same is true for a wide variety of penalty enhancement provisions, ranging from hate crime statutes and laws that impose severe penalties for attacks on "especially vulnerable victims" (like homeless people), to laws that increase the sentences of repeat offenders. In every case, efficient (or socially optimal) penalty escalation provisions perform two functions : 1) they decrease non-compliance with legal prohibitions on activities considered to be extremely dangerous or extremely harmful, and 2) with regard to hate crimes, penalty enhancement provisions decrease the avoidance costs of individuals or groups disproportionately impacted by historical discrimination and violence.[104] They are also expressive, often serving as a statutory symbol of a community's moral outrage and sense of fairness.[105]

[103] Polinsky and Shavell (1998, 3) ("In the standard model of public enforcement of law, sanctions are evaluated in terms of their deterrent benefits and their costs...").

[104] Dharmapala and Garoupa (2002, 4).

[105] See, e.g., Dau-Schmidt (1990) (suggests that penalty enhancement provisions might be expressive). For more on the role of fairness in the economic study of public law enforcement, see Swygert and Yanes (1998). For more on the role of fairness in the economic study of public law enforcement, see Swygert and Yanes (1998).

Traditional economics would measure the efficiency of symmetric hackbacks conceptualized as penalty enhancements primarily in terms of deterrence and cost, and secondarily in terms of expressiveness and fairness. But there is another way to measure the efficiency of penalty-enhancing symmetric hackbacks: opportunities. Economics often defines efficiency in terms of avoiding wasteful or duplicative uses of resources.[106] One could argue, for example, that policies encouraging the creation of cryptocurrencies like Bitcoin that consume vast quantities of energy performing complex 'proof of work' calculations are 'inefficient.' The same is arguably true for policies that encourage people to travel using methods that consume vast amount of energy (airplanes) instead of methods that use relatively little energy (bicycles).[107] Irene von Staveren has argued that Paretian efficiency is more likely to produce 'social waste,' in the form of "perverse preferences or false beliefs", than policies that increase both the quality and quantity of meaningful opportunities available to people. Opportunity-based policies, in other words, are 'less wasteful,' and therefore more efficient, than Pareto-based policies.[108] Waste directly impacts opportunity fulfillment. Some theorists have argued that hate crime laws—classic penalty enhancement mechanisms—not only reduce (or perhaps more accurately, shift) the avoidance costs shouldered by members of minority groups, but also minimize "the social waste [that] occurs because avoidance involves foregoing opportunities for profitable transactions with [members of the dominant group."[109] To the extent, therefore, that symmetrical hackbacks under zone 2 operate as penalty enhancement mechanisms, they efficiently reduce the social waste of lost opportunities endured by the capability-deficient victims of 'particularly dangerous' cyber attacks.'

References

Amsterdam, A. 1960. The Void for Vagueness Doctrine in the Supreme Court. *University of Pennsylvania Law Review* 108: 67–116.

Bitman. 2015. Bittman v Fox, 107 F.Supp 3d. 896.

[106] See, e.g., McCormick (1986) and Van Staveren (2007, 15) (briefly describes the role of the minimization of waste in the history of institutional economics).

[107] Id. at 7.

[108] Id. at 14.

[109] Dharmapala and Garoupa (2002, 4).

CFAA. 1986. Computer Fraud and Abuse Act of 1986, Pub L. No 99-474, 100 Stat. 1213 (1986).

CFAA. 2012. Section 1030 (a)(5), Supp. V 1988.

CFAAReport. 1984. House Committee Report on the Computer Fraud and Abuse Act, HR Rep 98-894.

CFAASuppI. 1984. 18 USC 1030a, Supp II 1984 CFAASuppI.

CFAASuppII. 1986. CFAA of 1986, Pub L. No 99-474, 100 Stat. 1213, Supp II 1986.

Chip. 2006. V Chip Corp. v Lee. 438 F Supp 2d 1087 (ND Cal 2006).

Citrin. 2006. IAC LLC v. Citrin, 440 F3d 418 (7th circ 2006).

Cordes, J. 2011. *An Overview of the Economics of Cybersecurity and Cybersecurity Policy*. Report GW-CSPRI-2011-6, June 1.

Creative. 2004. Creative Computing v. GetLoaded.com, LLC, 386 F3d 930 (9th circ, 2004).

CSR. 2020. Brief Amici Curiae of Computer Security Researchers, Electronic Frontier Foundation, Center for Democracy & Technology, Bugcrowd, Rapid7, Scythe, and Tenable in Support of Petitioner. https://www.eff.org/files/42020/07/08/19-783_eff_security_researchers_amici_brief_.pdf.

Dau-Schmidt, K. 1990. An Economic Analysis of the Criminal Law as a Preference-Shaping Policy. *Duke Law Journal*: 1–38.

Davis. 2019. US v Davis, 139 S. Ct. 2319, 2325 (2019) Davis. 2019.

Dharmapala, D., and Garoupa, N. 2002. *Penalty Enhancement for Hate Crimes: An Economic Analysis*. Economics Working Papers, 200212.

DOJCFAA. 2022. *Department of Justice Guidelines*, 9-48-000—Computer Fraud and Abuse Act, May 19.

Drew. 2009. United States v. Drew, 259 F.R.D. 449 (C.D. Cal. 2009).

EEA. 1996. Economic Espionage Act of 1996, Section 201 (1)(B)(ii), 110 Stat 3492.

EF. 2001. EF Cultural Travel BV v. Explorica, Inc, 274 F3d 577 (1st Circuit, 2001).

Epic. 2020. *Brief of Amici Curiae Electronic Privacy Information Center (Epic) and Fifteen Technical Experts and Legal Scholars in Support of Respondent.* On Writ of Certiorari to the United States Court of Appeals for the Eleventh Circuit Van Buren v. US.

Explorica. 2001. EF Cultural Travel NV v. Explorica, Inc., 274 F.3d 577, 582, n. 10 (1st Circ 2001).

Guy, S. 2013. Aaron Swartz 'Was Killed by the Government' Father Says at Funeral. *Chicago Sun Times*, January 15.

Heart. 2020. *Brief of Karen Heart and Anthony Volini of Ciplit as Amici Curiae in Support of Respondent*. On Writ of Certiorari to the United States Court of Appeals for the Eleventh Circuit, Van Buren v. US.

Hoffman. 1982. *Village of Hoffman Estates v Flipside*. Hoffman Estates, Inc., 455 US 489.

Intake. 2014. Memorandum from US Attorney to US attorney and Assistant Attorney Generals for the Criminal and National Security Divisions, "Intake and Charging Policy for Computer Crime Matters", September 22.
John. 2010. *US v John*, 597 F.3d 263 (5th circ, 2010).
Johnson. 2015. Johnson v United States, 135 S. Ct 2551, 2556 (2015).
Jones. 2001. US v Jones, 597 F3d 263 (5th Circ 2001).
Katyal, N. 2001. Criminal Law in Cyberspace. *University of Pennsylvania Law Review* 149: 1003–1114.
Kluben. 2009. Kluben Skahan and Associates, Inc. v Cardogen, Clark and Assoc. Inc., No 08-CV-1529, 2009 WL 466812 at 8 (ND Ill. February 2009).
Lessig. 2013. "Prosecutor Is a Bully". *Bloombergview.com*, January 17.
LVRC. 2009. LVRC Holdings, LLC v. Brekka, 581 F.3d 1127 (9th Circ. 2009).
McCormick, K. 1986. Towards a Definition of Waste in Economics: A Neoinstitutional Approach. *Review of Social Economy* 44 (1): 80–92.
McDonnell. 2016. McDonnell v US, 136 S. Ct. 2355, 2372-3 (2016).
Merritt. 2015. Merritt Hawkins and Assoc. LLC. v. Gresham, 79 F.Supp. 3rd 625 (ND Tex 2015).
Morris. 1991. US v. Morris, 928 F2d 504 (2d Cir. 1991).
NosalI. 2012. (Nosal I) US. v. Nosal, 676 F 3d 854 (9th Cir. 2012) (en banc).
NosalII. 2016. US. v. Nosal, 828 F.3d 865 (9th Cir. 2016).
Phillips. 2007. US. v. Phillips, 477 F.3d 215 (5th Circ. 2007).
Pildes, R., and Sunstein, C. 1995. Reinventing the Regulatory State. *University of Chicago Law Review* 62: 1–129.
PoliceBuren. 2020. Brief of the Federal Law Enforcement Officers Association as Amicus Curiae in Support of Respondent, On Writ of Certiorari to the United States Court of Appeals for the Eleventh Circuit, Van Buren v. US.
Polinsky, M., and Shavell, S. 1998. The Fairness of Sanctions: Some Implications for optimal Enforcement Policy. *Olin Program in Law and Economics*. Working Paper 167, September 1998.
Rodriguez. 2010. US v. Rodriguez, 628 f3d 1258 (11th Circ. 2010).
Sandvig. 2020. Sandvig v Barr, No. 16-1368 (JDB), 2020 WL 1494065, D.D.C, March 27.
Stevens. 2010. US v Stevens, 558 US 460, 480 (2010).
Swygert, M., and Yanes, K. 1998. Focus on Fairness, Efficiency and the Law: The Integration of Fairness into Efficiency. *Washington Law Review* 73 (2): 249–328.
Teague. 2011. US v Teague, 646 F.3d 1119 (8th circ 2011).
VanBuren. 2021. Van Buren v. United States, 593 U.S ___ (2021), 141 S. Ct. 1648 (2021).
VanBurenBrief. 2020. Brief for petitioner Van Buren, REPLY BRIEF FOR PETITIONER, Van Buren v US, On Writ of Certiorari to the United States Court of Appeals for the Eleventh Circuit.

VanBurenGov. 2019. Brief for the United States, Van Buren Vs United States, No. 19-783, On Writ of Certiorari to the United States Court of Appeals for the Eleventh Circuit.

Van Staveren, I. 2007. *The Ethics of Efficiency*. SCEME Working Paper No. 18.

Voatz. 2020. Brief of Voatz, Inc., As Amicus Curiae in Support of Respondent in Affirmance of the Decision Below, https://www.supremecourt.gov/DocketPDF/219/19-783/153062/20200903122434600_Voatz%20Amicus%20Brief.pdf

WEC. 2012. WEC Carolina Energy Sols, LLC v. Miller, 687 F.3d 199 (4th Cir. 2012).

CHAPTER 8

Amplifying American Power in Cyberspace

1 The International Order

1.1 Gradual Decline of the Liberal International Order

Few uncertainties in the geo-strategic landscape are more important than the appropriate exercise of state power during a national security crisis. Along with a host of pressing uncertainties in the international environment—the rising significance of non-governmental organizations, blurred distinctions between armed international conflict, non-international conflict and belligerent occupation, instability in the European union—policy-makers are, for the first time since World War II, seriously considering the distinct possibility that the United States should no longer play the role of the world's policeman. The collapse of Europe after World War II made the United States the world dominant economic power, and with the subsequent collapse of the Soviet Union, the United States emerged as the only superpower, blessed with a formidable arsenal of military weapons, economic capability, and political capital. It was also geographically blessed, with large bodies of water to its west and east, and good relations with its neighbors to the north and south.[1]

With these advantages, the United States, along with its principal European allies—England and France—embarked on an aggressive campaign

[1] Wucker (2016).

A. J. Thomas, *The Open World, Hackbacks and Global Justice*, https://doi.org/10.1007/978-981-19-8132-6_8

to remake the world using the favorite tools of western liberal societies: democracy, the rule of law, free trade and human rights guarantees. These innovations enabled the United States to grow its economy and establish the basis for later globalization, maintain a substantial military presence in at least fifteen countries (including German, Japan and South Korea), redraw the political and geo-strategic boundaries of the Greater Middle East (at first with, and eventually without, the assistance of Britain after 1969),[2] establish the legitimacy and effectiveness of international institutions like the United Nations and the GATT, consistently challenge Soviet expansionism, and set the ground rules for participants in the 'nuclear club' (such as the NPT) and the global economic system.

1.2 Realism and Its Critics

Providing a compelling political, geo-strategic and moral justification for a post-World War II 'liberal hegemonic order' guaranteed principally by the United States, however, has consumed the time and attention of policy makers for decades. Within international relations theory, the main point of disagreement hinges on the distinction between realism and idealism. The centerpiece of realism is that the dominant concern of states is the pursuit of power, which is primarily defined in terms of national identity, military capacity, independence and the control of territory. States are selfish, prioritize their own national interests, and are particularly sensitive to developments that could destabilize their relationship with regional neighbors, both friend and foe.[3] Although there are many variations of realism, most variations are deeply suspicious of ideology as a motivator for (or explanation of) state action. Regardless of whether a state seeks to defend its interests or expand into new physical or geo-strategic territory, state action is motivated by the pursuit of power rather than an affinity for a particular set of ideologies or moral sensibilities. The most effective way to protect state power may be through international institutions in certain situations, and military might in others, but the pursuit of power best explains state behavior.

Within the idealist tradition, state action can be motivated by a variety of different concerns, including but not limited to, ideological affinity.

[2] Jeffrey (2016).

[3] Id.

Interwar utopian liberalism[4] largely explains the idealist view that states to perceive themselves as members of a 'society of states,' "whose norms they adhere to not because of ongoing self-interested calculations that it is good for them as individual states, but because they have internalized and identify with them."[5] Such a view does not necessarily reduce state action to the ideological sensibility of important figures within government,[6] but it does suggest that modern states, as collective entities that must balance national interests and regional constraints against the demands of an international system that rewards cooperation and punishes defectors, usually factor existing norms, shared values and ideas into their geo-strategic deliberations. The crucial question is whether states comply with these norms or shared ideas solely because they perceive them to be legitimate, or if some other factor is involved, such as the fear of sanctions for non-compliance, or a more realist calculation that compliance is the best way to advance that nations self-interest.

Institutionalist critics of realism suggest that states best express their allegiance to shared ideas and norms not through their interactions with other states, but instead through their cooperation with international institutions.[7] Self-interest and the desire to defend or acquire power do not vanish when states purport to act based on the perceived legitimacy of norms and ideas developed and amplified by international institutions, but they place those interests within a web of broader super-regional and super-national concerns that may require a state to temporarily assume a burden in order to protect vulnerable parties in another part of the world. Such a cosmopolitan view of state action, particularly in the context of human rights and the 'responsibility to protect,' is based on the compelling idea that states should act both to protect their own citizens and the citizens of other countries where possible, not to enhance their

[4] See eg Heinze (2008, 111–114) and Ashworth (2006, 293–298).

[5] Wendt (1999, 242–3).

[6] Rao (2011, 207–8).

[7] Rao (2011, 209).

own power either regionally or internationally, but because providing assistance is the morally correct thing to do.[8]

1.3 *Primacy vs. Restraint*

America's role in the liberal international order is framed by two broad traditions: primacy and restraint. The latter suggests that America refrain from being the world's policeman, avoid massive military deployments in all but the most important areas during times of crisis, and cease democracy promotion efforts,[9] whereas the former paints the United States as the dominant military power and principal diplomatic, economic and political guarantor of the liberal economic order.[10]

Neo-conservatives and liberal interventionists tend to subscribe, either openly or quietly, to primacy.[11] Neo-conservatives, for example, believe that the United States is the sole super-power in terms of economic, political and military clout,[12] and it should not relinquish that role even when it is politically inconvenient or financially taxing. In practice, the neo-conservative view requires an aggressive and unmistakable US military presence in regional arenas, active democracy promotion efforts because democracy is morally desirable and encourages political and economic stability,[13] and, if necessary, unilateral military action to protect American interests or personnel threatened abroad.[14] Unlike the neo-conservatives, liberal interventionists believe that American primacy is necessary in order to prevent widespread suffering and human rights violations. Instead of asserting American military dominance as a means to maintain our status as the world's superior power or global hegemon, liberal interventionists

[8] Rao (2011, 210).

[9] Gholz, Press and Sapolsky (1997).

[10] Huntington (1993), Haass (2001), Walt (2018, 36), and Wohlforth (2008).

[11] Idealists of both the neo-conservative and liberal internationalist varieties used to dominate the foreign policy establishment. Toft (2017).

[12] Mearsheimer (2011) (neo-conservatives share with liberal interventionists the view that the United States is basically a 'global hegemon') and Jervis (2009).

[13] Kirchick (2017) (neo-conservative penchant for democracy promotion, so central to the decisions of Pres. Bush after 9/11, reflect America's historic responsibility to help beleaguered countries realize the 'dreams' of the people).

[14] Hoffman (2013) (distinguishes between neo-conservative unilateralists and (isolationist) sovereigntist unilateralists).

would ensure the pre-eminence of American values on the world stage by engaging in what has been described as 'armed philanthropy.'[15]

Proponents of restraint, on the other hand, tend to be realists who believe that states are primarily concerned about the pursuit of power. There are many forms of restraint. Neo-isolationists, for example, want America to withdraw from its post-World War II role as the world's policeman, and walk away from most of its current military and diplomatic obligations. Selective engagers believe that there are some regions in the world that demand special attention by the military community—mainly the Middle East, East Asia and Europe. In those regions, the United States must maintain an active military and diplomatic presence in order to ensure the liberal international order and deter aggressive regional hegemons like China, Russia or Iran. Such a presence will usually prevent the outbreak of war, and it is in our best interest to discourage conflicts, but that interest does not extend to democracy promotion—a mainstay of neoconservatism and, to a lesser extent, liberal interventionism. Off-shore balancers believe that America should devote its considerable military and diplomatic resources to fostering a balance between competing regional powers, rather than directly asserting American power in order to maintain regional stability.[16] Off-shore balancers are prepared to accept a world in which the United States avoids democracy-promotion, Russia 'takes the lead' in Syria, active diplomatic relations with Iran are pursued, and the United States plays a less active role in NATO. Allies that fail to provide their 'fair share' of the costs of collective security are labeled 'free-riders,' and run the risk of the United States drastically reducing its traditional role in the maintenance of vital security interests.[17]

2 Strategic Responses to Cyber Infiltrations

2.1 *The Authoritarian Advantage in Cyberspace*

After several years of ascendancy, the liberal international order is now entertaining a serious challenge from authoritarian regimes. And the challenge is most acute in an arena perfectly designed to level differences

[15] Hoffman (2013, 25).

[16] Mearscheimer and Walt (2016) and Posen (2014).

[17] Cordesman (2016), Hoffman (2016, 28), and Wright (2016) (free-riding is a particular concern for all realists).

between small or less advanced states, and large, highly advanced states: cyberspace. Even though the Internet has made it more difficult than ever for authoritarian regimes to hide their human rights abuses, corruption and frequent rule of law violations, studies have shown that authoritarianism has not only increased in the past 30 years, but that illiberal regimes now enjoy what can be described as an 'authoritarian advantage' in their competition with liberal, democratic societies. A strong increase in autocratization, "an umbrella term consisting of backsliding in democracy as well as democratic breakdown and autocratic regression"[18] has also been observed.

Evidence of an authoritarian advantage in the modern world falls into three categories. The first is that authoritarian regimes seem to respond better to emergencies. The idea that top down, hierarchical and largely unaccountable political system are better suited to handle emergencies than legislative systems is well established, even in a country as committed to horizontal, representative democratic rule as the United States. Even scholars deeply concerned about the tyrannical, anti-majoritarian implications of uncharted executive unilateralism have acknowledged the functional advantage of presidential lawmaking during wartime, an advantage that the founders also recognized and enshrined in the US Constitution.[19] Evidence for an authoritarian advantage, as well as creeping

[18] Maerz et al. (2020, 2).

[19] Arguments in favor of increased Presidential authority during war and/or national security crises are usually based on functionalism. The basic argument is that Presidents should be given as much leeway and flexibility as possible to conduct wars, evaluate threats and construct security arrangements with international partners, regardless of how much war-making and budgetary power is given to the President by the Constitution. Stronger functionalist defenses of wartime executive power are not constrained by time and place. Under this view, contemporary understandings of presidential power during wartime are the result of a dramatic shift in the balance of power between Congress and the President that started during the New Deal. See eg Schlesinger (1973), Posner and Vermeule (2007, 205) (emphasizes the superiority of executive decision-making during war-time in the current political climate). For a critique of functionalism and executive unilateralism during war time, see Strauss (2011, 55) (rejects extreme functionalist model, citing Manning (2011, 1987). The appeal of broad executive wartime authority based on the inherent superiority of Presidential expertise during national security crises is well documented. Presidents have traditionally enjoyed considerable authority to respond to imminent or actual attacks without additional Congressional authorization. See eg. Curtiss (1936, 319) ("The President is the sole organ of the nation in its external relations, and its sole representative with foreign nations."). and Egan (1988, 530) ("[U]nless Congress specifically has provided otherwise, courts traditionally have been reluctant to

autocratization, is strongest during emergencies like market crashes and public health disasters.[20] Many experts believe the autocratic regimes like China, Singapore and Vietnam responded better to the COVID crisis than Western democracies.[21] Trends that the champions of democracy find so troubling—lack of political accountability, human rights or due process guarantees, extreme surveillance, and state censorship—made it easier for autocratic regimes to institute draconian measures like brutal quarantines and imprisonment for failure to undergo mandatory testing and vaccination that quickly controlled the spread of the Covid virus.

Experts provide different explanations for public tolerance of repression during crises,[22] but it seems clear that a society capable of convincing it's people to tolerate repression is much more likely to control the spread of a contagious disease than other countries. The advantage that democracies have over authoritarian societies—providing healthcare options, durable pricing and access to a wide variety of medicines and services—seems limited to ordinary times. Even though the authoritarian advantage during health crisis does not necessarily translate into fewer deaths as a percentage of the population once demographic trends are factored into the equation,[23] autocratic regimes are much less likely to experience the widespread vaccine revolts that plagued so many democratic societies,

intrude upon the authority of the Executive in military and national security affairs.") Both the Commander-In-Chief Clause and settled legal precedent seem to provide the president with inherent power to initiate military action in self-defense, and that power has been used repeatedly by Presidents since the beginning of the republic. See eg Constitution (1776, Art II, § 2, cl 2); Hamdi (2004, 579–85) (characterizes the President as "Commander in Chief of the Armed Forces".), and Federalist74 (1961, 500).

[20] Diamond (2022, 186).

[21] Maerz et al. (2020, 1).

[22] There are two broad explanations for this phenomenon, and the second directly relates to the authoritarian advantage. The first is that people living in a democratic societies recognize that democratic institutions like representation, accountability and dissent work best during ordinary times. When crises occur, the same people are willing to tolerate a temporary diminution in democratic rights as long as minimum rights protections remain in place. The second explanation is that people living in autocratic societies are already used to repression. When extreme repression occurs during crises, the same people are not only prepared, but they may not demand democratic guarantees when the crisis is over.

[23] Maerz et al. (2020, 1).

and autocratization is still more likely to occur during health crises in democratic societies.[24]

Authoritarian regimes in the modern world also seem to enjoy an advantage over democratic societies in the area of development. In terms of developing their own countries, autocratic regimes that enjoy "a higher level of insulation from particularistic demands then do democratic regimes"[25] can avoid the collective action problem that forces governments to prioritize the short-term preferences of interest groups over the long-term decisions required to "ensure efficient economic performance."[26] In order to be successful, however, politically unaccountable autocratic societies have to respond to economic advice from without, which may not be possible because studies have shown that many authoritarian regimes have limited access to outside information. Authoritarian regimes have also used their economic muscle to increase their power in Third World countries. Under the Belt and Road Initiative, for example, China has constructed "roads, bridges, ports and telecommunications networks across Africa, Asia and Latin America."[27] Russia, for its part, "provided roughly a dozen African autocracies with formal military assistance and carried out social media disinformation campaigns to help them maintain their rule."[28] But these development deals, implemented with little input from the citizens of the autocratic state, and often executed with remarkable precision and swiftness, come with "strings attached." Beneficiaries are expected to enter into unfavorable debt arrangements, install monitoring equipment, and refrain from criticizing the questionable policies and practices of their benefactors. The UN Resolution condemning Russia's invasion of Ukraine, for example, was only supported by 54 African states, a bare majority of the total.[29]

Experts have also considered the distinct possibility that autocratic nations also enjoy certain advantages over liberal, democratic societies in cyberspace. Computer technology already has a pronounced leveling effect because it is relatively low cost, and widely available. For a variety

[24] Maerz et al. (2020, 2).

[25] Maravall (2008, 17–18).

[26] Id.

[27] Diamond (2022, 192).

[28] Id. at 188.

[29] Id. at 186.

of reasons, computers and the Internet can decrease the power differential of small states relative to larger, wealthier states, as well as the power differential between states and non-state actors.[30] Authoritarian states like North Korea, Iran, Russia and China have taken maximum advantage of the leveling effects of computer technology to dramatically increase their power over world affairs, even though three of the four have comparatively weak economies,[31] and all labor under a constant barrage of sanctions because of their repressive policies.[32] Authoritarian nations have also contributed, disproportionately, to the gradual normalization of hacking,[33] at least in part because so much illegal hacking originates either from their shores or from software that their operatives stole from the west. Even though the Snowden affair[34] revealed the extent of American hacking activity on allies and rivals, most cyber attacks come from China, Iran, Russia or North Korea.[35] Some of the most devastating cyber attacks, including the WannaCry attack launched by North Korea and NotPetya attack launched by Russia, came from modified versions of American software ('Eternal Blue') that were stolen from NSA databases and re-launched by Russian operatives.[36]

Autocratic nations in cyberspace also enjoy an advantage because they can engage in information operations, implement Internet kill switches,[37] election tampering strategic surveillance and espionage at a reasonably low cost and with low political visibility, secure in the knowledge that most of these hybrid, gray operations will fall below the international law definition full-scale armed conflict. Each successful cyber operation marginally increases influence of authoritarian regimes, even when their military and economic initiatives flounder. Some experts believe that

[30] Govella (2019, 41).

[31] Alperovitch (2022, 48).

[32] Id. at 49.

[33] Potter (2016).

[34] The world community was stunned to discover that the United States has been collecting telephone meta-data about all Americans since May 2006 from Edward Snowden. Greenwald (2013).

[35] Potter (2016) and Alperovitch (2022, 47).

[36] Gordon and Rosenbach (2022, 17).

[37] Manjikan (2021, 466).

the gradual normalization of hacking has actually increased the vulnerability of authoritarian regimes that hack frequently,[38] but authoritarian regimes still have a strong incentive to launch frequent cyber attacks because attacks are relatively inexpensive, amplify their geostrategic and economic interests, and perhaps most importantly, do not require any political accountability. Internally, authoritarian regimes spend an enormous amount of time suppressing political dissent and censoring information, often through the control of computer resources and Internet access. Externally, authoritarian regimes have found ways to decrease the reputational costs of being habitual rule of law violators by claiming to be democratic and in favor of open market rules even though they are not.[39] They also lessen the economic impact of sanctions by using dark money cryptocurrency transactions. All of these actions reduce the political, economic and strategic costs of not being constrained by, or beholden to, domestic political opinion, international law or human rights norms.

2.2 *Democratic Guarantees, Human Rights and the Open World*

Unlike autocratic regimes, United States and its principal allies labor under the honorable burden of protecting democracy, human rights and the rule of law. Protecting these values is usually considered a privilege, rather than a burden, of international leadership, but regardless of how it is characterized, America's relatively firm commitment to the rule of law and democratic accountability places constraints on American actions that authoritarian regimes rarely factor into their strategic calculus.

After the establishment of international institutions like the UN and World Bank, the normalization of rule of law guarantees and the protection of human rights, no objective is more valuable to the liberal international order then the promotion of democracy as an aspirational ideal. That statement remains true even though there are many examples of liberal democratic nations entering into alliances with autocratic regimes for strategic gain after World War II, and rogue autocratic nations like China and Russia successfully entering the global marketplace even

[38] Potter (2016).

[39] Gordon and Rosenbach (2022, 19). These claims may not be completely credible, but they are enough to avoid complete estrangement from legitimate international institutions.

though democratic guarantees, the rule of law and political accountability mean almost nothing to them. In spite of this checkered history, democracy remains an aspirational ideal powerful enough to incentivize emerging, new states to institute democratic reforms as a condition of receiving financial or developmental assistance from the World Bank or the IMF.[40] More established liberal nations have advanced the democracy promotion narrative into other ways: (1) piggy backing the establishment of democratic reforms onto humanitarian intervention campaigns in war ravaged countries, and (2) encouraging states to join pro-democracy global alliances in order to encourage the development of democratic reforms within the country, and counterbalance authoritarian regimes.

In cyberspace, democracy promotion by liberal states, *writ small*, often involves the defense of democratic political conduct online, such as voting, canvassing and political advertising. Liberal nations have also conducted online campaigns to counter illiberal nations or actors that undermine democratic processes, as well as tacitly support cyber rebellions that challenge the legitimacy of autocratic regimes.[41] The most visible and effective democracy promotion initiatives, however, link democracy and the protection of open world values. In fact, openness is considered the principal advantage that liberal societies enjoy over authoritarian regimes in cyberspace.[42] To support this view, open world zealots frequently rely on two simple dichotomies: (1) democracies always support openness and authoritarian regimes always support closeness, and (2) societies have to choose between cyber security and openness. Evidence for the first assertion seems to be everywhere: autocratic regimes institute, with little resistance from their citizens, data nationalization regimes that cordon off their nations' internet from the rest of the world[43]; severe censorship of internet traffic[44]; the distribution of massive amounts of misinformation online to undermine democratic movements and erode public

[40] Some developing countries are more receptive to the possibility of financial assistance than the moral desirability of democratic reform.

[41] Flynn (2018, 110).

[42] Id. at 109.

[43] Gordon and Rosenbach (2022, 20).

[44] Govella (2019, 40).

confidence in Internet freedom,[45] and resistance to regional data protection regimes like the GDPR. Liberal democratic support for openness, in the public and private sector, also seems to be everywhere, operating across all of the five open zones. In addition to being the counter narrative to closeness, openness also appears to be the counter narrative to calls for increased cyber security. Cyber security, according to the critics, signals implicit support for militarism (i.e., that the cyber world is incapable of policing itself without military or law enforcement intervention, which is exactly what authoritarian states believe),[46] and is based on the assumption that securing the integrity of the Internet necessarily decreases Internet freedom.[47]

While it is difficult to deny that autocratic regimes view Internet freedom and accountability with deep suspicion, and prioritize policies that discourage political dissent, undermine Internet access and balkanize data, it is entirely wrong to believe that any increases in cyber security necessarily decrease Internet freedom, or that a more granular understanding of the open world necessarily increases the power and influence of authoritarianism. Liberal democratic societies will not increase their power or push their advantage in cyberspace by relying on simple dichotomies. The structural complexities open wheel discourse described in Chapter 1 resists by bimodality. Furthermore, the mere suggestion that criminal and competitive skeptics are deliberately or subconsciously trying to undermine genuine, good faith internet freedom makes as much sense as the suggestion that anti-fraud or anti-discrimination public accommodation statutes are an attempt to undermine the marketplace. In the absence of affective and continuous self-policing by market stakeholders, which really happens, genuine, good faith participation in the economic market place is simply not possible without an anti-discrimination statute. In order to protect free markets, Internet access, open borders, educational opportunities and the global commons, the United States must 'command the middle,' or the gray, contested, skeptical but optimistic areas between open world zealotry and open world opportunism.

[45] Diamond (2022, 191).

[46] Flynn (2018, 108).

[47] Id. at 114.

2.3 *Capabilities and Strategic Discourse*

The power and influence of nation states is frequently considered a function of their 'capability,[48] particularly with regard to their kinetic military and offensive/defensive of cyber weapons.[49] We consider whether Americans military and cyber capabilities are strong enough to defend the homeland and our allies while deterring our enemies,[50] whether allies' capabilities compliment our own, and whether our military and cyber capabilities have outpaced the evolution of international law. But very few people have analyzed state power using capabilities theory. Capability theories have certainly examined the extent to which states adopt policies that make it difficult for people to take advantage of available opportunities, reward or punishment capability—deficient people who try to act responsibly in spite of the difficult situations they must confront every day, or incentivizes people to pursue life plans and entitlements oriented almost exclusively around the pursuit of wealth, primary goods and formal rights. But few capabilities theorists actually argue that nation-states might be capability-deficient, particularly a nation-state as powerful as the United States. There are two notable exceptions to the trend: just war theory and humanitarian intervention.

Like Sen, Nussbaum examined the gap between functionings and actual capabilities, particularly with regard to development, gender disparities and poverty, but Nussbaum's formulation of capabilities has played a role in larger debates about just war theory, national security and collective self-defense. Most people believe that the prima facie right to national self-defense flows from two sources: (1) state sovereignty, and (2) in the same way that individuals have a fundamental right to defend themselves, states have a similar obligation to protect citizens from hostile external threats. The latter is both paternalistic, comparable to the way that parents protect their children,[51] and delegative because citizens in most countries delegate authority to use wide-scale force to the state. Even if we temporarily bracket the fact that democratic and illiberal authoritarian societies may define sovereignty and political consent differently, sovereignty-based and delegative justifications for national self-defense

[48] Mazarr (2022, 52).

[49] Gordon and Rosenbach (2022, 12).

[50] Id. at 18.

[51] Baker and Roberts (2007, 25).

are still popular and ground many formulations of just war theory. The central challenge for international relations and international law experts, however, is that it is difficult to reconcile using military force in national self-defense based on state sovereignty with humanitarian intervention, another national justification for the use of military force that appears to undermine state sovereignty.

Theorists have tried to resolve the apparent conflict between sovereignty as the basis of using military force in self-defense and the critique of sovereignty as the basis for using military force in humanitarian interventions by pointing to a key difference between Sen and Nussbaum, which is that Nussbaum "[articulated] capability as a basis for constitutional principles that citizens have a right to demand from their government."[52] Among those demands are what Nussbaum described as 'central capabilities,' including life, bodily integrity and integrity, senses, emotions, practical reason, affiliation and control their environment.[53] Ensuring these capabilities is a responsibility of modern governance, regardless of its ideological orientation or administrative architecture. From these basic premises, we can imagine three understandings of the relationship between national self-defense, capabilities and humanitarian intervention.

Under the strong view, state A can use force to alleviate a humanitarian disaster in state B if proof of that disaster is accompanied by an actual strike on state A by state B. This strategy may, or may not, trigger a formal declaration of war, violate the rules of armed conflict, or operate alongside a 'belligerent occupation' prior to intervention.[54] Under an intermediate view, state A can use force (in self-defense) against state B if state B's negligence is causing a humanitarian disaster in state A and state B. Iraq's decision to intervene in Syria's civil war after ISIL eroded the border between Iraq and Syria, leading to a widespread humanitarian disaster in both countries, follows the logic of the intermediate view. Under the weak view, state A loses the right to use self-defense against other states when it violates the integrity and dignity of its own people, leading to a

[52] Id. at 28.

[53] Id. at 29.

[54] Rotunda (2009, 459) (discusses early distinction between enemy belligerents and enemy combatants).

widespread humanitarian disaster within its own borders.[55] Three themes unite the strong, intermediate and weak views: (1) a government's failure to ensure basic capabilities leads to gross atrocities, widespread poverty or rampant violence, (2) there's a pressing need for some kind of intervention but the offending state resists and (3) "the offending state cannot claim that a humanitarian intervention ... is a violation of its sovereignty, and it cannot claim the right to national defense."[56]

2.4 *State Power and VRN*

The idea that a state should be entitled, in the name of collective self-defense, to violate the sovereignty of another state that cannot secure the basic capabilities of its' own people, and that such a violation of sovereignty should be considered a humanitarian intervention rather than an invasion, is far from controversial. Most people would characterize a military operation conducted by state A into a neighboring country, state B, that cannot enforce its borders, resulting in a significant threat to the security and stability of state A, as a legitimate just war predicated on national self-defense rather than military or strategic ambition. It is also entirely possible to characterize state A as 'capability deficient' and a 'retaliatory actor,' the former because state A might have faced situational constraints that forced it to enter into the territory of its neighbor as opposed to pursuing other diplomatic or policy solutions based on its wealth, endowment of primary goods or access to formal rights under international law, and the latter because state A's violation of state B's sovereignty was done in retaliation for state B's inability to prevent its internal difficulties from spreading over into neighboring states. Capability analysis informs both sides of a humanitarian intervention: states that can't satisfy the basic capability needs of their citizens may lose control of their borders and become unstable, and states may launch military operations into neighboring states that cannot control their own borders or prevent internal difficulties from affecting other states because diplomatic opportunities typically available to them due to their wealth, primary goods endowment or access to formal rights simply don't exist. The latter scenario, most importantly, does not depend on the size or

[55] See eg Deeks (2012).

[56] Baker and Roberts (2007, 35).

power of the state that "intervenes" for humanitarian reasons. All states, regardless of their size, geo-strategic status or location, may find it difficult, at least temporarily, to take advantage of diplomatic opportunities to resolve territorial disputes.

If we can characterize the geo-strategic constraints that shape state action as a 'capability deficiency,' then we can certainly apply VRN to America's response to cyber attacks by other nation states or non-state proxies. Conflicts between states are roughly comparable to disputes between persons or organization, only in this case, the conflict must be resolved by international law and transnational moral norms. The permissibility or impermissibility of international cyber attacks and hack-backs done by state actors would be determined by geostrategic capability deficits that force nations to either hack or hack back against other state in the name of collective self-defense. Those hack backs would be subsequently assessed based on a nations' willingness to observe due diligence in their hackback (i.e. attribution, necessity, proportionality). Both the presumptive and desert rules could also protect territorial or economic entitlements in nation to nation disputes that turn into cyber attacks or hackbacks.

Critics may object to applying VRN to US sponsored cyber attacks and retaliatory hackbacks for two reasons. The first reason is a general reluctance to equate any geostrategic burden caused by America's near total commitment to political accountability, rule of law and human rights to the burden faced by public or private entities with cognitive/disability based capability deficits, existential capability deficits or even situationally constrained capability deficits. Once again, we are confronted with what appears to be a stark contrast between a capability deficit caused by dire necessity and a capability deficit caused by a privileged choice. Any reluctance, however, to compare the plight of a starving person in India or an abused married woman in Oklahoma to the geostrategic burdens placed on the United States is a failure of imagination, not a symptom of the limits of capability theory. Both structurally and substantively, capability theory is bi-directional; it works from the top down and from the bottom up. The ideological or hypothetical advantages of entities entitled to capability based protections are less important than whether an entity suffers from a significant capability constraint in its' actual, daily life that prevents it from taking advantage of opportunities based on wealth, endowment of primary goods or rights. Even powerful nation states experience this difficulty periodically.

Another potential objection speaks to the implied limits of VRN's desert rule. Under zone two, defendants must absorb the total cost of any damages caused by the plaintiffs retaliatory action. Under zone four, plaintiffs and defendants must split the cost of any damages caused by the plaintiffs retaliation. Even if two nations agree to have their entitlement disputes evaluated using the presumptive rule by an international body with jurisdiction, it might be very difficult for that international body to enforce judgments mandated by the desert rule. It is much more likely that states with a permissible claim to either cyber attack or hackback under zone two or zone four will pursue sanctions against the other party. This is exactly what happened after the US sanctioned North Korea in 2014 for the Sony picture attack, and the US sanctioned Russia, removed Russian diplomats, and closed Russian facilities in the United States after Russia interfered in the 2016 US presidential election.[57] But United States must also engage in what one commentator has described as "hard-nosed diplomacy" in cyber attacks and retaliatory hackback cases,[58] which is a gentle way of saying that it must force nations that launch cyber attacks against United States to 'make a deal,' agreeing to pay all of the cost of damages if the United States chooses to retaliate after a particularly dangerous, arguably permissible cyberattack (zone 2), or share the cost of damages caused by US retaliation if the initial attack is clearly impermissible, but the US' retaliatory attack was not done with 'due diligence' (zone 4).

The difficulty of securing these kinds of deals will undoubtedly depend on the nature of the initial cyber attack, our relations with the attacking nation, the manner in which negotiation takes place, and our bargaining position. In some cases, a quiet, quid pro quo agreement leading to an arrangement that splits the cost of retaliation might be effective. In other case, the United States might have to use a more public, cyber shaming approach, and in extreme cases, the United States might be forced to choose between shaming an attacking nation that refuses to share the cost of retaliation when appropriate and a full scale proportional, kinetic response. Using Obama's quiet deal with Xi Jiping in 2015 after China was caught stealing trade secrets as a baseline,[59] perhaps China might

[57] Gordon and Rosenbach (2022, 4).

[58] Alperovitch (2022, 46).

[59] Alperovitch (2022, 44).

agree to a cyber attack/retaliatory deal consistent with VRN if the United States calls off the trade war and temporally suspends sanctions; North Korea might agree if the United States pledges development assistance; Iran might agree if the United States fully resuscitates the JCPOA. The United States may even nudge parties into a cyber-attack/retaliatory deal if they publicly and loudly link the pursuit of cyber attack/retaliatory arrangements, the attacks of specific, named states and the protection of democracy.[60] But regardless of the context, the United States must defend its claim to engage in permissible retaliatory hackbacks given the increasing frequency of attacks, not because of its technological sophistication or desire to compete with other great nations, but because it operates under specific geostrategic constraints/capability deficits, and those constraints demand the use of retaliatory hackbacks as an 'adaptive preference' in an increasingly complex global environment.

References

Affairs. 1776. U.S. Constitution Article II, Section 2, cl 2.

Alperovitch, D. 2022. The Case for Cyber-Realism: Geopolitical Problems Don't Have Technical Solutions. *Foreign Policy* 101 (1, January/February): 44–51.

Ashworth, L. 2006. Where are the Idealists in Interwar International Relations. *Review of International Studies* 32: 291–308.

Baker, D., and D. Roberts. 2007. Extending Just War Theory: The Jus Ad Bellum and the Capabilities Approach to Armed Conflict. *Scientia Militaria, South African Journal of Military Studies* 35 (2): 21–38.

Cordesman, A. 2016. Anthony Cordesman. *Is There An Obama Doctrine?* March 10.

Curtiss. 1936. U.S. v Curtiss Wright Export Corp., 299 US 304, 319.

Deeks, A. 2012. 'Unwilling or Unable': Toward a Normative Framework for Extraterritorial Self-Defense. *Virginia Journal of International Law* 52: 483–531.

Diamond, L. 2022. All Democracy Is Global: Why America Can't Shrink from the Fight for Freedom. *Foreign Affairs* 101 (5): 182–198.

Egan. 1988. Dep't of the Navy v. Egan, 484 U.S. 518, 530.

Federalist74. 1961. Federalist 74 (Hamilton). In *The Federalist*, ed. Jacob E. Cooke. Wesleyan.

Flynn, M. 2018. Cyber Rebellions. *Journal of International Affairs* 71 (1.5): 107–114.

[60] Diamond (2022, 185).

Gholz, E., D. Press, and H. Sapolsky. 1997. Come Home America: The Strategy of Restraint in the Face of Temptation. *International Security* 21 (4): 5–48.

Gordon, S., and E. Rosenbach. 2022. America's Cyber-Reckoning: How to Fix a Failing Strategy. *Foreign Policy* 101 (1): 10–21.

Govella, K. 2019. Technology and Tension in the Global Commons. *Fletcher Security Review*: 38–44.

Greenwald, G. 2013. NSA Collects Phone Records of Millions of Verizon Customers Daily. *Guardian* (London), June 5.

Hamdi. 2004. Hamdi v. Rumsfeld, 542 U.S. 507, 579–585.

Haas, M. 2001. Prospect Theory and the Cuban Missile Crisis. *International Studies Quarterly* 45: 241–254.

Heinze, E. 2008. The New Utopianism: Liberalism, American Foreign Policy, and the War in Iraq. Journal of International Political Theory 4 (1): 105–125.

Hoffman. F. 2013. *Forward Partnership: A Sustainable American Strategy*. FPRI, Winter, 20–40 (reprint from *Foreign Policy*, January 2013).

Hoffman. F. G. 2016. Retreats Ashore: The Flaws of Offshore Balancing. Foreign Policy Research Institute (FPRI), July 5.

Huntington, S. 1993. Why International Primacy Matters. *International Security* 17 (4): 68–83.

Jeffrey, J. 2016. US Military Engagement in the Broader Middle East. *Washington Institute for Near East Policy, Policy Focus*, April.

Jervis, R. 2009. Unipolarity: A Structural Perspective. *World Politics* 61 (1): 197–198.

Kirchick, J. 2017. America in Retreat: Why Neo-Isolationism Exploded Under Obama and What We Can Do About it. *DailyBeast.com*, July 12.

Maerz, S., Anna Luhrmann, Jean Lachapelle, and Amanda Edgell. 2020. Worth the Sacrifice? Illiberal and Authoritarian Practices During Covid-19. Working Paper (Series 2020: 110). The Varieties of Democracy Institute.

Manjikan, M. 2021. The United States: A Declining Hegemon in Cyberspace? In *Routledge Companion to Global Cyber-Security Strategy*, ed. Scott Romaniuk and Mary Manjikan. London: Routledge.

Manning, J.F. 2011. Separation of Powers as Ordinary Interpretation. *Harvard Law Review* 124: 1939–2041.

Maravall, J. 2008. The Myth of Authoritarian Advantage. *Journal of Democracy* 5 (4): 17–31.

Mazarr, M. 2022. What Makes a Power Great: The Real Drivers of Rise and Fall. *Foreign Affairs* 110 (4): 52–63.

Mearsheimer, J. 2011. Imperial by Design. *The National Interest* 111: 16–34.

Mearscheimer, J., and S. Walt. 2016. The Case for Offshore Balancing. *Foreign Affairs* 95 (4): 60–83.

Posen, B. 2014. *Restraint: A New Foundation for US Grand Strategy*. Ithaca, NY: Cornell University Press.

Posner, E., and A. Vermeule. 2007. *Terror in the Balance: Security, Liberty and the Courts*. New York: Oxford University Press.
Potter, R. 2016. Cyber Warfare as a Double-Edged Sword for Authoritarian States. *RealClear Defense.com*, August 2.
Rao, N. 2011. Public Choice and International Law Compliance: The Executive Branch Is a 'They Not an It.' *Minnesota Law Review* 96: 194–280.
Rotunda, K. 2009. A Compartive Historical Analysis of War Time Procedural Protections and Presidential Powers: From the Civil War to the War on Terror. *Chapman Law Review* 12: 449–487.
Schlesinger, A. 1973. *The Imperial Presidency*. Boston: Houghton Mifflin.
Strauss, P.L. 2011. A Softer Formalism. *Harvard Law Review* 124: 55–61.
Toft, M. 2017. Why Is America Addicted to Foreign Interventions? *The National Interest*, December 20.
Walt, S. 2018. *The Hell of Good Intentions: America's Foreign Policy Elite and the Decline of US Primacy*. New York: Farrar, Strauss and Giroux.
Wendt, A. 1999. *Social Theory of International Politics*. Cambridge: Cambridge University Press.
Wohlforth, W. 2008. *World Out of Balance: International Relations and the Challenge of American Primacy*. Princeton: Princeton University Press.
Wright, T. 2016. Realism's Utopia: What Academic Realism Almost Never Works in Washington. *International Affairs*, January 13.
Wucker, M. 2016. Move over Black Swan, Here Comes Gray Rhino. *Carnegie Council*, June 14.

CHAPTER 9

Capabilities, Populism and Presidential Leadership

1 Morality and The Critique of Neoliberalism

In the last chapter, I considered two possible objections to incorporating VRN into American foreign policy generally, and America's retaliatory hackback policy more specifically: (1) the difficulty of conceptualizing the United States, one of the worlds' remaining superpowers, as capability deficient, and (2) the difficulty of enforcing the imposition of costs for damages caused by retaliatory action against either the attacker or the retaliator. For a wide variety of reasons, I rejected both objections, but there is a third possible objection that deserves careful consideration. One could argue, quite convincingly, that it will be impossible to incorporate VRN into America's active defense calculus, particularly in regard to the moral and legal permissibility/impermissibility of specific cyber attacks and hackbacks, without addressing the role of morality in American foreign policy. Standing alone, VRN is already deeply moral because it posits a theory of human flourishing and well-being that is, notwithstanding the overlap problem described in Chapter 4, significantly different from the theory of human flourishing embraced by Rawlsians, utilitarians and proponents of wealth maximization. VRN's presumptive rule extends special protection to individuals who commit cyber attacks or retaliatory hackbacks because their capability deficiencies deprive them of the freedom to flourish and lead a good life, not to people who launch cyber attacks and hackbacks because they could not pursue their

A. J. Thomas, *The Open World, Hackbacks and Global Justice*,
https://doi.org/10.1007/978-981-19-8132-6_9

own version of the good life without primary goods, rights or wealth. Furthermore the legal and moral protection provided by VRN extends to capability-deficient individuals, entities or nation-states.

Characterizing the United States as a capability-deficient nation is also deeply moral because it suggests that America's collective well-being can't be reduced, in the current geostrategic climate, to America's primary goods, rights or national wealth. In many ways, incorporating VRN into American foreign policy requires the moral certainty to not only level with Americans about what the United States can and cannot do in the world beyond its shores, but also convince the American people that America's foreign policy decisions must be based on something other than the need to maintain its competitive edge or America's military, economic or technological prowess. The same logic applies, with particular urgency, to cyber attacks and hackbacks performed by the American government. It should be possible for the United States to justify active measures programs on the basis of more than great power competition, technological capacity and simple revenge.

In order to incorporate the full substantive, moral message of VRN into American cyberspace policy, policy-makers must prioritize two cybersecurity goals that reflect the capability approach to human flourishing: (1) promote liberal democracy as a response to the authoritarian advantage in cyberspace and (2) reject neoliberal defenses of zealous open world views that place primary blame for cyber security breaches on victims rather than perpetrators, and implicitly normalize hacking. If a capability-deficient American government pursues retaliatory hackbacks for explicitly moral reasons, however, then it must do so without recreating the historical excesses of American exceptionalism. We also need to develop a common consensus about what kind of American president is more likely to sympathize with, and successfully execute, a capability-based retaliatory agenda in cyberspace. The last two sections of this chapter argue that presidents with a particular understanding of neo-populism (or left-variant populism), like Joseph Biden, are more likely to embrace and implement a capability-based retaliatory hackback agenda then "right populist" presidents like Donald Trump.

In the last chapter, I described the advantages that authoritarian regimes enjoy in cyberspace, as well as a comparable set of disadvantages that liberal democratic societies generally, and the United States more specifically, must endure in the current environment. These two trends have slowly transformed the United States into a 'capability-deficient' nation-state. Capability theory, in this sense, is part of a broader narrative of national vulnerability that persists in spite of America's wealth, endowment of primary goods and formal rights as a sovereign power. But capability theory also provides a partial moral justification from linking American foreign policy goals to ongoing critiques of economic neoliberalism.[1] Neoliberalism[2] is the intellectual descendent of a cluster of economic ideas generally associated with the Chicago school of economics and Friedrich Hayek in the 1930s.[3] It is now generally considered to be the cause of a "wave of market de-regulation, privatization, and welfare state withdrawal that swept the first, second and third world."[4] It was also the dominant economic orientation of the Clinton, Bush and Obama administrations. In addition to encouraging privatization and the de-regulation of markets, proponents of economic neoliberalism also tend to exhibit two tendencies: (1) a preference for expertise and technocratic rule over democratic governance, and (2) support for ethics of 'responsibilization' and 'self-care.'[5] The former stems from "a widespread collapse of faith in government's ability to deliver on democratic demands,"[6] whereas the latter combines the self-sufficiency ethos so central to 'homo-economicus' with a desire to blame the troubled beneficiaries of social welfare programs for causing their own problems.[7]

Sen is both a critic of, and partially indebted to, neoliberalism. On the one hand, capability theory is a clear counter narrative to wealth maximization, market fetishism, welfarism and utility, four ideas that most neoliberals endorse.[8] Sen also prefers democratic participation over expert-driven, technocratic rule because the former encourages collective decision-making and human development over the naked pursuit of

[1] Economic neoliberalism is distinguishable from, but not entirely inconsistent with, foreign policy neoliberalism, which encourages alliance building, soft power and multilateral cooperation. De Castro (2011, 2).

[2] Brand and Sekler (2009, 6) (distinguishes between neoliberalism as an intellectual movement, an elitist anti-Fordist critique and a political phenomenon).

[3] Venugopal (2015, 168).

[4] Id.

[5] Pyysiainen et al. (2017, 217–19).

[6] Singh and Purdy (2014, 7).

[7] Pyysiainen et al. (2017, 218).

[8] Laruffa (2018, 175).

economic growth.[9] Critics contend, however, that in spite of Sen's aversion to neoliberalism principles, the liberty he seeks for capability-deficient people is limited to capitalism. As one commentator observed,

> Sen, as a supporter of the free market, attributes cases of injustice and poverty to societies not having completed development into capitalism. This is as far as Sen goes with regard to his critique of capitalism: attributing flaws to pre-capitalist … arrangements, while maintaining the importance of individual freedoms.[10]

A fully postcolonial version of capability theory may provide a more compelling critique of neoliberalism,[11] but capability theory as it currently exists can still link American foreign policy to the promotion of democracy policies that limit the power of technocratic elites.

Neoliberals also tend to support responsibilization regimes that blame victims for their suffering, dependency and weak moral character. By contrast, Sen believes that capability-deficient people lack meaningful freedom, and that both the state and the individual share responsibility for human flourishing.[12] Any responsibilization program that assigns blame for the failure to take advantage of opportunities solely at the feet of victims is morally toxic, disempowering and ultimately doomed to fail because it under-theorizes the complexities of capability deficiency. The same logic applies to responsibilization regimes in cyberspace. In spite of their constant critiques of the hegemonic impact of 'Big Tech' firms like Apple, Microsoft, Facebook and Google, some open world zealots and cyber libertarians embrace the central memes of neoliberalism: technocratic expertise, the 'mantra of self-care' and a completely unregulated Internet.[13]

Given the hyper-competitive character of cyberspace, societies should also, according to the same cyber libertarians, accept the near inevitability of cyber attacks, and assign primary responsibility for cyber security to victims, regardless of who they are, or the intentions/expertise of the hackers. A capability-based defense of hackbacks would reject neoliberal

[9] Id.

[10] Comiling and Sanchez (2014, 11).

[11] Id. at 15.

[12] Sen (1999a, 288).

[13] Jacob (2017, 13).

critiques of democratic participation, and neo-liberal defenses of responsibilization regimes that needlessly blame the victims of cyber attacks for their own suffering.

Full incorporating the capability critique of neoliberalism into active defense policy will require two things: (1) acceptance of a moral justification for American foreign policy generally and American cyberspace more specifically, and (2) suitable presidential leadership. The former requires American decision-makers to revisit past, and at times painful, instances of exceptionalism, when Americans were convinced that they could do no wrong in the global world. The historical record is filled with regrettable examples of exceptionalism when the United States planned and/or executed coups in other countries to protect its own interests, the most familiar examples being Iraq, Chile, Iran, Nicaragua, Panama, the Congo and almost Cuba. There are also well intentioned but poorly executed examples of American exceptionalism, such as Somalia, Rwanda and Libya. Ultimately, decision-makers will have to justify American foreign policy programs on something other than the naked self-interest calculus favored by realists, including programs that purport to promote democracy and protect people in cyberspace.[14] Convincing the American people that the United States should shoulder another burden, even though it is laboring under the dual burden of not only being the only military hegemon but also, as a member of the democracy club, self-limiting its options due to its professed if imperfect commitment to political accountability, human rights and the rule of law, will be challenging.[15]

This leads to the second requirement: presidential leadership. In order to implement the capability-based active defense regime, the American president must link vulnerabilities that ordinary people experience in cyberspace to the vulnerabilities that America faces in cyberspace. The kind of president most likely to accomplish this objective in the current political climate must have populist or quasi-populist, rather than neoliberal, sensibilities, but there are malevolent and benevolent forms of populism. America has seen its share of populous presidents, including James Polk and Andrew Jackson in the nineteenth century[16]

[14] May (2021).

[15] Id.

[16] Andrew Jackson, an unapologetic supporter of states rights and slavery who was extremely critical of concentrated federal power in all forms, billed himself as a populist leader so beloved by the majority that he was one of only three Democrats to win

and Teddy Roosevelt in the twentieth century.[17] The twenty-first century has witnessed two presidents with populist sensibilities, and their presidencies have proceeded and very different directions. Populism, however, is the last step of the journey to presidential support for capability-based active defense strategy. The journey starts with an examination of institutional limits on presidential power in cyberspace.

2 Cyberspace Challenges to Presidential Leadership

Cyber security, international relations and international law played an outsized role in the American presidential campaign of 2016. To an unprecedented degree, a foreign power—Russia—sought to tip the scales of an American presidential election against former Secretary of State Hillary Clinton, and in favor of Donald Trump. Putin and his operatives manipulated electoral officials, obscured legal responsibility for political disinformation campaigns and facilitated the financing of hackers. Intelligence and law-enforcement agencies discovered that the Russian government had employed the expertise of individuals with murky professional affiliations—both public and private—to hack into the mail systems of the two major American political parties, plant fake news on public

a popular vote in three consecutive elections. Jackson's popularity and populist sensibility shaped his overarching critique of government-created monopolies and special state-created charters that appeared to operate for the benefit of elites. Hartz (1948, 289–320). The result was Jackson's decision to defund the Second Bank of the United States, arguably the first major clash between the President and Congress in the early republic. Jackson's decision to defund the Second Bank of the United States increased his popularity with the voting public, further stoked popular distrust for pro-corporate interests in Congress and eventually lead to the election of President Van Buren, another anti-monopolist proponent of Jacksonian democracy.

[17] Theodore Roosevelt translated his personal popularity, populism and an appreciation for the institutional uniqueness of the Presidency into an aggressive understanding of executive unilateralism. Roosevelt (1920, 464). Roosevelt's perception of executive power was fueled by Progressive sympathies for the plight of ordinary individuals, the poor and the most vulnerable members of society, but Roosevelt, unlike Jackson, came from a privileged background. He also came of age politically during a time when the power of the federal government was well established because of the Civil War and the gradual growth of the administrative state. As a result, Roosevelt's ability to translate his popularity into unfettered executive authority was much more limited than that of Jackson. Roosevelt's background and extensive experience in politics before assuming the Presidency made him more of an institutionalist than Jackson.

websites and facilitate the transfer of stolen documents to Wikileaks. Individuals with strong ties to the Trump campaign and the Russian government communicated frequently before and after the election, both online and in person.

Throughout the campaign, intelligence experts and political decision-makers worried about the possibility that Russia might launch a major cyber-attack on election software systems, and use that software to compromise the operation of actual voting machines. The careful use of anonymizers and fake ISPs made it difficult for the US government to attribute full responsibility for any attacks to rogue blackhat actors, the Russian government itself or some combination of the two. For many national security and foreign policy experts, Russia's systematic hacking attempts during an election almost constituted an act of war, pushing the boundaries of armed conflict into new, and potentially dangerous, territory. Putin's 'weaponization' of cyberspace, and Donald Trump's suspiciously timid response to hacking allegations, lead to the biggest American political scandal since the Iran-Contra Affair, intensifying ongoing tensions between two powerful nations.

Like many national security challenges in the post-9/11 era, Russia's cyber-based efforts to subvert the American electoral process forced policy-makers to confront new challenges on many fronts: counter-terrorist targeting, persistent turmoil and instability across the Middle East and Northern Africa, long-standing sectarian and religious conflicts that seem to resist democratic or economic resolution, heightened fear-mongering by media-savvy political elites and transnational crime. It also forced decision-makers to re-examine the limits of American power in the global community. After the collapse of the Soviet Union, the United States became the world's sole superpower, blessed with geographic advantages, superior military capacity, unmatched economic power and unquestioned control over a liberal international order that protected global trade, encouraged multilateral relations, discouraged nuclear proliferation to unstable or unaccountable regime regimes, and preserved American hegemonic control in Europe, Asia, the Middle East, Africa and Latin America.

Liberal internationalism, interventionism and neoconservatism were the dominant credos in two successive administrations—Bush and Clinton—resulting in a series of military and political operations that aggressively projected American values and ideals into virtually every regional conflict across the globe. Critics during this period began to characterize America as an 'empire' with boundless self-confidence, unchallenged transformative aspirations on the world stage and an extremely elastic understanding of its own 'national interest.' American policies toward Bosnia, Kosovo and Rwanda, however, began to undermine the appeal of liberal interventionism and internationalism during the Clinton Administration, resulting in a second Bush Administration that called, prior to 9/11, for a more 'modest,' less interventionist, American foreign policy, but the attacks of 9/11 brought an end to early Bush-era critiques of interventionism. Neoconservatives simultaneously championed a policy of aggressive military interventionism to counter-terrorist threats overseas, and an equally aggressive critique of liberal internationalist norms that placed American military and diplomatic policy at the mercy of customary international law, United Nations directives and multilateral organizations like the International Criminal Court. Realist critics of interventionism generally, and the disastrous invasion and subsequent occupation of Iraq more specifically, also proposed numerous justifications for American 'retrenchment' and disengagement from various regional conflict zones during the Bush and Obama administrations.

The chain of events leading up to the Iraq War in 2003, and the presidential election of 2016 exposed fundamental fissures in the American presidential decision-making process, fissures that America's enemies in cyberspace eagerly exploited. Presidents exercise considerable rule-making authority over strategic, military, diplomatic and cyber affairs, typically through executive orders, National Security Decision Directives (NSSD), or Presidential Policy Directives (PPD). Presidents also work with Congress, the diplomatic and military establishment, partners in the private sector and allies to craft America's cyberspace policy, but the biggest shifts in cyberspace policy tend to coincide with changes in Presidential administrations. Each president has a different understanding of America's power in the global world, and the legitimizing effects of international law.

3 Economic Espionage vs. Armed Conflict

During the Obama Administration, two pieces of legislation were instrumental in ramping up America's response to theft of intellectual property or trade secrets. The first was the Economic Espionage Act (EEA), which was passed in 1996.[18] The EEA authorized prosecutions against any individual who "[intended or knew] "that the offense [would] benefit any foreign government, foreign instrumentality, or foreign agent ....".[19] Attributing responsibility for international cyber-espionage and trade secret theft was always a challenge, but prosecutors in the Obama Justice department managed to overcome that hurdle, only to discover that the real challenge was service of process and jurisdiction, not attribution. Even if they could identify the overseas cyber hacker responsible for a particular instance of IP theft, courts wouldn't permit service of process, even for hackers linked to a company operating in the United States.[20] Chinese hackers, for example, indicted under the EEA during the Obama administration, but they weren't prosecuted.[21] Section 1637 of the National Defense Authorization Act of 2015 also provided President Obama with new authority, under the International Emergency Economic Powers Act,[22] to prohibit suspected international IP thieves from transferring information through the US financial system.[23]

As the Obama administration developed new tools to counter international IP theft and cyber-espionage, another massive hack-related crisis loomed on the horizon: cyber-attacks on critical infrastructure. The first major, international cyber attack on critical infrastructure took place in 2008. Iran announced[24] in 2010 that someone had infiltrated their nuclear software systems and successfully compromised the operation of their fuel subterfuges, significantly slowing down their nuclear enrichment program. After extensive analysis, computer forensics experts and systems programmers discovered that Iran had been infected by the 'Stuxnet' virus, a clever piece of deliberately destructive code engineered by private hackers in both the United States and Europe (among others), working in both the public and private sectors.[25] The primary beneficiary of the hack was Israel, which had been repeatedly threatened with nuclear/military

[18] EEA (1996).

[19] Id.

[20] Pangang (2012).

[21] Ellings (2014).

[22] IEEP (2014).

[23] NDAA (2017).

[24] Stuxnet (2010).

[25] Neuman (2012) and Nakashima and Warrick (2012).

destruction by pro-Palestinian activists as well as the Iranian government itself.[26]

International condemnation of the Stuxnet attack, subsequent major cyber attacks engineered by Iran against critical infrastructures sites in Saudi Arabia and Qatari,[27] and earlier DDoS infrastructure attacks on Georgia and Estonia engineered by Russia, was swift, but the Obama administration's response was slower than expected, for two reasons. The first was that Obama's foreign policy neoliberalism (as opposed to economic neoliberalism) lead him to pursue soft power, diplomatic and retorsive solutions. The United States was also still trying to disentangle itself from two expensive, long-term wars in Afghanistan and Iraq. The second reason was that it wasn't entirely clear if attacks like Stuxnet were acts of war under international law. A few authors of the first Tallinn Manual (Independent Group of Experts, or IGE), for example, did not believe that Stuxnet "was sufficient to reach the level of armed attack."[28] Snowden's hack of the NSA, and Russia's cyber attack on Estonia and Georgia also failed to meet the requirements of an armed attack under the Talinn Manual.[29] Without that determination, it wasn't clear if the United States could look to Article 51 of the UN Charter for authority to respond based on its "inherent right of … self-defense,"[30] or take advantage of unilateral 'retorsion' powers established under international law that permit nations to retaliate against perceived wrongdoers.[31]

In spite of uncertainty about the legality of certain kinds of international cyber attacks, the Obama Administration, led by the National Security Council, authored a report that established firm procedural limits on any retaliatory action by the US government after a cyber-attack.

[26] Gordon and Rosenbach (2022, 12).

[27] Id.

[28] Yoo (2015, 12) Yoo , Christopher S., "Cyber Espionage or Cyber War?: International Law, Domestic Law, and Self-Protective Measures" (2015). Faculty Scholarship Paper 1540, p. 12.

[29] Id.

[30] UNCharter51 (1945).

[31] Gabcikovo (1997) ("… a state victim of a violation of an international obligation by another state may resort to countermeasures that might otherwise be unlawful ….") and Zoller (1984). For a critique of retorsion that is not based on necessity or conducted proportionally, see Setear (1997, 75–76).

Presidential Policy Directive #20 (PPD-20),[32] issued in 2012, required a thorough risk assessment for most active offensive operations,[33] interagency vetting, the approval of both the President and National Security Advisor, and operations had to be consistent with American values.[34] Before receiving approval, operation coordinators had to determine if conventional forces might be more effective, identify potential spillover effects on allies, affected nation-states and private firms,[35] and prioritize non-lethal tactics to avoid cyber escalations.[36] Such caution[37] was not surprising considering Obama's intellectual, political[38] and strategic[39] sensibilities. But that caution sparked a fundamental re-assessment of America's active defense program in the next administration.

4 Trump's 'Right-Populism'

4.1 'America-First' Realism

In 2016, America elected the first arguably populist president in the twenty-first century: Donald Trump. Suddenly, we entered a period in which the President of the United States regularly characterized any piece of information with which he disagreed as 'fake news,' and announced major governmental policy changes in increasingly incoherent pre-dawn tweets.[40] The rhetorical (as opposed to intellectual) centerpiece of the Trump administration was an ongoing, relentless critique of economic and foreign policy neo-liberalism. With regard to the former, Trump professed his desire to put the needs of people, usually working class and middle class people, over the needs of the marketplace. Like all populists,

[32] PPD20 (2012).

[33] Id. at Section VI (risk assessments).

[34] Id. at Section III ("consistent with applicable values, principles and norms of state behavior that the US promotes domestically and internationally…").

[35] Id. at Section III (obtain consent from affected parties).

[36] Id at Section V (must consider other alternatives, preference for non-lethal solutions).

[37] This was a common critique of the Obama Administration. See Youssef and Harris (2016).

[38] Sanger (2012, 265) (White House was "absolutely allergic to talking about our cyber offensive capabilities.").

[39] Murdoch (2013).

[40] Ptweet (2017) and Bromwich (2017).

he also pledged to implement policies that were sensitive to commonsensical perceptions held by ordinary people, as opposed to 'elites.' Foreign policy objectives took a dramatic turn from the Obama administration: he pledged to pull the United States out of bilateral and multilateral agreements, ranging from the Paris Climate Accords to NATO, that appeared to reflect the best interests of our partners and allies, rather than the American people.

Trump's daily pronouncements and bloviations about everything, including cyberspace policy, were an incoherent stew, alternating wildly between reflexive hyper-masculine posturing and obsequious appeasement of Putin and his Russian handlers. The intellectual legacy and strategic objectives of the people working in the Trump administration, however, were more complex and coherent. The Trump administration's approach to foreign policy flowed from three sources. The first was Trump's approach to deal-making. As a private citizen, Trump was famous for striking favorable deals after threatening to withdraw entirely from the negotiation process. Saber-rattling, symbolic threats and brinksmanship were part of his deal-making arsenal.[41] Trump was also notoriously transactional and impulsive, often more concerned about the appearance of winning and saving face than actually mastering the details of any agreement or securing a reasonable, durable accommodation with bargaining partners. A negotiating strategy based on brinksmanship and unpredictability,[42] combined with the belief that Obama 'left many deals on the table,' produced a checkered diplomatic record for Trump[43]: failure to convince Mexico for pay for a new border wall; Iran openly bragging about restarting their nuclear enrichment program after the UN withdrew from the JCPOA; and unremarkable and inconclusive trade agreements with China as trade tariffs remained in effect. To be fair, all states take

[41] Downie (2015) ("[W]hen negotiating a deal in the private sector, Trump's ideas makes some sense: playing chicken with someone can work as a strategy to force them to accept your terms. That's game theory 101. But that brinksmanship can only be pulled off in a private real estate deal; it's impossible as President of the United States... with an economic catastrophe approaching.").

[42] For a limited defense of Trump's unpredictability, see Dueck (2018) ("Being unpredictable at the operational level is a plus. Even peacetime unpredictability designed to throw adversaries off balance can be advantageous. Unpredictability with allies is more of a mixed bag. It has secured certain limited concessions for the US over the past two years, but at the same time carries undeniable risks.").

[43] Warford (2019) (criticizes Trump and Theresa May for their failed brinksmanship).

advantage of opportunities,[44] and many states and political communities have either fomented, or take advantage of, chaos in order to secure their borders.[45] Putin, for example, took advantage of unrest in Ukraine to annex the Crimea, securing a vital seaport for the Kremlin. ISIS took advantage of instability in Iraq and Syria to establish the beginnings of a new Islamic caliphate in the Levant. The United States took advantage of instability in Central America to engineer a coup in Panama, securing control of the Panama Canal construction project. But Trump's opportunism was not limited to traditional statecraft.

Trumpians also took advantage of ongoing critiques of the liberal international order, ranging from America's protector role dating back to World War II, to the inadequacies of international institutions and liberal constitutionalism.[46] Populist critiques of the liberal international order, from the left and the right, systematically undermined the legitimacy and effectiveness of international norms-setting bodies, including cyber-related institutions and consortia.

Opportunists in the Trump administration also took advantage of the growing ubiquity of state-sponsored hackbacks. The open trade war with China made it easier to acknowledge that as China hacked US facilities, the United States aggressively hacked China back. The United States had to acknowledge placing code in Russian facilities after discovering that Russia attempted to infiltrate American plants and election sites. As a result, the 'cold cyber-war' with our geostrategic rivals, already well underway before 2016, was thoroughly normalized by Trump. Many people in democratic societies viewed Trump's opportunistic cyber saber-rattling as just one of many bargaining tactics available to the American government, rather than an incendiary prelude to war.[47]

[44] Owen (2000, 51) ("states respond to opportunities as well as threats."), Zakaria (1999, 21) ("all states are opportunistic, and will expand power if they can do so a relatively little cost.") and Saideman (2001).

[45] Snyder (2013, 4) ("[T]he solution to security problems at the 'turbulent frontier' of the empire is to conquer still more territory in order to punish or prevent harassment by contiguous barbarians.").

[46] Scheppele (2019, 315) ("Populists exploit the debates that liberal constitutionalists have with each other over normative foundations. They mirror and mimic the language of constitutional liberalism in order to undermine it in practice.").

[47] Warford (2019) ("Somehow we have managed to produce a situation where democratically elected politicians are threatening substantial to their own countries as a bargaining tactic.").

The third contributing factor to Trumpianism was the composition of Trump's governing circle. Trump's 'American First' philosophy was a shotgun marriage between neo-isolationism, neoconservatism[48] and conservative nationalism.[49] The neo-isolationist and conservative nationalist wing consisted of an ad hoc, if historically familiar,[50] mixture of partial military retrenchment,[51] the reprioritization of domestic policies, the systematic re-examination of ongoing treaty obligations, frequent critiques of free-riding by our allies, a potential trade-war against perceived economic competitors, nativism, and gratuitous saber-rattling that provokes unpredictable illiberal nations (except Russia, in Trump's case) and irritates allies. The neoconservative wing called for dramatically increased military spending and a renewed emphasis on great power competition as the central foreign policy challenge currently facing the United States. Instead of eliminating, as the neoconservatives urged after 9/11, the 'Axis of Evil' and violent Islamist jihadis in the Middle East, neoconservatives wanted the United States to focus on deterring, pre-empting and out-competing its principal rivals: China, Iran and Russia.

4.2 *Cyber-Brinkmanship*

In spite of their quasi-nihilistic sensibilities, Trumpian cyber-opportunists lived in a world that rewarded people who supported the open world ethos, and punished people who opposed the open world ethos. Investments, academic talent, trade opportunities, favorable media coverage and social capital tend to flow toward states that espouse support for open world values. Opportunists in the Trump administration, some motivated by fears of unchecked neoliberalism, navigated this unstable terrain through mixed messaging, unapologetically adopting anti-open world policies at in one forum, while voicing support for open world goals in another forum. Trump, for example, repeatedly justified withdrawing

[48] Larson (2018) (Trump's reflexive hostility toward Iran, and willingness to take risks, reflects the influence of hardliners in his administration).

[49] Dueck (2019).

[50] Kitfield (2016).

[51] Finer and Malley (2018).

from international agreements that protected open world expectations[52] because he found them 'unfavorable' to American interests. In response to the inevitable criticism, he then claimed that he was merely trying to 'fix' a flawed document.

Trump's opportunistic *penchant* for unpredictability, or recklessly 'blowing things up,' directly affected open world governance questions. Trump supporters often proposed solutions to difficult political, economic, social or strategic problems that harmed open world expectations more than necessary. They signaled their opportunism through an extreme tolerance for escalation risks in cyberspace. Brandon and Valeriano uncovered this tendency in their study of Trump's active defense policy. In a 2018 report issued by the US Cyber Command, the government apparently endorsed open world values when it criticized Russia and China for threatening the US military and the open world. The United States, unlike its Russian rival, supported the entire global infrastructure and (the) open exchange of information.[53] America's response to intense great power competition in the cyber realm, as announced in the report and elsewhere, was a new policy called 'persistent action,' defined as "the continuing ability to anticipate the adversary's vulnerabilities, and formulate and execute cyberspace operations or.. adversary course of action under determined conditions."[54] Persistent action would "delay, degrade or deny adversary aggression,"[55] ensuring that the United States can command control of commons areas[56] through deterrence, and if necessary, "and acceleration of cyber offense."[57]

But Brandon and Valeriano took issue with Trump's ideological and administrative approach to cyber security and active defense. Instead of the careful risk assessment, interagency cooperation and covert signaling that Obama adopted in response to potential escalations in cyberspace,[58] Trump pushed for pre-emption and the policy of giving agencies broad

[52] Wu and Yoo (2007) (internet does not favor one application over another).

[53] Valeriano and Jensen (2019, 5).

[54] Id. at 6.

[55] Id. at 6.

[56] Id. at 6.

[57] Id. at 6.

[58] Id. at 9.

discretion to engage in aggressive active defense measures with little interagency coordination.[59] According to Brandon and Valeriano, Trump's policies flew in the face of compelling evidence that retaliations to cyber attacks tend to be proportional, that states employ cyber attacks precisely because they want to avoid kinetic conflicts and that a permanently pre-emptive active defense posture is likely to convince enemies that any cyber attack is merely designed to "prepare the environment ... for an offensive strike."[60] As a result, the environment created by persistent action was likely to be "misinterpreted and lead to inadvertent escalation."[61]

While it is entirely possible that the Trump administration was either unaware of evidence suggesting that retaliatory cyber attacks tend to be proportional and lean toward less invasive tactics, or was aware of the research but chose to adopt an aggressive, pre-emptive active defense posture anyway because they don't believe that it will necessarily lead to escalation. But it is also entirely possible that there were people in Trump's national security circle who were aware of the increased risk that comes with a pre-emptive active defense posture, and chose to adopt it anyway. During the Cold War, American planners embraced the doctrine of 'mutually assured destruction' to deter the Soviet Union from launching nuclear weapons[62] even though the strategy made the United States look impulsive and irrational. Opportunists in the Trump administration might have been prepared to accept persistent actions' heightened escalation risk, along with the paralyzing effect an actual escalation might have on open world expectations, in order to prevent rivals like China or Russia from gaining an insurmountable strategic or military advantage in cyberspace.

[59] Id. at 6–7.

[60] Id. at 12.

[61] Id. at 12.

[62] Warford (2019) ("There is a third way to make threats credible: create the risk of an accident. Thomas Schelling ... described handcuffng yourself to your opponent the cavorting to the edge of a cliff. You aren't suicidal, but you are willing to create the risk of an accident. If your counterpart fears that risk more than you, you may extract concessions.").

4.3 *Overmatch and Under-Target*

Historically, populists tend to be isolationist, and at least rhetorically, Trumps' right-variant populism seemed to demand American withdrawal from a wide range of bilateral and multilateral agreements. But the brinksmanship in Trumps' pronouncements and, to a lesser extent, his policies advanced a goal shared by right-populists and neoconservatives: dramatically increasing America's military and cyber capacity in an increasingly competitive global environment. The intellectual inspiration for this shared goal is the concept of 'overmatch.' Overmatch ensures that America "retains military and technological superiority over Russia and China (and all other potential challenges) for as far into the future as we can see."[63] It was a response to the competitive gains of our allies and rivals,[64] as well as a response to a military depleted by two wars in Iraq and Afghanistan.[65] It also seemed to serve other purposes, such as reinforcing the insatiable budgetary demands of neoconservatives in the military establishment (quite possibly ensuring their loyalty in a relatively isolationist administration) and calming the fears of right-populists who were extremely paranoid about the apparently porous condition of America's borders.

In retrospect, overmatch was a perfect illustration of the deep hypocrisy in Trumps' right-variant populism: in the same way that Trump talked like a populist but governed like a plutocrat, overmatch was pushed primarily by elites in the defense establishment, even though it virtually guaranteed that "poor and working people... [would] be denied essential services to pay for a new military buildup."[66] The hyper-competitive global environment that neoliberals, isolationists and neoconservatives considered inevitable spawned a new arms race during the Trump administration, an arms race conducted both on the ground and in cyberspace.

As neoconservatives in the Department of Defense did everything they could to ensure that America continued to 'overmatch' its principal rivals, Trump signed a flurry of cyberspace directives that betrayed

[63] Klare (2018).

[64] Id.

[65] Id.

[66] Id. at 10.

a lingering fascination with neoliberal obsessions like 'voluntary' agreements. Executive Order 13800, for example, created a new set of procedures that protected critical infrastructure,[67] public/private partnerships and information sharing. The same reforms also bolstered supply chain management, facilitated dismantling botnets, disseminated best practice guidelines and encouraged innovation through 'trade-related engagement.'[68] All of these reforms took advantage of "voluntary, non-binding norms" in order to "preserve America's overmatch."[69]

Ultimately, it would have been very difficult for the Trump administration to develop a capability-based active defense program for four reasons. First, the overmatch concept exposed a fundamental conflict between Trump's stated support for programs that help 'ordinary working-class' people and his less visible support for expensive military programs developed and championed by elites. Second, it would've been difficult for Trump to promote an active defense program designed to protect vulnerable, capability-deficient victims in cyberspace (even if the US is not one of them) while embracing a divisive, racist political agenda that scapegoated a wide swath of the American public.

Third, Trump's moral credibility outside of his narrow political base was practically non-existent.[70] It would have been difficult, if not impossible, for him to argue that American foreign policy should be based on strong moral values and morally defensible permissibility thresholds for hackbacks, rather than naked, strategic self-interest or technological capacity. Fourth, Trumps' refusal to criticize Putin would have undermined any administration support for serious democracy promotion efforts. At least on paper, the Trump administration devoted a considerable amount of energy to the protection of critical infrastructure, but Trump's refusal to personally criticize Russia after they attacked America's political infrastructure in 2016 and 2020 also left the impression that he was really only interested in infrastructure protection when it did not directly impact his political relationship with Putin. The toxic combination of right-populism, neoconservatism and surreptitious neoliberalism

[67] Atkinson (2020, 9).

[68] Id. at 16.

[69] Id.

[70] Kupchan and Trubowitz (2021, 99) ("The Trump era.. seriously compromised American moral authority.").

would have made it very difficult for Trump to convince the American people of the rightness of a capability-based active defense program.

5 Biden's New Internationalism

5.1 *Realism, Progressive Idealism and Neo-Populism*

As President Biden entered the Oval Office in 2021, America was just beginning to comprehend the scope of the January 6 insurrection on the US Capital, an insurrection wholeheartedly encouraged, and partially funded by, Bidens' immediate predecessor, Donald Trump. Biden had many political tools at his disposal—democratic control over both houses of Congress, clear victories in both the electoral college and popular vote, and overwhelming international support—but he also brought a new set of intellectual tools into the White House. Bidens' 'new internationalism,' if the term existed at such an early point in his presidency, is a curious combination of strategic realism, progressive idealism and neo-populism.

Biden's worldview, first and foremost, is informed by realism,[71] which encourages states to act in their own best interest, assume that the global environment is fundamentally anarchic with "no overarching authority to place limits on the actors behavior"[72] and forge strong alliances to counter balance competing regional powers.[73] Even with a rival is determined to undermine America as Russia, Biden is determined to keep diplomatic channels open in order to maintain a minimum level of trust, gauge the adversaries' willingness to negotiate or use force, and preserve both the legitimacy and effectiveness of international law.[74] Biden is also absolutely determined to resuscitate alliances, partnerships and strategic arrangements that George Bush and Trump systematically shredded.[75]

[71] Kassab (2014, 70) (defines defensive realism).

[72] Id. at 60.

[73] Craig and Valeriano (2018, 1) (distinguishes offensive and defensive realism) and Kegley and Raymond (2021).

[74] Biden and Carpenter (2018, 56).

[75] Neoconservatives in the Bush Administration downplayed alliances because they knew that they could not muster enough international support for the invasion of Iraq in 2003, and believed that alliance politics should not supersede sovereign claims based on self-defense. Right-populists, on the other hand, downplayed international alliances

Bidens' determination to rebuild alliances is also a pragmatic response to a dramatically changed geostrategic landscape in which the United States is no longer the only superpower, having squandered that lofty perch after two prolonged, expensive wars in Afghanistan and Iraq. The United States is still the dominant military power, but it shares the economic world stage with a number of other countries, including most notably, China, one of its chief political and strategic rivals.

Bidens' new internationalism is also informed by 'progressive idealism.' Idealism, the counter narrative to realism, assumes that states make foreign policy decisions primarily on the basis of ideological commitment, not strategic interest. It is a Manichean, black-and-white view of the world that divides states into two clear categories: good versus evil. Neoconservatives in the Bush administration used idealism to generate the infamous "axis of evil" (Iraq, Afghanistan, Iran),[76] as well as, to a lesser extent, justify entering Syria. Liberal interventionists in the Clinton administration used idealism to justify humanitarian operations in Kosovo, Somalia and Rwanda. Progressive idealists in the Biden administration, however, are not interested in massive humanitarian interventions or regime changes. On the contrary, progressive idealists are pursuing a less ambitious goal: the promotion and protection of technocratic liberal democracies.[77] As Biden stated in an article he wrote before he became president, Russia considers liberal democracy the "greatest external threat to its survival."[78] Shoring up democracies worldwide is undoubtably controversial, and may even precipitate another Cold War,[79] but it serves the dual strategic purpose of not only counterbalancing authoritarian states (or pro-autocratization movements within liberal states), but also providing a moral framework for distinguishing between American policies and the policies of its rivals. Cynics will argue Bidens' democracy promotion campaign[80] is just a clever way to distract attention from the

because they believed that those alliances undermine the economic interest of working-class and middle-class Americans, and wanted to help Putin continue to undermine liberal democratic states.

[76] Brands (2016, 107) (discusses the "Manichean ethos of Bush's post-9/11 statecraft..").

[77] Wright (2020).

[78] Rodrik and Walt (2022, 145).

[79] Diamond (2022, 186).

[80] Liptak (2021).

historical excesses of American exceptionalism,[81] but it prevents liberal democratic states from endlessly questioning their own motives.

Promoting democracy also plays well to a domestic audience, which leads to the third prong of Bidens' 'new internationalism': neo-populism. Populism is a flexible, "thin-centered ideology."[82] Historically, there have been left populists and right populists.[83] Trump's America first ideology was a right populist variant, steeped in anti-elitism, anti-globalism and overt racism. Trumps' philosophy nurtured an extraordinary tolerance for foreign interference in American political processes, and an equally perplexing tolerance for massive tax cuts for wealthy Americans. All of these policy choices led Biden to accuse Trump of being a phony populist, suggesting that Biden was either a populist himself, or that he understood the true character of populism better than Trump. Only the judgment of history can confirm the latter, but the former is decidedly correct. Rhetorically, Bidens' populism avoids scapegoating and extreme divisiveness, two hallmarks of Trumpianism. Biden's economic policies, ranging from direct transfers under the Care Act[84] to quasi-protectionist policies like tariffs, 'buy-American' procurement guidelines and support for a weaker dollar, are also a marked departure from the neoliberal policies of Clinton, Bush and Obama.[85]

5.2 *Constraints and Cyberspace*

Bidens' cyberspace policies are quasi-populist, and that quasi-populism lays the groundwork for what he perceives to be the main geostrategic vulnerabilities in American society. In their everyday lives, many people living in the United States feel vulnerable in four ways: (1) participating in trade agreements that undermine domestic job security; (2) interference

[81] Wright (2020).

[82] Pozen (2018, 524).

[83] Pozen (2018, 525).

[84] Liptak (2021) (the Affordable Care Act was based on Senator Mitt Romney's neoliberal, market based insurance program).

[85] Roubini (2021). Obama's foreign policy combined neoliberalism and defensive realism. The former requires alliance building, and diplomatic, rather than military, solutions to global crises like terrorism or cyber threats. De Castro (2011, 3). The latter is comparable in many respects to offshore balancing or selective engagement. See Juneau (2020) and Kassab (2014, 69).

with political processes by foreign powers; (3) cyber attacks that erode critical infrastructure, create supply chain problems and contribute to the spread of transnational criminality and terrorism; and (4) global impacts that undermine domestic climate and public health priorities. Eliminating foreign interference, particularly by Russia and China, is a key foreign policy objective with tremendous domestic appeal.[86]

Biden has also taken advantage of the perception, held by many 'ordinary' Americans, that defensive measures are not enough to deter cyber attacks. After meeting with Putin during the first few weeks of his presidency, for example, Biden reportedly told Putin that the United States was prepared to "respond in kind" to Russian cyber attacks.[87] After the Solar Winds and Colonial Pipeline attacks, Biden administration officials announced that they were considering a 'retaliatory response' targeting the suspected perpetrators. The popular sentiment that defensive measures are insufficient is shared by members of Congress, the military and technology leaders in the private sector. Numerous bills calling for an amendment to the CFAA authorizing private sector retaliatory hackbacks[88] also enjoy public support. The bills have yet to be passed,[89] but neo-populists in the Biden administration probably believe that the bills have stalled not because the public no longer believes that defensive measures are insufficient, but because 'elites' in the private sector, academic community and open world community have aggressively criticized both the motivation for, and possible effects of, legal authorizations for retaliatory hackbacks.

The strategic and technological superiority of cyber offense over cyber defense seems to confirm the common perception that in spite of law-enforcement efforts, and the proliferation of cyber hardening software, cyber hacks continue to increase. That common perception, along with realist fears about basically anarchic domains that resist control by any overarching international body on a daily basis, is slowly pushing

[86] Rodrik and Walt (2022, 145).

[87] Liptak (2021).

[88] ACDC (2017) and SCAROA (2022).

[89] The mere suggestion that the ACDC Act and SCAROA stalled in Congress because they lack popular support is absurd. Most Americans support abortion rights, but Congress failed to codify Roe. Most Americans also support basic gun control legislation, but the recent gun control law is the only major piece of gun control legislation to pass Congress since the Brady bill.

states toward anticipatory self-defense, which includes active defensive measures.[90] To be fair, there is also elite support for the perception that defensive measures are not enough to determine cyber attacks.[91] But elite support for the view that defense mechanisms are insufficient in cyberspace does not diminish the significance of two additional facts: (1) there is considerable non-elite support for the view that defensive measures are insufficient in the cyberspace, and (2) that support might be particularly appealing to a populist president.

Biden's sympathy for the vulnerability that most Americans feel about cyberspace in terms of the limits of defensive measures is largely consistent with his appreciation of American vulnerability in cyberspace. The former might be primarily populist while the latter might be primarily strategic, but both represent the Achilles' heel of American global leadership. American adherence to openness, the rule of law and human rights might be its greatest strength in the long term, but in the short term, adherence to those principles has made America vulnerable in cyberspace.

From the standpoint of capability theory, American vulnerability in cyberspace seems to persist even though America has tremendous wealth, a considerable endowment of primary goods and vast legal rights as a sovereign power. For the foreseeable future, the United States, under Bidens leadership, will be 'situationally constrained' in cyberspace, forced to develop the strategic, technological and military equivalent of 'adaptive preferences' like 'over-the-horizon' initiatives, dependence on transnational law enforcement, cyberspace specific bilateral agreements, non-interventionist democracy promotion efforts and public/private sector retaliatory hackbacks. These facts alone move America's foreign policy into the scope of VRN. If we can accept that America and its peoples are 'capability—deficient' under VRN, then the next question is how Biden, the realist with neo-populist sensibilities, will define the permissibility of specific cyber attacks and retaliatory hackbacks.

5.3 Permissibility Thresholds

Three weeks after the 2020 presidential election, the FireEye cyber security firm reported the theft of its 'Red Team' software. It was later

[90] Kegley and Raymond (2021) and Craig and Valeriano (2018, 6).

[91] Craig and Valeriano (2018, 5).

discovered that malware code called Sunburst was developed using the RedTeam software, and was found on computers loaded with software from a company called Solarwinds after those computers were updated with Orion software. Hundreds of thousands of companies and government agencies used Solarwinds products. Eventually, a hacking group affiliated with the Russian SVR called Cozy Bear (APT 29) was blamed for the attack. Mike Pompeo, Secretary of State for the Trump administration, criticized the Russian government for orchestrating the Solarwinds attack, but President Trump insisted that the Chinese, rather than the Russians, were responsible. Even though the malware loaded during the Orion Code update infected hundreds of thousands of computers and servers, there were no reports of damage to either hardware or software. No files were deleted, renamed or relocated to other directories or servers, and no hardware was shut down or inaccessible. But the malware did allow hackers to roam around systems, access files and map the system's computational and network architecture.

Experts began to consider the possibility that the Solarwinds attack, widespread in its impact on critical infrastructure in both the public and private sector, might constitute an act of war against the United States under international law. Senators Durbin and Rubio concluded that the Solarwinds attack was an act of war, and President-elect Biden immediately expressed outrage. Experts had to determine if the attack could be attributed to a state (Russia), constituted a breach owed to another state, as required by Articles 2, 4 and 8 of the Articles on State Responsibility, and constituted an 'act of force' as required by Article 2 (4) of the UN Charter. Without any physical damage or functional incapacity, however, experts could not satisfy the 'act of force' requirement.[92] It was also difficult to prove that the Solarwinds attacks constituted an 'act of war' executed by one state against another state or a non-state actor[93] or an impermissible intervention into the internal affairs of another country.[94]

On May 7th, 2021, the nation heard about a ransomware attack on Colonial Pipeline, a fuel line that serves roughly half of the eastern United States. The suspected attacker, Darkside, demanded a ransom of almost 4.5 million dollars (in Bitcoin) and Colonial paid the ransom. Fuel

[92] Tallinn20 (2013), Rule 69.

[93] Tallinn20 (2013), Rules 82 and 83.

[94] Tallinn20 (2013), Rule 66.

services from the pipeline were temporarily shut down, crippling gas deliveries across many states. After the ransom was paid, Colonial was given an IT tool and eventually service was restored. On June 7th, the DOJ announced the recovery of roughly half of the ransom, and on May 9th, President Biden signed Executive Order 14028 to, among other things, increase cyber security standards across the government, and enhance information sharing. DOJ never disclosed how it managed to obtain the private key necessary to retrieve part of Colonial's ransom payment from Darkside's bitcoin wallet.

At least publicly, Biden's reaction to the Colonial Pipeline attack was retorsive, but that public response might have involved secretive countermeasures. After the Colonial attack, Biden gave a speech on 13 May 2021 in which he stated that the Colonial pipeline hackers were living in Russia, that the Russian government was involved, that he was "in direct communication with Moscow about the imperative for responsible countries to take decisive action against these ransomware attacks" and that his administration had decided to "pursue a measure to disrupt their ability to operate." The use of the term 'measure' might have been a signal that the United States not only considered the Colonial attack an 'unlawful act' (as opposed to merely an unfriendly act) because the ransomware attack caused damage to critical infrastructure, but also that the United States was prepared to use 'countermeasures,' as defined by international law, to identify and punish the hackers responsible for the Colonial attack.

Some commentators urged Biden to be more transparent about his intention to use hackback countermeasures. Instead of making any further announcements about his administration plans to possibly launch a retaliatory hackback against Russia or the Darkside hacker group (a for-profit ransomware group operating within Russia), however, Biden announced, on 11 May 2021, a series of reforms to establish public/private information-sharing arrangements, unlock potential supply chain problems caused by the attack and enhance existing protections of critical infrastructure. Eventually, in February 2022, we learned that the Russian government, using information provided by the United States, had arrested members of the ransomware group Revil, and discovered that one of the arrestees was responsible for the Colonial attack. The administration attributed the successful operation to enhanced diplomatic relations between the United States and Russia.

As I mentioned before, it is not uncommon for law enforcement officials or private 'greyhat' programmers to use hackback strategies like breaking into a C2 message network, or breaking into a server to retrieve a private key for accessing a ransomware wallet, in order to destroy a ransomware operation. The real question is whether the US government will ever publicly acknowledge that it not only uses the aforementioned 'cyber countermeasures' to thwart rogue, illiberal regimes, but that doing so is morally justifiable. Besides the general problem of not wanting to reveal technological secrets, and risk escalating a cyber conflict into kinetic conflict by publicly (as opposed to privately) shaming the rogue nation ultimately responsible for a cyber attack, the Biden administration is probably already wargaming its response to a particular constellation of facts:

1. an extremely dangerous attack on infrastructure or a military installation owned by the federal government
2. the attack is deliberately widespread and damaging
3. the attack happens repeatedly
4. attribution is clear—it was directly perpetrated by a rogue, authoritarian nation
5. the attacking nation acknowledges its role in the attack and announces its intention to launch further attacks on the same facility
6. retorsive strategies are unlikely to be successful as a deterrent
7. potential spillover effects (including third-party damages) can be contained using due diligence

These are the same triggering conditions flagged by the DOJ in its post-Van Buren CFAA domestic law guidelines, and zone 2 of VRN. In an actual conflict situation, the aforementioned conditions might be so disturbing and incendiary that the US government will feel compelled to publicly acknowledge that it intends to respond with a retaliatory hackback.

5.4 *Acknowledgement, Vulnerability and Credibility*

Public acknowledgement, as opposed to an internal Administration decision to seek legal authorization or treat a retaliatory hackback as a countermeasure instead of a retorsive act because of the nature of the initial attack, is important because it will provide a sitting president with an opportunity to do something that no other president has done:

explicitly link America's active defense program to a moral agenda with potentially bipartisan appeal. The President would have to juggle three balls at once: (1) allay public fears of down-stream spillover effects or third-party damage; (2) ensure the American people that in all but the most severe circumstances, a retaliatory hackback response to a cyber attack of the aforementioned variety is preferable to the use of military force; and (3) emphasize that even though the primary purpose of a retaliatory hackback would be national self-defense, a secondary purpose would be the protection of increasingly vulnerable, similarly situated democratic nations across the globe. More than anything else, the third point would reinforce "... a broader mission to help America rediscover its' 'higher purpose' as a benign superpower."[95]

This three-pronged justification for a retaliatory hackback, launched by the American government in 'zone 2' circumstances, generates a familiar contradiction between, on the one hand, an America that continues to labor under a 'disadvantage' against authoritarian nations in cyberspace and, on the other hand, an America that retains its status as the best positioned, most 'capable,' 'protector' of vulnerable, liberal democratic nations. From the standpoint of capabilities theory, however, the contradiction is an illusion. In the same way that a wealthy, abused married woman with a formidable endowment of primary goods and access to lawyers may believe that she must remain in her home to protect her children, a nation-state with wealth, sovereign rights and generous national endowments may find its options limited in crisis situations because of its strategic circumstances or ideological commitments. The real question is not, from the standpoint of capabilities theory, whether it is possible to compare the situational constraints placed on burdened individuals with the situational constraints placed on nation-states. On the contrary, the real question is whether effective presidential leadership can bridge the gap between the disadvantages experienced by individuals and the disadvantages experienced by states in order to justify American foreign policy goals in the current geostrategic environment.

One of the most pressing foreign policy goals of the Biden administration is to re-establish international alliances and relationships shattered by Trump. Biden has stated repeatedly that America is "ready to

[95] May (2021, 6).

lead the world."[96] Any initiative to establish a capability-based active defense program must be able to demonstrate that America is willing to align national security and foreign policy goals, which includes an openly acknowledged retaliatory hackback program, with the political, economic and emotional priorities of ordinary, working-class ad middle-class Americans. This is the populist advantage over neoliberalism or neoconservativism: a willingness to openly, unapologetically and consistently align foreign policy and national security goals with the interests of ordinary, 'non-elite' Americans. Commentators have already proposed a number of administrative changes that Biden can make in order to align working-class goals with foreign policy goals, including making the Secretary of Labor "a permanent member of the National Security Council," and deepening ties between the NSC or State Depart and domestic policy offices like the "National Economic Council and the Domestic Policy Council."[97]

Biden must establish and normalize administrative pathways between the national security and domestic policy staffs in order to ensure that Washington "aggressively [enforces] US domestic trade laws," and consistently "[promotes] the interests of American workers."[98] The principal ideological advantage held by a neo-populist president faced with pressing national security and foreign policies in the current environment is that capability theory is extremely sensitive to the actual constraints experienced by ordinary people in their daily lives. It protects the interests, and prioritizes the experiences, of capability-deficient people, groups and nation-states. VRN extends that protection to capability-deficient entities who hackback in good faith, using due diligence. But neo-populist presidents determined to align working-class, 'non-elite' domestic goals with foreign policy goals must also possess tremendous moral integrity, and speak with moral clarity. Integrity and clarity are two qualities that Trump lacked, but Biden appears to possess both, personally and politically. As a result, he is better positioned to forge "a new internationalist consensus by making significant investments in the domestic economy,"[99] and incorporate capability theory in America's active defense strategy in the current

[96] Kupchan and Trubowitz (2021, 92).

[97] Id. at 96–97.

[98] Id. at 97.

[99] Id. at 98.

political environment, than Trump ever was. Promoting democracy at home and abroad, while avoiding responsibilization regimes that shift most of the blame for cyber-attacks onto victims rather than perpetrators (the technological equivalent of populist scapegoating), will reinforce America's standing in cyberspace and the open world. A capability-based retaliatory hackback program that advances national interests and a compelling moral agenda could become something that the American government is actually prepared to openly acknowledge.

References

ACDC. 2017. Active Cyber Defense Certainty Act, H.R. 4036, 115th Congress.

Atkinson, W. 2020. A Review of the Trump Administrations' National Cyber Strategy: Need for Renewal and Rethinking of the Public/Private Partnership in US National Security Policy. *Active Measures Journal.*

Biden, J., and M. Carpenter. 2018. How to Stand Up to the Kremlin. *Foreign Affairs* 97 (1): 44–57.

Brand, U., and N. Sekler. 2009. Postneoliberalism: Catch-All Word or Valuable Analytical and Political Concept? *Development Dialogue* 51: 5–14.

Brands, H. 2016. Barack Obama and the Dilemmas of American Grand Strategy. *The Washington Quarterly* 39 (4): 101–125.

Bromwich, J. 2017. How US Military Policy in Transgender Personnel Changed Under Obama. *N.Y. Times*, July 26.

Craig, A., and B. Valeriano. 2018. Realism and Cyber Conflict: Security in the Digital Age. In *Realism in Practice: An Appraisal.* Berlin: Springer.

De Castro, R. 2011. The Obama Administration's (Neoliberal) Reengagement Policy in East Asia: Implications for U.D.-China Relations in the Twentieth Century. *Issues and Studies* 47 (2): 1–44.

Comiling, K., and R. Sanchez. 2014. A Post Colonial Critique of Amartya Sen's Capability Approach. *Perspectives in the Arts and Humanities, Asia* 4 (1): 1–26.

Diamond, L. 2022. All Democracy Is Global: Why America Can't Shrink from the Fight for Freedom. *Foreign Affairs* 101 (5): 182–198.

Downie, J. 2015. What Donald Trump Doesn't Understand About Dealmaking. *Washingtonpost.Com*, October 26.

Dueck, C. 2018. Trump's Strategic Unpredictability, Its Pros and Cons. *NationalReview.com*, December 28.

Dueck, C. 2019. What's Conservative American Nationalism? *NationalReview.com*, October 21.

EEA. 1996. Economic Espionage Act of 1996 (October 11, 1996), and 18 U.S.C. § 1831 (2013).

Ellings, R. 2014. Five Chinese Military Officers Indicted. Now What? *Nat'l Bureau Asian Research*, May 22. http://nbr.org/downloads/pdfs/ETA/Ellings_FiveIndicted_052214.pdf.

Finer, J., and R. Malley. 2018. Trump Is Right to Seek an End to America's Wars. *NY Times.com*, January 8.

Gabcikovo. 1997. *Gabcikovo-Nagymaros Project*. Hung.

Gordon, S., and E. Rosenbach. 2022. America's Cyber-Reckoning: How to Fix a Failing Strategy. *Foreign Policy* 101 (1): 10–21.

Hartz, L. 1948. *Economic Policy and Democratic Thought*. Ann Arbor: University of Michigan Press.

IEEP. 2014. 50 U.S.C. §§ 1701–1707 (2014) (?)International Emergency Economic Powers Act 50, U.S.C. §§ 1701–1707 (2014). IEEP.

Jacob, A. 2017. *Cyber-Physical Systems, or the Cultural logic of Techno-Liberalism*. Master Thesis. McMaster University.

Juneau, T. 2020. Obama, Biden and Realism in the Middle East. *ResponsibleStatecraft.com*, July 15.

Kassab, H. 2014. In Search of Cyber Stability: International Relations, Mutually Assured Destruction and the Age of Cyber Warfare. In *Cyberspace and International Relations*, ed. J.F. Kremer and B. Muller. Berlin: Springer-Verlag.

Kegley, C., and G. Raymond, 2021. Realism in the Age of Cyber Warfare. *Ethics and International Affairs*, April 26.

Kitfield, J. 2016. Trump's 'America First' Neo-Isolationism. *Yahoo.com*, April 26.

Klare, M. 2018. 'Overmatch' Is Overkill: The Pentagon's Strategy for Global Supremacy Could Bankrupt the Country and Spark a World War. *TheNation.com*, December 20.

Kupchan, C., and P. Trubowitz. 2021. The Home Front: Why an Internationalist Foreign Policy Needs a Stronger Domestic Foundation. *Foreign Affairs* 100 (3): 91–101.

Larson, D. 2018. The Perils of Trump's Iran Obsession. *The American Conservative*, August 13.

Laruffa, F. 2018. Towards a Post Neoliberal Society? Social Investment Versus Capability Approach. *Momentum Quarterly* 7 (4): 171–187.

Liptak, K. 2021. Biden Repeats Call to NATO Allies to Defend Democracy Against Autocracy. *CNN.com*, June 14.

May, R. 2021. *Moralism and American Foreign Policy: Biden's Greatest Challenge to Revising US Primacy*. Atlas Institute of International Affairs Blog.

Murdoch, S. 2013. US Plans to Respond to Chinese Hacking. *The Australian*, February 21.

Nakashima, E., and J. Warrick. 2012. Stuxnet Was Work of U.S. and Israeli Experts, Officials Say. *WASHPOST.com*, June 3.

NDAA. 2017. National Defense Authorization Act of 2015, Pub. L. No. 113-291, § 1637, 128 Stat. 3292.

Neuman, S. 2012. As the Worm Turns: Cybersecurity Expert Tracks Blowback from Stuxnet. *NPR.com*, June 1. http://www.npr.org/blogs/thetwo-way/2012/06/01/154162121/as-the-worm-turns-cybersecurity-expert-tracks-blowback-from-stuxnet.

Owen, J. 2000. *Liberal Peace, Liberal War: American Politics and International Security*. Ithaca, NY: Cornell University Press.

Pangang. 2012. United States v. Pangang Grp. Co., Ltd., 879 F. Supp. 2d 1052, 1056 (N.D. Cal. 2012).

Pozen, D. 2018. David Pozen, Populist Constitutions. *University of Chicago Law Review* 85: 521–543.

PPD20. 2012. United States Government. Presidential Policy Directive 20/PPD-20. *US Cyber Operations Policy*, October. https://fas.org/irp/offdocs/ppd/ppd-20.pdf.

Ptweet. 2017. Presidential Tweet on July 26, at 5:55AM.

Pyysiainen, J., D. Harpin, and A. Guilfoyle. 2017. Neo-Liberal Governance and Responsibilization of Agents: Reassessing the Mechanisms of Responsibility-Shift in Neo-Liberal Discursive Environments. *Distinkton: Journal of Social Theory* 18 (2): 1–21.

Rodrik, D., and S. Walt. 2022. How to Build a Better Order: Limiting Great Power Rivalry in an Anarchic World. *Foreign Affairs* 101 (5): 142–155.

Roosevelt, T. 1920. *Theodore Roosevelt: An Autobiography*. New York: Charles Scribner's Sons.

Roubini, N. 2021. Biden's Neo-Populist Economic Agenda. *Projectsyndicate.org*, August 3.

Saideman, S. 2001. *The Ties That Divide: Ethnic Politics, Foreign Policy and International Conflict*. New York: Columbia University Press.

Sanger, D. 2012. *Confront and Conceal: Obama's Secret Wars and Surprising Use of American Power*. New York: Penguin Random House.

SCAROA. 2022. Study on Cyber-Attack Response Options Act, S.2292-117th Congress (2021–22).

Scheppele, K. 2019. The Opportunism of Populists and the Defense of Constitutional Liberalism. *German Law Journal* 20 (3): 314–331.

Sen, A. 1999a. *Development and Freedom*. Oxford: Oxford University Press.

Sen, A. 1999b. Equality of What? In *Equal Freedom*, ed. Stephen Darwall. Ann Arbor: University of Michigan Press.

Setear, J. 1997. Responses to Breach of a Treaty and Rationalist International Relations Theory: The Rules of Release and Remediation in the Law of Treaties and the Law of State Responsibility. *Virginia Law Review* 83 (1): 1–126.

Singh, D., and J. Purdy. 2014. Introduction: Law and Neo-liberalism. *Law and Contemporary Problems* 77 (4): 1–23.

Snyder, J. 2013. *Myths of Empire: Domestic Politics and International Ambitions*. Ithaca, NY: Cornell University Press.

Stuxnet. 2010. Iran Confirms Stuxnet Worm Halted Centrifuges. *CBSNEWS.com*, November 29. http://www.cbsnews.com/stories/2010/11/29/world/main7100197.shtml.

Tallinn20. 2013. *The Tallinn Manual on the International Law Applicable to Cyber Warfare*, ed. Michael N. Schmitt et al. Cambridge: Cambridge University Press.

UNCharter51. 1945. *Charter of the United Nations*. San Francisco.

Valeriano, B., and B. Jensen. 2019. The Myth of Cyber Offense: The Case for Restraint. Cato Institute Policy Analysis, No. 862, January 15.

Venugopal, R. 2015. Neoliberalism as a Concept. *Economy and Society* 44 (2): 165–187.

Warford, T. 2019. Trump, May and the Art of Political Brinksmanship. *Financialtimes.Com*, January 11.

Wright, R. 2020. Robert Wright, Biden's Foreign Policy Team Is Full of Idealists Who Keep Getting People Killed. *Washingtonpost.com*, December 15.

Wu, T., and C. Yoo. 2007. Keeping the Internet Neutral?: Tim Wu and Christopher Yoo Debate. *Federal Communications Law Journal* 59 (3): 575–604.

Yoo, C. 2015. Cyber Espionage or Cyber War?: International Law, Domestic Law, and Self-Protective Measures. Faculty Scholarship. At Penn Carey Law 1540.

Youssef, M., and S. Harris. 2016. Obama Officials Wonder: Why Won't the Boss Stand Up to Putin? *The Daily Beast*, September 29.

Zakaria, F. 1999. *From Wealth to Power: The Unusual Origins of America's World Role*. Princeton: Princeton University Press.

Zoller, E. 1984. *Peacetime Unilateral Remedies: An Analysis of Countermeasures*. New York: Transnational Publishers.

CHAPTER 10

Conclusion

Open world expectations concerning the free flow of information, persons and goods occupy an uncomfortable space between cyber victims and accused hackers. Well-conceived hackback regulations protect open world expectations from the potential chaos of hackbacks by balancing the entirely reasonable demands of private firms and governmental institutions that want to retaliate against attackers against the equally reasonable demands of law enforcement agencies, international institutions and self-governing cyber communities that want to punish or deter attackers without resorting to retaliation. The former signals state support for competitive resolve and retributive justice, while the latter signals state support for social trust.

Even though the liberal international order legitimized open world expectations, and created international institutions to protect them, many decision-makers firmly believe that urgent public requests for hackback authorizations are a sign that the liberal international order has run out of steam as an effective enforcement mechanism in cyberspace because the norms produced by the liberal international order failed to deter bad actors from seeking, and eventually gaining, unauthorized access to other people's computers. Self-help measures that alleviate public reservations about the security and stability of cyberspace may offer alternative conflict resolution mechanisms that standards-setting, norms-producing international institutions cannot provide. As Congress and various state

A. J. Thomas, *The Open World, Hackbacks and Global Justice*,
https://doi.org/10.1007/978-981-19-8132-6_10

legislatures propose new hackback authorization statutes, it is important to remember that these statutes are likely to be drafted by zealots, criminal skeptics and competitive skeptics determined to protect reasonable open world expectations from misconduct, cyber escalations and manipulation by allies and enemies abroad. But opportunists are also likely to participate in the legislative process, equally determined to exploit fissures in the open world ethos for personal or national gain. In their reckless hands, retaliatory hackbacks could become the errant virtual missile that turns into a full-scale kinetic war.[1]

Almost entirely missing from most discussions of retaliatory hackbacks, however, is any consideration of global justice. Comparatively few national security, technology or open world experts ever consider whether a particular critique or defense of retaliatory hackbacks, undertaken by either private or public sector actors, actually promotes global justice. When the victim of a cyber attack decides to hackback, as opposed to contact law enforcement, directly sue the suspected attacker or harden his security apparatus, is the decision motivated by the absence of wealth, primary goods or former rights? Or is it possible for someone to possess all of these things, and still conclude that a retaliatory hackback is the most desirable and efficient response to a cyber attack? Conceptualizing the decision to hackback as a choice between wealth, primary goods/rights, efficiency or something else immediately transports active defense discourse into a heavily contested philosophical arena dominated by means -ends calculations, complicated assessments of personal, organizational and nation-state vulnerability, socio-economic equality, Pareto-optimality, inter-personal comparisons, cost-benefit analysis and distributive justice. The typical question is whether personal decisions, government policies or disputes over entitlements should be based on fairness (Rawls' 'justice as fairness') or utility (wealth maximization, efficiency, welfare). But there are many decisions and policies that should not be reduced to the choice between fairness and utility, and the decision to hackback is one of them.

Many victims of cyber attacks choose to hackback even if they possess primary goods, formal rights and wealth, three attributes that usually drive people towards "reasonable" options like law enforcement, private suits or hardening strategies. And yet they either cannot or will not to take

[1] Clayton (2013) (reporting cyber attacks on American gas pipelines and citing US suspicions of Chinese military involvement).

advantage of reasonable options to secure justice from a wrong committed in cyberspace, in the same way that a poor person without skills might starve in a city awash with development money and legal guarantees, or an abused married woman with children might remain in the house because she is afraid of what her husband will do to the children if she leaves. The best explanation for the behavior of these victims is capability theory, not fairness or utility. Capability theory explains the constraints that people face in their daily lives, constraints that make it difficult for them to take advantage of available opportunities, even if they have wealth, primary goods or formal rights. The justice that they seek (either consciously or unconsciously), and the justice they deserve in the cyber world, depends on their ability to immediately retrieve stolen information and efficiently deter future attacks. Regardless of whether the cyber attack took the form of ransomware, spear phishing emails, web scraping, DDoS or 'insider' access with authorized credentials but an improper purpose, capability deficient people, institutions and nation-states may choose to retaliate after the attack.

The central challenge faced by decision-makers is whether the unique injustices faced by capability deficient victims should be treated differently from the justice claims of other victims, particularly when the cyber world is experiencing an ever increasing number of attacks. The latter dilemma has generated a wide variety of potential solutions. Some people have suggested the algorithmic equivalent of Israel's 'Iron Dome' or Reagan's 'Star Wars'—a massive, anti-virus software protection system implemented at the national or even regional level.[2] Others suggest that we not only lower our expectations of cyber security in the open world but also place most of the blame for repeated attacks on victims rather than perpetrators. The third alternative is a 'meta-regime' capable of resolving conflicts between rival parties who attack and counterattack against each other in cyberspace on a semi-regular basis. Something similar has been proposed to guide the resolution of disputes between rival nations when conflict resolution mechanisms established by the liberal international order lose their effectiveness.[3]

[2] Kassab (2014, 71–73).

[3] Rodrik and Walt (2022, 144) (the proposed meta-regime is a "device for guiding a process through which rival states or even adversaries could seek agreement or accommodation on a host of issues. When they don't agree, as will often be the case, adopting the framework can still enhance communication among them, clarifying ways they disagreed,

In many ways, VRN belongs in the third category, but there are four important differences: (1) it is limited to permissible or impermissible cyber attacks and retaliatory hackbacks, (2) it extends legal and moral protection to cyber attacks or hackbacks engineered by capability deficient parties, (3) it rewards good faith capability-deficient parties who hack-back using "due diligence" (i.e., reasonable attribution, proportionality, necessity) and (4) it punishes parties that engineer particularly dangerous attacks against capability-deficient individuals, organizations or nation-states by forcing the attacker to absorb the total cost of any damages caused by the victims retaliation.

The most controversial aspect of VRN is that it extends both moral and legal protection to all capability-deficient victims of cyber attacks, including nation states. Comparing the plight of people with disability/cognitive capability deficiencies or people with existential capability deficiencies to people with broad situational deficiencies is difficult for some people. Comparing the plight of anyone in the aforementioned categories to nation-states appears to be even more difficult. But the geo-strategic constraints faced by specific nation-states in cyberspace significantly limits the range of policy options available to decision-makers. Even states with a considerable endowment of primary goods, formal sovereignty rights and national wealth will find it difficult to take advantage of available diplomatic, military or technological opportunities.[4] Retaliation after a cyber attack might be the ultimate adaptive preference for liberal democratic states laboring under the 'burden' of their moral beliefs: adherence to the rule of law, international institutions, political accountability and human rights.

When cyber attacks engineered by illiberal, authoritarian regimes are 'particularly dangerous,' the United States is, under VRN, fully 'entitled' to hackback against the suspected attacker as long as the retaliation is done safely, not because the United States is technologically capable of doing so, or because hackbacks are necessary to maintain America's

and offer them incentives to avoid harm on others, even if they seek to protect their own interest.").

[4] Ikenberry (2022, 60) ("[O]ne of the United States' greatest strengths is its capacity to fail; as a liberal society, it can acknowledge its vulnerabilities and errors and seek to improve, a distinct advantage over its illiberal rivals in confronting crises and setbacks.").

competitive edge, but because the hackback serves a legitimate moral purpose. It is morally permissible in a way that hackbacks conducted by authoritarian regimes (or their proxies) are not. Not every president is capable of communicating the advantages of a capability-based hackback program effectively to the American people. Establishing protection for such a framework under existing domestic and international law will prove difficult enough. But President Joseph Biden, a realist with neo-populist sensibilities and a skeptical, but optimistic, understanding of the open world, may outperform his rivals, once again.[5]

References

Clayton, M. 2013. Exclusive: Cyberattack Leaves Natural Gas Pipelines Vulnerable to Sabotage. *Christian Science Monitor.com*, February 27.

Foroohar, R. 2022. After Neoliberalism. *Foreign Policy*. 101 (6): 134–145.

Ikenberry, G. 2022. Why American Power Endures. *Foreign Policy*. 101 (6): 56–73.

Kassab, H. 2014. In Search of Cyber Stability: International Relations, Mutually Assured Destruction and the Age of Cyber Warfare. In *Cyberspace and International Relations*, ed. J.F. Kremer and B. Muller. Berlin: Springer-Verlag.

Rodrik, D., and S. Walt. 2022. How to Build a Better Order: Limiting Great Power Rivalry in an Anarchic World. *Foreign Affairs* 101 (5): 142–155.

[5] Foroohar (2022, 141) ("The idea that the global economy must be put back in the service of national needs is gaining traction, but neither party has put forward a complete plan for how to do so (although the Biden administration has come the closest.").

References

Abbott, E., and O. Hetznel. 2010. *Homeland Security and Emergency Management: A Legal Guide for State and Local Governments*. Chicago: ABA.

ACDC. 2017. Active Cyber Defense Certainty Act, H.R. 4036, 115th Congress.

ACDR. 2019. *Active Cyber Defense Report*, July. http://www.ncsc.gov.uk/report/active-cyber-defence-report-2019.

Ackerman, B. 2011. Legal Acrobatics, Illegal War. *N.Y. Times*, June 21.

Adath, D. 2019. Science Under Maximum Pressure in Iran. *Thescientist.Com*, September 3.

Adler, M. 2005. Against 'Individual Risk': A Sympathetic Critique of Risk Assessment. *University of Pennsylvania Law Review* 153: 1121–1250. http://papers.ssrn.com/abstract=487123.

Adler, M., and R. Posner. 1999. Re-thinking Cost-Benefit Analysis. *Yale Law Journal* 109: 165–247.

Affairs. 1776a. U.S. Constitution Article I, Section 8.

Affairs. 1776b. U.S. Constitution Article II, Section 2, cl 2.

Aghaei-Forushman, V., and A. Zineir-Heywood. 2013. On Evaluating IP Traceback Schemes: A Practical Approach. *IEEE Security and Privacy Workshop*: 127–134.

Alperovitch, D. 2022. The Case for Cyber-Realism: Geopolitical Problems Don't Have Technical Solutions. *Foreign Policy* 101 (1, January/February): 44–51.

Amadae, S. 2004. Amartya Sen's Rationality and Freedom. *Economics and Philosophy*. 20 (2), 381–389.

Ambastha, M. 2019. Taking a Hard Look at the Vulnerability Equities Process and Its National Security Implications. *Berkeley Technology Law Journal (BTLJ Blog)*, April 22.

A. J. Thomas, *The Open World, Hackbacks and Global Justice*,
https://doi.org/10.1007/978-981-19-8132-6

Amsterdam, A. 1960. The Void for Vagueness Doctrine in the Supreme Court. *University of Pennsylvania Law Review* 108: 67–116.

Andropopolis, H. 1994. *The Laws of War: Constraints in Warfare in the Western World*. New Haven: Yale University Press.

Armeur, E., S. Gamb, and A. Ho. 2012. Maintaining Sovereignty over Personal Data on Social Networking Sites. In *Managing Privacy Through Accountability*, ed. Calla Iffen, Inga Kroener, Daniel Heybard, and Hector Postigo. London: Palgrave Macmillan.

Ashburton, L. 1858. Letter from Lord Ashburton to Daniel Webster, US Secy of State (July 28, 1842). British and Foreign State Papers 1841-2. http://www.yale.edu/lawweb/avalon/diplomacy/britain/br-1842d.htm.

Ashworth, L. 2006. Where are the Idealists in Interwar International Relations. *Review of International Studies* 32: 291–308.

Atapour-Abarghouei, A., Stephen Bonner, and Andrew Stephen McGough. 2019. Volenti Non Fit Injuria L Ransomware and Its Victims. arxiv:1911,08364v1 [cs.CR], November 10.

Atkinson, W. 2020. A Review of the Trump Administrations' National Cyber Strategy: Need for Renewal and Rethinking of the Public/Private Partnership in US National Security Policy. *Active Measures Journal*.

Bacon, D. 2008. *Illegal People: How Globalization Creates Migration and Criminalizes Immigration*. Boston: Beacon Press.

Baker, D., and D. Roberts. 2007. Extending Just War Theory: The Jus Ad Bellum and the Capabilities Approach to Armed Conflict. *Scientia Militaria, South African Journal of Military Studies* 35 (2): 21–38.

Barbour, S. 2010. *Does Illegal Immigration Harm Society?* San Diego: Reference Point Press.

Barnett, R. 2003. The Original Meaning of the Necessary and Proper Clause. *University of Pennsylvania Journal of Constitutional Law* 6: 183–227.

Barreme. 1804. Little v. Barreme, 6 U.S. (2 Cranch) 170 (1804).

Becker, G. 1996. *Accounting for Tastes*. Cambridge: Harvard University Press.

Benedek, W., Veronica Bauer, and Matthias Ketterman. 2008. *Internet Governance and the Information Society: Global Perspectives and European Dimensions*. New Jersey: Eleven International Publishing.

Bentham, J. 1843. Principles of Death Law. In *J. Bentham's Works*, Part II, Bk. 1, ed. J. Bowring. Edinburgh.

Bentham, J. 2003. *An Introduction to the Principle of Morals and Legislation* (reproduced in Darwall, 2001).

Berlin, I. 1961. *Two Concepts Of Liberty*. New York: Clarendon Press.

Bertachini, E. 2012. *Cultural Commons: A New Perspective on the Production and Evolution of Cultures*. Edward Elgar Publishing.

Bethlehem, D. 2012. Self-Defense Against an Imminent or Actual Armed Attack by Nonstate Actors. *American Journal of International Law* 106: 769–803.

Biden, J., and M. Carpenter. 2018. How to Stand Up to the Kremlin. *Foreign Affairs* 97 (1): 44–57.

Bitman. 2015 Bittman v Fox, 107 F.Supp 3d. 896 (2015).

Bleichrodt, H., and Wakker, P. 2015. Regret Theory : A Bold Alternative to the Alternatives. *The Economic Journal.* 125(3), 493–532.

Bobbitt, P. 2008. *Terror and Consent: The Wars for the Twenty-First Century.* New York: Alfred A. Knopf.

Boose, S. 2012. Blackhat Sources: 36% of Internet Security Experts Have Engaged in Retaliatory Hacking. *BusinessWire.com*, July 26.

Borders. 2019. *Open Borders: In Defense of Free Movement*, ed. Reece Jones. Athens: University of Georgia Press.

Bothe, M. 2003. Terrorism and the Legality of Pre-Emptive Force. *European Journal of International Affairs* 14: 227–231.

Boyce, R., and R. Perkins. 1989. *Criminal Law and Procedure*, 7th ed. Foundation Press.

Bracknell, R. 2007. Trust Not Their Presents, Nor Admit the Horse: Countering the Technically-Based Espionage Threat. *Roger Williams University Law Review* 12 (3): 832–851.

Bradley, C., and M. Flaherty. 2004. Executive Power Essentialism and Foreign Affairs. *Michigan Law Review* 102: 545–580.

Braine, D., and M. Clark. 1972. Varieties of Necessity. *Proceedings of the Aristotelian Society* (Supplementary Volumes) 46: 139–187.

Brand, U., and N. Sekler. 2009. Postneoliberalism: Catch-All Word or Valuable Analytical and Political Concept? *Development Dialogue* 51: 5–14.

Brands, H. 2016. Barack Obama and the Dilemmas of American Grand Strategy. *The Washington Quarterly* 39 (4): 101–125.

Brkan, M. 2017. The Courts of Justice on EU: Data Privacy and Data Protection—Judge-Made Law as a Leitmotif on Fundamental Rights Protection. In *Courts, Privacy and Data Protection in the Digital Environment*, ed. Maja Brkan and Evangelia Psychogropoulo. Cheltenham: Edqard Elgar Publishing.

Brock, G. and Bighouse, H. 2005. *The Political Philosophy of Cosmopolitanism.* (New York, Cambridge University Press)

Bromwich, J. 2017. How US Military Policy in Transgender Personnel Changed Under Obama. *N.Y. Times*, July 26.

Bruch, E. 2007. Open or Closed: Balancing Border Policy with Human Rights. *Kentucky Law Journal* 96: 197–224.

Burk, D. 2000. The Trouble with Trespass. *Journal of Small and Emerging Business* 4: 27–58.

Burnyeat, M. 1983. Can the Skeptic Live His Skepticism? In *The Skeptical Tradition*, ed. Myles F. Burnyeat. Berkeley.

Calabresi, G., and A. Melamed. 1972. Property Rules, Liability Rules and Inalienability: One View of the Cathedral. *Harvard Law Review* 85: 1089–1097.

Calandrillo, P. 2001. Responsible Regulation: A Sensible Cost-Benefit, Risk Versus Risk Approach to Federal Health and Safety Regulation. *Boston University Law Review* 81: 957–992.

Campbell. 2000. Campbell v. Clinton, 203 F.3d 19, 20 (D.C. Cir. 2000).

Carter, S. 1983. The Political Aspects of Judicial Power: Some Notes on the Presidential Immunity Decision. *University of Pennsylvania Law Review* 131: 1341–1384.

Carvalhos, W. 2019. Deforestation Control in the Brazilian Amazon: A Conservation Struggle Being Lost as Agreements and Regulations Are Subverted and Bypassed. *Perspectives in Ecology and Conservation* 17 (3): 122–130.

CFAA. 1986. Computer Fraud and Abuse Act of 1986, Pub L. No 99-474, 100 Stat. 1213 (1986).

CFAA. 2012. Section 1030 (a)(5), Supp. V 1988.

CFAA. 2018. 18 USC 1030(f) (2018). Computer Fraud and Abuse Act of 1986, Pub. L. No. 99-474, § 2(c), 100 Stat. 1213 (1986).

CFAAReport. 1984. House Committee Report on the Computer Fraud and Abuse Act, HR Rep 98-894 (1984).

CFAASuppI. 1984. 18 USC 1030a, Supp II 1984 CFAASuppI (1984).

CFAASuppII. 1986. CFAA of 1986, Pub L. No 99-474, 100 Stat. 1213, Supp II 1986.

Chip. 2006. V Chip Corp. v Lee. 438 F Supp 2d 1087 (ND Cal 2006).

Chong, D. 2000. *Rational Lives*. Cambridge: Harvard University Press.

CIC. 1776. U.S. CONST. art. II, § 2, cl. 1.

CIRA. 2021. The Cyber Incident Reporting Act of 2021, S.3875, September 2021, 117th Congress, 1st Session.

Citrin. 2006. IAC LLC v. Citrin, 440 F3d 418 (7th circ 2006).

Clancy, E. 1998. The Tragedy of the Global Commons. *Indiana Journal of Global Legal Studies* 5 (2): 601–610.

Clapper. 2013. Clapper v Amnesty International USA, 133 S. Ct. 1138 (2013).

Clausing, K. 2019. Open : The Progressive Case for Free Trade, Immigration, and Global Capital (Harvard University Press).

Clayton, M. 2013. Exclusive: Cyberattack Leaves Natural Gas Pipelines Vulnerable to Sabotage. *Christian Science Monitor.com*, February 27.

Coase, R. 1960. The Problem of Social Cost. *Journal of Law and Economics*. 3: 1–44.

Coats, D. 2019. Director of National Intelligence, Senate Select Committee on Intelligence. *Statement for the Record: Worldwide Threat Assessment of the US Intelligence Community*, January 29.

Cohen, J. 1998. Copyright and the Jurisprudence of Self-Help. *Berkeley Technology Law Journal* 13: 1089–2020.
Cohen, Z., and A. Marquard. 2019. Intelligence Warns China Is Using Student Spies to Steal Secrets. *CNN.com/politics*, February 2.
Comiling, K., and R. Sanchez. 2014. A Post Colonial Critique of Amartya Sen's Capability Approach. *Perspectives in the Arts and Humanities, Asia* 4 (1): 1–26.
Compuserve. 1997. Compuserve, Inc. v Cyber Promotions, Inc., 962 F.Supp. 1015.
Connoll, T., and Zeelenberg, M. 2002. Regret in Decision Making. *Current Direction in Psychological Science.* 11(6), 212–216.
Connolly, T., and Butler, D.J. 2006. Regret in Economic and Psychological Theories of Choice. *Journal of Behavioral Decision-making.* 19(2), 139–154.
Constitution. 1776a. U.S. CONST. art. 1, § 10, cl. 3.
Constitution. 1776b, U.S. CONST, art. I, § 8, cl. 10.
Constitution. 1776c. U.S. CONST, art. I, § 8, cl. 15.
Convention of the Sea. 1982. *Third United Nations Conference on the Law of the Sea, United Nations Convention on the Law of the Sea.* Opened for signature December 10, 1982, Art. 136, 21 ILM.
Cooter, R. 1988. Expressive Law and Economics. *Journal of Legal Studies* 27: 585–604.
Cooter, R. 2000. *The Strategic Constitution*, 129–134. Princeton: Princeton University Press.
Copes, H. 2003. Societal Attachment, Offending Frequency, and Techniques of Neutralization. *Deviant Behavior: An Interdisciplinary Journal* 24: 101–110.
Cordes, J. 2011. An Overview of the Economics of Cybersecurity and Cybersecurity Policy. *Report GW-CSPRI-2011-6.* June 1, 2011.
Cordesman, A. 2016. Anthony Cordesman. *Is There An Obama Doctrine?* March 10.
Corn, G.S. 1998. Presidential War Power: Do the Courts Offer Any Answers? *Military Law Review* 157: 180, 210.
Cortes, S. 2015. MLAT Jiu-Jitsu and Jur: Mutual Legal Assistance Treaties in Surveillance. *Richmond Journal of Law and Technology* 22 (1): 2–19.
Craig, A., and B. Valeriano. 2018. Realism and Cyber Conflict: Security in the Digital Age. In *Realism in Practice: An Appraisal.* Berlin: Springer.
Creative. 2004. Creative Computing v. GetLoaded.com, LLC, 386 F3d 930 (9th circ, 2004).
Crimm, N. 2005. Post-September 11 Fortified Anti-Terrorism Measures Compel Heightened Due Diligence. *Pace Law Review* 25: 203–224.
CSR. 2020. Brief Amici Curiae of Computer Security Researchers, Electronic Frontier Foundation, Center for Democracy & Technology, Bugcrowd,

Rapid7, Scythe, and Tenable in Support of Petitioner. https://www.eff.org/files/42020/07/08/19-783_eff_security_researchers_amici_brief_.pdf.
Curtiss. 1936. U.S. v Curtiss Wright Export Corp., 299 US 304, 319.
Cyanamid. 1973. United States v. American Cyanamid Co., 354 F. Supp. 1202 (S.D.N.Y. 1973).
D'Amato, A. 1984. Is International Law Really Law? *Northwestern University Law Review* 79: 1293–1314. Faculty Working Paper 103.
D'Amato, C. 2014. Rawls and Capabilities. *Virginia Tech. Papers*. vtechworks.edu.lib.vt.edu.
Dagan, H., and M. Heller. 2001. The Liberal Commons. *Yale Law Journal* 110 (4): 549–612.
D'Agostino, S. 2022. An Outrage, Silence and Finally Acknowledgment of a Cyber Attack. *Inside Higher Education.com*, August 18.
Damrosch, L. 1997. Enforcing International Law Through Non-Forcible Measures. Working Paper Series.
Daniels, N. 2000. Reflective Equilibrium and Justice as Political. In *The Idea of a Political Liberalism: Essays on Rawls*, ed. Victoria Davion and Clark Wolf. Lanham: Rowman and Littlefield.
Darwall, S. 2001. *Consequentialism*. Oxford: Blackwell Publishing.
Dau-Schmidt, K. 1990. An Economic Analysis of the Criminal Law as a Preference-Shaping Policy. *Duke Law Journal*. 1–38.
Davidson, A. 2021. Guido Calabresi's 'Other Justice Reasons.' *University of Chicago Law Review* 88: 1625–1631.
Davis. 2019. US v Davis, 139 S. Ct. 2319, 2325 (2019).
De Castro, R. 2011. The Obama Administration's (Neoliberal) Reengagement Policy in East Asia: Implications for U.D.-China Relations in the Twentieth Century. *Issues and Studies* 47 (2): 1–44.
Deeks, A. 2012. 'Unwilling or Unable': Toward a Normative Framework for Extraterritorial Self-Defense. *Virginia Journal of International Law* 52: 483–531.
Deibert, R. 2013. *Black Code: Surveillance, Privacy and the Dark Side of the Internet*. Toronto: McClellan and Stewart.
Dharmapala, D. and Garoupa, N. 2002. Penalty Enhancement for Hate Crimes : An Economic Analysis. *Economics Working Papers*. 2002, 12.
Diamond, L. 2022. All Democracy Is Global: Why America Can Shrink from the Fight for Freedom. *Foreign Affairs* 101 (5): 182–198.
Dinstein, Y. 1992. *War, Aggression and Self-Defense*, 3rd ed. Princeton: Princeton University Press.
Dipshaw, R. 2018. A Federal Policy Loophole Is Supporting the Hack-for-Hire Market. Can It Be Closed? *Slate.com*, June 20.

Diveloh, A., and R. Dishpande. 2004. *The Global Market: Developing a Strategy to Manage Across Borders*, ed. John A. Diveloh and Rahit Dishpande. New York: Wiley.

DMCA. 2003. *Digital Millennium Copyright Act* (17 USC 512).

DODNIST. 2011. Department of Defense Strategy for Operating in Cyberspace, NIST : Computer Security Resource Ctr. 7 (July 2011). https://csrc.nist.gov/CSRC/media/projects/ISPAB/documents/DOD-Strategy-for-Operating-In-cyberspace.pdf

DODWar. 2016. *US Department of Defense, Law of War Manual* (June 2015, updated in 2016).

DOJCFAA. 2022. *Department of Justice Guidelines*, 9-48-000—Computer Fraud and Abuse Act, May 19.

Donoff, J. 1992. Reconciling International Trade with Preservation of 'The Global Commons': Can We Prosper and Protect? *Washington and Lee Law Review* 49: 1408–1422.

Douligeris, C., and A. Mitrokotsa. 2004. DDoS Attacks and Defense Mechanisms: Classification and State of the Art. *Computer Networks* 44: 643–667.

Downie, J. 2015. What Donald Trump Doesn't Understand About Dealmaking. *Washingtonpost.Com*, October 26.

Drew. 2009. United States v. Drew, 259 F.R.D. 449 (C.D. Cal. 2009).

DTSA. 2016. Public Law 114–153 (114th Congress) 130 Stat. 376.

Dueck, C. 2018. Trump's Strategic Unpredictability, Its Pros and Cons. *NationalReview.com*, December 28.

Dueck, C. 2019. What's Conservative American Nationalism? *NationalReview.com*, October 21.

Dunkley, G. 2000. *The Free Trade Adventure: The WTO, the Uruguay Round and Globalization: A Critique*. London: Zed Books.

Dworkin, R. 1981. What is Equality? Part 2 : Equality of Resources. *Philosophy and Public Affairs* 10 (4): 283–345.

eBay. 2000. eBay, Inc, v, Bidders Edge, 100 F. Supp. 2d 1058 (ND Cal. 2000).

EEA. 1996. Economic Espionage Act of 1996, Section 201 (1)(B)(ii), 110 Stat 3492.

EEA. 2013. Economic Espionage Act of 1996 (October 11, 1996) See 18 U.S.C. § 1831.

EF. 2001. EF Cultural Travel BV v. Explorica, Inc, 274 F3d 577 (1st Circuit, 2001).

Egan. 1988. Dep't of the Navy v. Egan, 484 U.S. 518, 530.

Eichensehr, K. 2018. Today's Revolution: Cybersecurity and the International Order—A Review of Lucas Kello's the Virtual Weapon and International Order. *LawfareBlog.com*, February 8.

Eisenberg, M. 1999. Corporate Law and Social Norms. *Columbia Law Review* 99: 1253–1286.

Elgin, B., D. Lawrence, and M. Riley. 2012. Coke Gets Hacked and Doesn't Tell Anyone. *Bloomberg.com*, November 4.

Elkin-Koren, N. 2005. What Contracts Cannot Do : The Limits of Private Order Facilitating a Creative Commons. *Fordham Law Review*. 74(2): 375–422.

Ellickson, R. 1992. *Order Without Law*. New York: Oxford University Press.

Ellings, R. 2014. Five Chinese Military Officers Indicted. Now What? *Nat'l Bureau Asian Research*, May 22. http://nbr.org/downloads/pdfs/ETA/Ellings_FiveIndicted_052214.pdf.

Elster, J. 1989. *Nuts and Bolts for the Social Sciences*. Princeton: Princeton University Press.

Emory. 1994. Emory v. State, 647 A.2d 1243, 1254 (Md. Ct. Spec. App. 1994).

Enhanced. 2000. 18 U.S.C. $924©(1)(2000). Act of November 13, 1998, Pub. L. No. 105-386, $1, 112 Stat. 3469, 3469-70.

Epic. 2020. *Brief of Amici Curiae Electronic Privacy Information Center (Epic) and Fifteen Technical Experts and Legal Scholars in Support of Respondent*. On Writ of Certiorari to the United States Court of Appeals for the Eleventh Circuit Van Buren v. US.

Epstein, R. 2005. The Theory and Practice of Self-Help. *Journal of Economics and Policy* 1: 1–26.

Ericksen, T. 2007. *Globalization: The Key Concepts*. Oxford: Berg.

Explorica. 2001. EF Cultural Travel NV v. Explorica, Inc., 274 F.3d 577, 582, n. 10 (1st Circ 2001).

Export. 2020. Export Administration Regulations, 15 C.F.R. 703, et. seq.

Fairchild, A. 2013. *Science at the Border, Immigrant Media Inspection and the Modern Industrial Work Force*. Baltimore: John Hopkins University Press.

Farivar, C. 2012. CISPA Advances in House, as EFF Decries Bill's Revisions. *Ars Technica.com*, April 26. http://arstechnica.com/tech-policy/2012/04/cispa-advances-in-house-as-eff-decries-bills-revisions/.

Farzarnegan, M. 2012. Dark Side of Trade in Iran: Evidence from a Structural Equation Model. In *Illicit Trade and the Global Economy*, ed. Claudia Costa Sorti and Paul De Grauwe. Cambridge: MIT Press.

FBI. 2018. *FBI Report*, March 23.

FBIReportB. 2022. DOJ/FBI Report, March 9.

Federalist74. 1961. Federalist 74 (Hamilton). In *The Federalist*, ed. Jacob E. Cooke. Wesleyan.

Fehige, C. 2000. Justice Beyond Desires Norman Daniels, Reflective Equilibrium and Justice as Political. In *The Idea of a Political Liberalism: Essays on Rawls*, ed. Victoria Davion and Clark Wolf. Lanham: Rowman and Littlefield.

Feldman, A. 1998. Kaldor-Hicks Compensation. In *The New Palgrave Dictionary of Economics and the Law*, ed. Peter Newman. London: Springer.

Fennell, L. 2009. Commons, Anticommons and Semicommons. Reprinted in *Research Handbook on the Economics of Property Law*, ed. Kenneth Ayotte and Henry E. Smith. Cheltenham: Edward Elgar.

Finer, J., and R. Malley. 2018. Trump Is Right to Seek an End to America's Wars. *NY Times.com*, January 8.

Finkelstein, C. 2002. Death and Retribution. *Criminal Justice Ethics* 21 (2): 12–21.

Finnemore, M., and K. Sikkink. 1998. International Norm Dynamics and Political Change. *International Organization* 52: 887–898.

Finnis, J. 1998. On the Practical Meaning of Secularism. *Notre Dame Law Review* 73: 491–526.

Fisher, L. 1995. *Presidential War Power*. Cambridge: Harvard University Press.

Fisher, L., and D. Adler. 1998. The War Powers Resolution: Time to Say Goodbye. *Political Science Quarterly* 113: 1–14.

Flynn, M. 2018. Cyber Rebellions. *Journal of International Affairs* 71 (1.5): 107–114.

Ford, C. 1995. War Powers as We Live Them: Congressional-Executive Bargaining Under the Shadow of the War Powers Resolution. *Journal of Law and Policy* 11: 609–652.

Foster, J. 2010. Freedom, Opportunity and Well-Being. *OPHI Working Paper No. 175*. Oxford: Policy and Human Development Initiatives, April.

Foucault, M. 1977. *Discipline and Punish*, trans. Alan Sheridan. New York: Pantheon Books.

Foucault, M. 1991. Governmentality. In *The Foucault Effect: Studies in Governmentality*. New York: Harvester Wheatshaf.

Franchin, M., and E. Vida. 2019. Myths and Images in Global Climate Governance, Conceptualization and the Case of Brazil (1989–2019). *Revista Brasiliera De Politica International* 62 (2): 1–25.

Freeman, S. 2006. Frontiers of Justice: The Capabilities Approach vs Contractarianism. *Texas Law Review* 85 (2): 385–430.

French, P. 2001. *The Virtues of Vengeance*. Oxford: Oxford University Press.

Fretsedenas, P. 2012. *The Immigration Crucible: Transforming Race, Nation and the Limits of Law*. New York: Columbia University Press.

Fumerton, R., and K. Kress. 2001. Causation and the Law : Pre-emption, Lawful Sufficiency and Causal Sufficiency. *Law and Contemporary Problems* 64 (4): 83–90.

Gabcikovo. 1997. *Gabcikovo-Nagymaros Project*. Hung.

Gaillet, L., and L. Gigliemo. 2014. *Scholarly Publication in a Changing Academic Landscape: Models for Success*. New York: Springer.

Gandhi, D. 2017. Legal Hack Back: Balancing Risks and Benefits. *Information and Privacy Law Group (Au Law Blog)*, November 7.

Gasper, D. 2007. Adding Links, Adding Persons, and Adding Structures : Using Sen's Frameworks – A Response to Qizilbash, *Feminist Economics*. 13 (1) : 67–85.

Geers, K. 2011. *Strategic Cyber Security Blog*.

Georgia. 2016. Georgia Bill SB 315. Section (b.1)(2)(C).

Gholz, E., D. Press, and H. Sapolsky. 1997. Come Home America: The Strategy of Restraint in the Face of Temptation. *International Security* 21 (4): 5–48.

Gilman, B. 1992. *Response to Rep Benjamin Gilman and Senator Jesse Helms from Assistant Secretary of State Wendy Sherman*, July 21.

Gilroy, A. 2011. Access to Broadband Networks. *Net Neutrality Debate*. CRS 7-5700, R40616, March 15.

Girel, M. 2013. From Doubt to Its Social Articulation: Pragmatist Insights. In *Pragmatism and the Social Dimension of Doubt: Fresh Perspectives*, vol. 2.

Glen, C. 2017. *Controlling Cyberspace: The Politics of Internet Governance and Regulation*. Santa Barbara: ABC-CLIO.

Glennon, M. 1995. Too Far Apart: Repeal the War Powers Resolution. *University of Miami Law Review* 50: 17–35.

Goodman, M. 2015. *Future Crimes: Inside the Digital Underground and the Battle for Our Connected World*. New York: Random House.

Google. 2018. *Letter from Google and Microsoft to Governor Nathan Deal*, April 16.

Gordon, S., and E. Rosenbach. 2022. America's Cyber-Reckoning: How to Fix a Failing Strategy. *Foreign Policy* 101 (1): 10–21.

Gottschalk, M. 2006. *The Prison and the Gallows: The Politics of Mass Incarceration in America*.

Govella, K. 2019. Technology and Tension in the Global Commons. *Fletcher Security Review*: 38–44.

Gray, C. 2013. *Making Strategic Sense of Cyber Power: Why the Sky Is Not Falling*. Lulu.com.

Greenwald, G. 2013. NSA Collects Phone Records of Millions of Verizon Customers Daily. *Guardian* (London), June 5.

Greenwald, G., and E. MacAskill. 2013. NSA Prism Program Taps into User Data of Apple, Google and Others. *Guardian* (London), June 6.

Grewal, D., and Purdy, J. 2014. Introduction: Law and Neo-liberalism. *Law and Contemporary Problems* 77 (4): 1–23.

Gross, M. 2011a. Fed. Res. Div., Library of Congress. *Transnational Activities of Chinese Crime Organizations*.

Gross, M. 2011b. Enter the Cyber-Dragon. *VanityFair.com*, 9.

GTPB. 2016. Game Theory: Prison Breakthrough. *The Economist*. 5/12/2016.

Gundert, L., S. Chohan, and G. Lesnevich. 2018. Iran's Hacker Hierarchy Exposed: How the Islamic Republic of Iran Uses Contractors and Universities to Conduct Cyber Operations. *Recorded Future*. CTA_2018-0509.

Gupta, S. 1997. *The Political Economy of Globalization*. Berlin: Springer Science and Business Media.

Guy, S. 2013. Aaron Swartz 'Was Killed by the Government' Father Says at Funeral. *Chicago Sun Times*, January 15.

Ha. 1999. Ha v State, 892 p.2d 174, 191 (Alaska App.).

Haas, M. 2001. Prospect Theory and the Cuban Missile Crisis. *International Studies Quarterly* 45: 241–254.

Habermas, J. 1992. Citizenship and National Identity: Some Reflections on the Future of Europe. *Praxis International*. 12: 1–19.

Haenssgen, M., and P. Ariana. 2018. The Place of Technology in the Capability Approach. *Oxford Development Studies* 46 (1): 98–112.

Hafner, K. 1999. *Where Wizards Stay Up Late: The Origins of the Internet*. New York: Simon and Schuster.

Haggart, B. 2014. *Copyfight: The Global Politics of Digital Copyright Reform*. London: University of Toronto Press.

Haidt, J. 2007. When Morality Opposes Justice. *Social Justice Research* 20 (1): 98–108.

Hallaert, J., and L. Munro. 2009. Binding Constraints to Trade Expansion: Aid for Trade Objectives and Diagnostics Tools. OECD Trade Policy Working Paper, No. 94, December 2.

Hamdi. 2004. Hamdi v. Rumsfeld, 542 U.S. 507, 579–585.

Hannon, M. 2019. Skepticism: Impractical, Therefore Implausible. *Philosophical Issues*.

Hardin, G. 1968. The Tragedy of the Commons. *Science* 162: 1243–1252.

Hardin, R. 2003. *Liberalism, Constitutionalism and Democracy*. Oxford: Oxford University Press.

Harmelin. 1991. Harmelin v. Michigan, 501 US 957.

Hartz, L. 1948. *Economic Policy and Democratic Thought*. Ann Arbor: University of Michigan Press.

Hathaway, O., Rebecca Crootof, Philip Levitz, Haley Nix, Aileen Nowlan, William Perdue, and Julia Spiegel. 2012. The Law of Cyber-Attack. *California Law Review* 100: 817–856.

Hazlett, T. 2011. *The Fallacy of New Neutrality*. Encounter Books.

Heart. 2020. *Brief of Karen Heart and Anthony Volini of Ciplit as Amici Curiae in Support of Respondent*. On Writ of Certiorari to the United States Court of Appeals for the Eleventh Circuit, Van Buren v. US.

Heinze, E. 2008. The New Utopianism: Liberalism, American Foreign Policy, and the War in Iraq. Journal of International Political Theory 4 (1): 105–125.

Hekmat, F. 1994. The Origins and Evolution of Lex Talionis in the Abrahamic Faiths.

Heller, M., and R. Eisenberg. 2002. Can Patents Deter Innovation? The Anti-Commons in Biomedical Research. In *Perspectives on Property Law*, 3rd ed.,

ed. Robert C. Ellickson, Carol M. Rose, and Bruce A. Ackerman. New York: Aspen Law and Business.
Hendrich, K. 2001. The German States in Comparative Perspective in the Year 2000. In *Leviathan Transformed: Seven National States in the New Century*, ed. Theodore Kaplow, 51–55. Montreal: McGill-Queens University Press.
Hernon, P., R. Dugan, and J. Shuler. 2003. *US Government on the Web: Getting the Information You Need*. Englewood: Libraries Unlimited.
Hicks, J. 1939. The Foundations of Welfare Economics. *Economic Journal* 49: 696–724.
Hill, C. 1998. How Investors React to Political Risk. *Duke Journal of Comparative and International Law* 8: 283–312.
Himma, K. 2007. The Ethics of Hacking Back: Active Response to Computer Intrusion. In *Internet Security: Hacking, Counterhacking and Society*, ed. Kenneth Eimar Himma. Sudbury: Jones and Bartlett Publisher.
Hinduja, S. 2007. Neutralization Theory and Online Software Piracy: An Empirical Analysis. *Ethics and Information Technology* 9: 187–197.
Hobson, J. 1997. *The Wealth of States: A Comparative Sociology of Institutional and Political Change*. Cambridge: Cambridge University Press.
Hoffman. 1982. *Village of Hoffman Estates v Flipside*. Hoffman Estates, Inc., 455 US 489.
Hoffman. F. 2013. *Forward Partnership: A Sustainable American Strategy*. FPRI, Winter, 20–40 (reprint from *Foreign Policy*, January 2013).
Hoffman. F. G. 2016. Retreats Ashore: The Flaws of Offshore Balancing. Foreign Policy Research Institute (FPRI), July 5.
Hoffman, W., and S. Nyikos. 2008. Governing Private Sector Self-Help in Cyberspace: Analogies from the Physical World. *Carnegie Endowment for International Peace*.
Hohfeld, W. 1913. Some Fundamental Legal Conceptions as Applied in Judicial Reasoning. *Yale Law Journal* 23: 16–38.
Honeypots. 2012. European Union Agency for Network and Info. Sec., Proactive Detection of Security Incidents Ii: Honeypots 17.
Hope, M. 2019. The Brazilian Development Agenda Drives Amazon Devastation. *The Lancet/planetary Health* 3: 1–5.
Hsu, S. 2005. What Is a Tragedy and the Commons?: Over-Fishing and the Campaign Spending Problem. *Albany Law Review* 69: 75–94.
Hume, D. 1965. *A Treatise on Human Nature*, ed. L.A. Selby-Bigge. Oxford: Clarendon Press.
Hume, D. 1978. *An Enquiry Concerning Human Understanding*, ed. P.H. Nidditch. Oxford: Oxford University Press.
Humphrey, S. 1995. Regret Aversion or Event-Splitting Effects? More Evidence Under Risk and Uncertainty. *Journal of Risk and Uncertainty* 11: 263–274.

Huntington, S. 1993. Why International Primacy Matters. *International Security* 17 (4): 68–83.
Ides, A. 1984. Congress, Constitutional Responsibility and the War Power. *Loyola Law Review* 17: 599–652.
IEEP. 2014. 50 U.S.C. §§ 1701–1707 (2014) (?)International Emergency Economic Powers Act 50, U.S.C. §§ 1701–1707 (2014). IEEP.
IEPCapabilities. 2022. Global Ethics: Capabilities Approach. *Internet Encyclopedia of Philosophy*. ISSN 2161-0002. Last visited, 11/22/2022. https://iep.utm.edu/ge-capab/.
Ikenberry, J. 2011. The Future of the Liberal World Order. *Foreign Affairs* 90 (3): 56–68.
ILCDraft. 2000a. ILC Draft Articles of Responsibility, Article 51.
ILCDraft. 2000b. ILC Draft Articles of Responsibility, Article 49.
Intake. 2014. Memorandum from US Attorney to US attorney and Assistant Attorney Generals for the Criminal and National Security Divisions, "Intake and Charging Policy for Computer Crime Matters", September 22.
Intel. 2003. Intel Corp. v. Hamidi, 30 Cal 4th 1342 (2003).
Irons, B., and C. Hepburn. 2007. Regret Theory and the tyranny of Choice. *The Economic Record 83* (261): 191–203.
Jacob, A. 2017. *Cyber-Physical Systems, or the Cultural logic of Techno-Liberalism*. Master Thesis. McMaster University.
Jaggar, A. 2006. Reasoning About Well-Being: Nussbaum's Methods of Justifying the Capabilities. *The Journal of Political Philosophy* 14 (3): 301–322.
Janes. 1993. State v Janes, 121 Wash. 2d 220, 241-2 (1993).
Jayawal, V., W. Yurcik, and D. Doss. 2002. Internet Hackback: Counter Attacks as Self Defense or Vigilantism? *Proceedings of the IEEE International Symposium on Technology and Society*, June.
Jeffrey, J. 2016. US Military Engagement in the Broader Middle East. *Washington Institute for Near East Policy, Policy Focus*, April.
Jervis, R. 2009. Unipolarity: A Structural Perspective. *World Politics* 61 (1): 197–198.
John. 2010. *US v John*, 597 F.3d 263 (5th circ, 2010).
Johnson. 2015. Johnson v United States, 135 S. Ct 2551, 2556 (2015).
Johnson, D., and M. Schmitt. 2021. Durand Johnson and Michael Schmitt, Responsibility for Proxy Cyber Operations Under International Law. *The Cyber Defense Review* (Fall): 15–30.
Jones. 2001. US v Jones, 597 F3d 263 (5th Circ 2001).
Joyner, C. 1986. Legal Implications of the Concept of Common Heritage of Mankind. *International and Comparative Law Quarterly* 35: 190–201.
JSR. 2019. *Fundamental Research Security*. JSR, December 19–21.
Jun, J., and N. Kostyuk. 2021. *The Pros and Cons of Mandating Reporting from Ransomware Victims*, November 1.

Juneau, T. 2020. Obama, Biden and Realism in the Middle East. *ResponsibleStatecraft.com*, July 15.

Jupp, B. 2001. *Divided by Information? The Digital Divide and the Use of the New Meritocracy*. London: Demos.

Kahan, D., and M. Nussbaum. 1996. Two Concepts of Emotion in Criminal Law. *Columbia Law Review* 96 (2): 269–375.

Kaldor, N. 1939. Welfare Propositions in Economics and Interpersonal Comparisons of Utility. *Economic Journal* 49 (195): 549–552.

Kam, S. 2004. Intel Corp. v. Hamidi: Trespass to Chattels and a Doctrine of Cyber-Nuisance. *Berkeley Technology Law Journal* 19 (1): 427–453.

Kanbur, R. 2016. Capability, Opportunity, Outcome ... And Equality. *Working Paper*. Charles H. Dyson School of Applied Economic and Management, Cornell University. WP 2016—05. March 2016.

Kant, I. 1784. Idea for a Universal History With a Cosmopolitan Purpose, in *Kant : Political Writings* (Cambridge Texts in the History of Political Thought, 2d. ed) (Cambridge, Cambridge University Press, 1991).

Kaplow, L. 2007. Louis Kaplow, Primary Goods, Capabilities ... or Well-Being. *Philosophical Review* 116 (4): 603–632.

Kaplow, L., and S. Shavell. 2002. *Fairness Versus Welfare*. Cambridge: Harvard University Press.

Kassab, H. 2014. In Search of Cyber Stability: International Relations, Mutually Assured Destruction and the Age of Cyber Warfare. In *Cyberspace and International Relations*, ed. J.F. Kremer and B. Muller. Berlin: Springer-Verlag.

Kassab, H., and J. Rosen. 2018. *Illicit Markets, Organized Crime and Global Security*. Cham: Spring.

Katyal, N. 2001. Criminal Law in Cyberspace. *University of Pennsylvania Law Review*. 149: 1003–1114.

Kegley, C., and G. Raymond. 2021. Realism in the Age of Cyber Warfare. *Ethics and International Affairs*, April 26.

Kellner, D. 1985. Critical Thought, Marxism and Modernity. *Studies in Soviet Thought* 44 (2): 144–148.

Kesan, J., and Carol Hayes. 2012. Mitigative Counterstriking: Self-Defense and Deterrence in Cyberspace. *Harvard Journal of Law and Technology* 25(2) : 429–543.

Kostyuk, N. 2022. The Digital Prisoner's Dilemma: Challenges and Opportunities for Cooperation. *Worldwide Cybersecurity Initiative*.

Kirchick, J. 2017. America in Retreat: Why Neo-Isolationism Exploded Under Obama and What We Can Do About It. *DailyBeast.com*, July 12.

Kitfield, J. 2016. Trump's 'America First' Neo-Isolationism. *Yahoo.com*, April 26.

Kivisto, P., and T. Faist. 2009. *Beyond a Border: The Causes and Consequences of Contemporary Immigration*. London: Sage Press.

Klare, M. 2018. 'Overmatch' Is Overkill: The Pentagon's Strategy for Global Supremacy Could Bankrupt the Country and Spark a World War. *TheNation.com*, December 20.

Kluben. 2009. Kluben Skahan and Associates, Inc. v Cardogen, Clark and Assoc. Inc., No 08-CV-1529, 2009 WL 466812 at 8 (ND Ill. February 2009).

Krauss, M. 1999. Property Rules vs. Liability Rules, Paper 3800 (GMU Law School), JEL Series. *International Encyclopedia of Law and Economics*.

Kreps, S., and J. Schneider. 2019. Escalation Firebreaks in the Cyber, Conventional and Nuclear Domains: Moving Beyond Effects-Based Logics. *Journal of Cybersecurity* 5 (1): 1–11.

Kumari, A., Md Zakirul Alam Bhuiyan, Jigyasa Namdeo, Shipra Kanaujia, Ruhul Amin, and Satyanarayana Vollala. 2019. Ransomware Attack Protection: A Cryptographic Approach. In *Security, Privacy, and Anonymity in Computation, Communication, and Storage*, ed. G. Wang et al., 15–25. Atlanta, GA: Springer.

Kundnani, H. 2017. What Is the Liberal International Order? German Marshall Fund Policy Essay, No. 17.

Kupchan, C., and P. Trubowitz. 2021. The Home Front: Why an Internationalist Foreign Policy Needs a Stronger Domestic Foundation. *Foreign Affairs* 100 (3): 91–101.

Kuran, T. 2013. Moral Overload and Its Alleviation. In *Economics, Values and Organization*, ed. Avner Ben-Her and Louis Putterman. New York: Cambridge University Press.

Kymlicka, W. 1995. *Multicultural Citizenship: A Liberal Theory of Minority Rights* (Oxford, Oxford University Press).

Laham, N. 2008. *Ronald Reagan and the Politics of Immigration Reform*. Greenwich: Greenwich Publishing.

Lal, D. 2004. *In Defense of Empires*. Washington, DC: AEI Press.

Lambrects, W., and S. Situba. 2016. *Last-Mile Internet Access: Emerging Energy Economies*. Berlin: Springer.

Larson, D. 2018. The Perils of Trump's Iran Obsession. *The American Conservative*, August 13.

Laruffa, F. 2018. Towards a Post Neoliberal Society? Social Investment Versus Capability Approach. *Momentum Quarterly* 7 (4): 171–187.

Law, J., and Martin, E. 2013. *A Dictionary of Law*, 7th ed. London: Oxford University Press.

Leigh, M. 1975. *Letter from State Department Legal Advisor Monroe Leigh and Dod General Counsel Martin Hoffman*, June 3. To House Foreign Affairs Committee Chair Clement Zablocki, and Office of Legal Counsel (Carter), 4a Op. O.L.C.

Lessig, L. 2002. The Architecture of Innovation. *Duke Law Journal* 51 (6): 1783–1801.

Lessig, L. 2013. "Prosecutor Is a Bully". *Bloombergview.com*, January 17.
Lessnoff, M. 1990. *Social Contract Theory*, ed. Michael Lessnoff. New York: New York University Press.
Levey, J. 2010. *Illegal Immigration and Amnesty: Open Borders and National Security*. New York: The Rosen Publishing Group.
Levy, J. 1996. Loss Aversion, Framing and Bargaining: The Implication of Prospect Theory for International Conflict. *International Political Science Review* 17 (9): 179–195.
Li, S., and I. Alon. 2020. China's Intellectual Property Provocation: A Political Economy View. *Journal of International Business Policy* 3: 60–72.
Lichtman, D. 2004. How the Law Responds to Self-Help (John M. Olin Program in Law and Economics). Working Paper No. 232.
Lin, H. 2010. Offensive Cyber Operations and the Use of Force. *Journal of National Security Law and Policy* 4: 63–86.
Lindner, R. 1939. An Experimental Study in Anticipation. *American Journal of Psychology* 51 (2): 253–260.
Liptak, K. 2021. Biden Repeats Call to NATO Allies to Defend Democracy Against Autocracy. *CNN.com*, June 14.
Lipton, R. 1987. Altruism and Sympathy in Human Ethics. *Austrakasian Journal of Philosophy* 65 (1): 18–32.
Lipton, R. 1990. Beyond Skepticism, Foundationalism and the New Fuzziness: The Role of Wide Reflective Equilibrium in Legal Theory. *Cornell Law Review* 75 (4): 810–877.
Lissner, R., and Mira Rapp Hooper. 2020. An Open World: How America Can Win The Content For the Twenty First Century Order (Yale University Press).
Low, L. 2003. Linda Low, Multilateralism, Regionalism, Bilateral and Cross-Regional Free Trade Agreements: All Paved with Good Intentions for Asean? *Asian Economic Journal* 17 (1): 65–86.
LVRC. 2009. LVRC Holdings, LLC v. Brekka, 581 F.3d 1127 (9th Circ. 2009).
MacCoun, R.J. 1993. Drugs and the Law: A Psychological Analysis of Drug Prohibition. *Psychological Bulletin* 113 (3): 497–512.
Maerz, S., Anna Luhrmann, Jean Lachapelle, and Amanda Edgell. 2020. Worth the Sacrifice? Illiberal and Authoritarian Practices During Covid-19. Working Paper (Series 2020: 110). The Varieties of Democracy Institute.
Manjikan, M. 2021. The United States: A Declining Hegemon in Cyberspace? In *Routledge Companion to Global Cyber-Security Strategy*, ed. Scott Romaniuk and Mary Manjikan. London: Routledge.
Manning, J.F. 2011. Separation of Powers as Ordinary Interpretation. *Harvard Law Review* 124: 1939–2041.
Mansell, R. 2002. From Digital Divides to Digital Entitlements in Knowledge Societies. *Current Sociology* 50 (3): 407–426.

Mantiw, J. 2014. *Drugs on the Dark Net: How Cryptomarkets Are Transforming Global Trade in Illicit Drugs*. Houndmills, Basingstoke, Hampshire, and New York: Palgrave Macmillan.
Manzanec, B., and B. Thayer. 2014. *Deterring Cyber Warfare: Bolstering Strategic Stability in Cyberspace*. Berlin: Springer.
Maranganti, A. 2012. Tragedy of the Commons Revisited (I): Granite Quarrying in Telangana. *Economic and Political Weekly* 47 (42): 10–13.
Maravall, J. 2008. The Myth of Authoritarian Advantage. *Journal of Democracy* 5 (4): 17–31.
Material. 2006. 18 U.S.C. § 2339A (2006) (terrorists) and § 2339B.
May, R. 2021. *Moralism and American Foreign Policy: Biden's Greatest Challenge to Revising US Primacy*. Atlas Institute of International Affairs Blog.
Maynz, R. 2002. National States and Global Governance, VII Inter-American Congress of Clad on State and Public Reform, October 8–11.
Mazarr, M. 2022. What Makes a Power Great: The Real Drivers of Rise and Fall. *Foreign Affairs* 110 (4): 52–63.
McAdams, R. 1997. The Origin, Development, and Regulation of Norms. *Michigan Law Review* 96 (2): 338–433.
McCormick, K. 1986. Towards a Definition of Waste in Economics: A Neoinstitutional Approach. *Review of Social Economy* 44 (1): 80–92.
McCormick, M. 2020. Value Beyond Truth-Value: A Practical Response to Skepticism. *Synthese* 198 (9): 8601–8619.
McCulloch. 1819. McCulloch v. Maryland, 17 U.S. (4 Wheat.) 316, 413–414, 420–421 (1819).
McDonnell. 2016. McDonnell v US, 136 S. Ct. 2355, 2372-3 (2016).
Mearsheimer, J. 2011. Imperial by Design. *The National Interest* 111: 16–34.
Mearsheimer, J., and S. Walt. 2016. The Case for Offshore Balancing. *Foreign Affairs* 95 (4): 70–83.
Merritt. 2015. Merritt Hawkins and Assoc. LLC. v. Gresham, 79 F.Supp. 3rd 625 (ND Tex 2015).
Meyer, T. 2017. Saving the Political Consensus in Favor of Free Trade. *Vanderbilt Law Review* 70 (3): 985–1026.
Michelman, F. 1982. Ethics, Economics and the Law of Property. In *Nomos Xxiv: Ethics, Economics and the Law*, ed. J.R. Pennock and J.W. Chapman, 3–40. New York: New York University Press.
Miller, H. 1934. Treaties and Other International Acts of the United States of America. vol. 4, ed. Hunter Miller. Government Printing Office.
Miller, D. 1999. *Bounded Citizenship in Cosmopolitan Citizenship* (eds. Kimberly Hutchings & Roland). (London, Palgrave Macmillan).
Miller, W. 2005. *Eye for an Eye*. New York: Cambridge University Press.

Mitsilegas, V. 2012. Immigration Control in an Era of Globalization: Vetting Foreigners, Weakening Citizens, Strengthening the State. *Indiana Journal of Global Legal Studies* 19 (1): 3–60.

Moloney, D. 2012. *National Insecurities: Immigrants and US Deportation Policy Since 1882*. Chapel Hill: University of Carolina Press.

Monten, J. 2005. The Roots of the Bush Doctrine: Power, Nationalism, and Democracy Promotion in U.S. Strategy. *International Security* 29 (4): 112–156.

Moon, I. 1994. Modeling Programmable Logic Controllers for Logic Verification. *IEEE Control Systems Magazine* 14 (2): 53–59.

Moore, R., and E. McMullan. 2009. Neutralizations and Rationalizatoins of Digital Piracy: A Quantitative Analysis of University Students. *International Journal of Cyber Criminology* 3 (1): 441–451.

Morris. 1991. US v. Morris, 928 F2d 504 (2d Cir. 1991).

MPC. 2005. *Use of Force in Self-Protection*. Model Penal Code (Thomson/West, 2005).

Murdoch, S. 2013. US Plans to Respond to Chinese Hacking. *The Australian*, February 21.

Murphy, J. 2003. *Getting Even: Forgiveness and Its Limits*. New York: Oxford University Press.

Nakashima, E., and J. Warrick. 2012. Stuxnet Was Work of U.S. and Israeli Experts, Officials Say. *WASHPOST.com*, June 3.

NCSS. 2016. National Cyber Security Strategy, 2016–2021, UK Government, 12. See also Gordon Carera, *UK Moves to Active Cyber Defence*, *BBCNews.com*, September 13.

NDAA. 2017. National Defense Authorization Act of 2015, Pub. L. No. 113-291, § 1637, 128 Stat. 3292.

Neal, P. 1997. *Liberalism and Its Discontents*. New York: New York University Press.

Neuman, S. 2012. As the Worm Turns: Cybersecurity Expert Tracks Blowback from Stuxnet. *NPR.com*, June 1. http://www.npr.org/blogs/thetwo-way/2012/06/01/154162121/as-the-worm-turns-cybersecurity-expert-tracks-blowback-from-stuxnet.

Nixon, R. 1973. Veto of the War Powers Resolution. *Public Papers* 5: 893 (October 24).

Nosal1. 2012. (Nosal I) US. v. Nosal, 676 F 3d 854 (9th Cir. 2012) (en banc).

Nozick, R. 1974. *Anarchy, State and Utopia*. New York: Basic Books.

Nussbaum, M. 2000. *Women and Human Development: The Capabilities Approach*. Cambridge: Cambridge University Press.

Nussbaum, M. 2009. *Frontiers of Justice: Disability, Nationality and Species Membership*. Cambridge: Harvard University Press, Belknap Press.

Oberdiek, J. 2004. The Ethics of Risk Regulation: Towards a Contractualist Re-orientation. *Rutgers Law Journal* 36: 199–205.
O'Brien, R. 2005. Amnesty and International Law. *Nordic Journal of International Law* 74 (2): 261–278.
O'Malley, P. 1996. Risk and Responsibility. In *Foucault and Political Reason: Liberalism, Neo-Liberalism and Rationalities of Government*. London: University College of London Press.
Olson, M. 1965. *The Logic of Collective Action*. Cambridge: Harvard University Press.
Oppenheim, L. 1912. *International Law: A Treatis*. Oxford: Oxford University Press.
Ostrom, E., Roy Gardner, and James Walker. 1984. *Rules, Games and Common Pool Resources*. Ann Arbor: The University of Michigan Press.
Owen, J. 2000. *Liberal Peace, Liberal War: American Politics and International Security*. Ithaca, NY: Cornell University Press.
Ozkaya, E., and R. Islam. 2019. *Inside the Dark Web*. Boca Raton: CRC Press.
Panaganya, A. 2019. *Free Trade and Prosperity: How Openness Helps the Developing Countries Grow Richer and Combat Poverty*. New York: Oxford University Press.
Pangang. 2012. United States v. Pangang Grp. Co., Ltd., 879 F. Supp. 2d 1052, 1056 (N.D. Cal. 2012).
Parfit, D. 2000. Equality or Priority. In *The Ideal of Equality*, ed. M. Clayton and A. William. Basingstoke: Macmillan Press.
Paasi, A. 2009. Bounded Spaces in a Borderless World? Border Studies, Power and the Anatomy of Territory. *Journal of Power* 2 (2): 213–224.
Paternoster, R. 1987. The Deterrent Effect of the Perceived Certainty and Severity of Punishment: A Review of the Evidence and Issues. *Justice Quarterly* 4 (2): 173–217.
Pathology. 1994. Pathologies of Rational Choice Theory: A Critique of Applications in Political Science. Edited by Donald Green and Ian Shapiro. New Haven: Yale University Press.
Pell, S., and J. Finocchiaro. 2017. The Ethical Imperative for a Vulnerability Equities Process and How the Common Vulnerability Scoring System Can Aid the Process. *Connecticut Law Review* 49 (5): 1549–1572.
Peoria. 1983. Gonzales v City of Peoria, 722 F.2d 468 (9th Cir. 1983).
Philippon, T. 2019. The Great Reversal : How America Gave Up on Free Markets (Harvard University Press.)
Phillips. 2007. US. v. Phillips, 477 F.3d 215 (5th Circ. 2007).
Pildes, R. and Sunstein, C. 1995. Reinventing the Regulatory State. Reinventing the Regulatory State. *University of Chicago Law Review*. 62 : 1–129.
Plott, C., and K. Zeiler. 2005. The Willingness to Pay-Willingness to Accept Gap and the Endowment Effect: Subject Misconceptions, and Experimental

Procedures for Eliciting Valuations: Reply. *American Economic Review* 101 (2): 1012–1028.
Pogge, T. 1992. Cosmopolitanism and Sovereignty. *Ethics*. 103(1) : 48–75.
Pogge, T. 2002. Can Capability Theory Be Justified? *Philosophical Topics* 30 (2): 167–228.
PoliceBuren. 2020. Brief of the Federal Law Enforcement Officers Association as Amicus Curiae in Support of Respondent, On Writ of Certiorari to the United States Court of Appeals for the Eleventh Circuit, Van Buren v. US.
Polinsky, M. and Shavell, S. 1998. The Fairness of Sanctions : Some Implications for optimal Enforcement Policy. Olin Program in Law and Economics, *Working Paper* 167. September 1998.
Posen, B. 2014. *Restraint: A New Foundation for US Grand Strategy*. Ithaca, NY: Cornell University Press.
Posner, R. 1981. *The Economics of Justice*. Cambridge: Harvard University Press.
Posner, R. 1992 [2001]. *Frontiers of Legal Theory*. Cambridge: Harvard University Press.
Posner, E. 1998. Symbols, Signals and Social Norms in Politics and the Law. *The Journal of Legal Studies* 27: 765–798.
Posner, E. 2000. *Law and Social Norms*. Cambridge: Harvard University Press.
Posner, R., and E. Rasmussen. 2000. Creating and Enforcing Norms, with Special Reference to Sanctions. Working Paper No. 96, March.
Posner, E., and A. Vermeule. 2007. *Terror in the Balance: Security, Liberty and the Courts*. New York: Oxford University Press.
Potter, R. 2016. Cyber Warfare as a Double-Edged Sword for Authoritarian States. *RealClear Defense.com*, August 2.
Pozen, D. 2018. David Pozen, Populist Constitutions. *University of Chicago Law Review* 85: 521–543.
PPD20. 2012. United States Government. Presidential Policy Directive 20/PPD-20. *US Cyber Operations Policy*, October. https://fas.org/irp/offdocs/ppd/ppd-20.pdf.
Prakash, S., and M. Ramsey. 2001. The Executive Power over Foreign Affairs. *Yale Law Journal* 111: 231–356.
Preemption. 2002. US Dept. of Justice, Office of Legal Counsel. *Memo for the Attorney General Re: Non-Preemption of the Authority of States and Local Law Enforcement Officers to Arrest Aliens for Immigration Violations*, April 3.
Prince, B. 2012. BlackHat: Hacking Back: The Best Defense May Not Be the Best Offense. *Security Week*, July 27.
Prize. 1863. The Prize Cases, 67 U.S. (2 Black) 635, 668 (1863).
Promoting. 2002. *Government's Role in Promoting the Future of the Telecommunications Industry and Broadband Development* (October 1), S. Hrg. 107-1148, Hearing Before the Committee on Commerce, Science and Transportation, US Senate, 107th Congress, Second Session.

Protocol_I. 1977a. *Protocol Additional to the Geneva Convention of 12 August 1949, and Relating to the Protection of Victims of International Armed Conflict*, opened for signature December 12th, Article 51(2), 1125 UNTS 3 (Protocol 1), Article 35(2) (rejects superfluous injury and unnecessary suffering).

Protocol_1. 1977b. *Protocol Additional to the Geneva Conventions of 12 August 1949, and Relating to the Protection of Victims of International Armed Conflicts (Protocol I)*, adopted June 8, 1977, art. 51(1), 1125 U.N.T.S. 3, 26, (lawful combatants); art. 51(3), 1125 U.N.T.S. at 26 (unlawful combatants).

Przeworski, A. 1991. *Democracy and the Market*. Cambridge: Cambridge University Press.

Ptweet. 2017. Presidential Tweet on July 26, at 5:55AM.

Pyysiainen, J., D. Harpin, and A. Guilfoyle. 2017. Neo-Liberal Governance and Responsibilization of Agents: Reassessing the Mechanisms of Responsibility-Shift in Neo-Liberal Discursive Environments. *Distinkton: Journal of Social Theory* 18 (2): 1–21.

Qizilbash, M. 2005. Incommensurability and the First Fundamental Welfare Theorem. *Oxford Economic Papers*. 57, 664–673.

Randall. 1976. US v. Randall, 104 Daily Wash. L. Rep. 2249, DC Superior Court.

Rao, N. 2011. Public Choice and International Law Compliance: The Executive Branch Is a 'They Not an It.' *Minnesota Law Review* 96: 194–280.

Rawls, J. 1972. *A Theory of Justice*. London: Oxford University Press.

Rawls, J. 1999a. *The Law of Peoples*. Cambridge: Harvard University Press.

Rawls, J. 1999b. Reply to Alexander and Musgrave. In *John Rawls, Collected Papers*, ed. Samuel Freeman. Cambridge: Harvard University Press.

Rawls, J. 1999c. A Kantian Conception of Equality. In *John Rawls, Collected Papers*, ed. Samuel Freeman. Cambridge: Harvard University Press.

Rawls, J. 2001. *Justice as Fairness: A Restatement*, ed. Erin Kelly. Cambridge: Harvard University Press.

Reddy, A., and A. Daoud. 2018. Entitlements and Capabilities. *SSRN Electronic Journal*. https://doi.org/10.21391.ssrn.320924.

Register. 2000. Register.com, Inc. v. Verio.com, Inc., 126 F. Supp. 2d 238 (SDNY 2000), aff'd 356 F.3d 393.

Resilience. 2014. *Ensuring the Security, Stability, Resilience and Freedom of the Global Internet, Hearing Before the Subcommittee on Communication and Technology of the Committee on Energy and Commerce*. House of Representatives, 113th Congress, Second Session, April 2.

Restatement_Agency. 1992. Section 112 of the Second restatement of Agency.

RestatementForeign. 1987. Restatement (Third) of the Foreign Relations Law of the United States, § 905.

Rheingold, H. 2002. *Smart Mobs: The Next Social Revolution*. Cambridge: Basic Books.

Richardson, M. 2013. CISPA Remains Fatally Flawed Secret Committee Markup. *ACLU.com*, April 12.

Rid, S. 2012. Kashmir: The Prisoner's Dilemma for India and Pakistan. *Peace Prints: South Asian Journal of Peacebuilding* 4 (2). http://www.wiscomp.org/peaceprints.htm

Rinard, S. 2019. Pragmatic Skepticism, Philosophy and Phenomenological Research. *Rutgers Epistemology Conference*.

Rivera, I. 2012. Global Age Cosmopolitanism. *International Journal of Business and Globalization* 9 (1): 90–105.

Roberts, A. and N. Lamp. 2022. *Six Faces of Globalization : Who Wins, Who Loses, and Why It Matters*. (Cambridge, Harvard University Press).

Robinson. 1992. Robinson v State, 417 S.E.2d 88, 91 (S.C. 1992).

Robinson, P., S. Garvey, and T. Ferrari, 2009. Measurable Provocation and Self-Defense: Recognizing the Distinction Between Act Reasonableness and Emotion Reasonableness and Criminal Law. In *Conversations*, ed. Paul Robinson, Stephen Garvey, and Timberway Ferrari. Oxford: Oxford University Press.

Rodrigues, A. 2020. If the Law Can Allow Takebacks, Shouldn't It Also Allow Hackbacks? *Marquette Intellectual Property Law Review* 24 (1): 1–13.

Rodriguez. 2010. US v. Rodriguez, 628 f3d 1258 (11th Circ. 2010).

Rodrik, D., and S. Walt. 2022. How to Build a Better Order: Limiting Great Power Rivalry in an Anarchic World. *Foreign Affairs* 101 (5): 142–155.

Roemer, J. 1996. *Theories of Distributive Justice*. Cambridge: Harvard University Press.

Roosevelt, T. 1920. *Theodore Roosevelt: An Autobiography*. New York: Charles Scribner's Sons.

Rorty, R. 1979. *Philosophy and the Mirror of Nature*. Princeton: Princeton University Press.

Rotunda, K. 2009. A Compartive Historical Analysis of War Time Procedural Protections and Presidential Powers: From the Civil War to the War on Terror. *Chapman Law Review* 12: 449–487.

Roubini, N. 2021. Biden's Neo-Populist Economic Agenda. *Projectsyndicate.org*, August 3.

Rowe, E. 2016. RATs, TRAPs, and Trade Secrets. *Boston College Law Review* 57: 381–426.

Rushkoff, B. 1984. Note, A Defense of the War Powers Resolution. *Yale Law Journal* 93 (7): 1330–1354.

Ruys, T. 2016. Sanctions, Retorsion and Counter Measures: Conceptualizing and Interpreting a Legal Framework. ssrn.com/abstract=2760853.

Saideman, S. 2001. *The Ties That Divide: Ethnic Politics, Foreign Policy and International Conflict*. New York: Columbia University Press.

Sales, N. 2013. Regulating Cyber-Security. *Northwestern University Law Review* 107 (4): 1503–1568.

Salter, M. 2016. *Crime, Justice and Social Media*. Hertfordshire: Taylor and Francis.

San, S., M. Bastug, and H. Basil. 2021. Crisis Management in Authoritarian Regimes: A Comparative Study of Covid-19 Responses in Turkey and Iran. *Global Public Health* 16 (4): 485–501.

Sandvig. 2020. Sandvig v Barr, No. 16-1368 (JDB), 2020 WL 1494065, D.D.C, March 27.

Sanger, D. 2012. *Confront and Conceal: Obama's Secret Wars and Surprising Use of American Power*. New York: Penguin Random House.

Sanger, D., and N. Perlroth. 2019. US Escalates Online Attacks in Russia's Power Grid. *N.Y. Times*, June 15.

Santana-Garcia. 2001. US v Santana-Garcia, 264 F.3d 1188 (10th Circ. 2001).

Savage, C. 2011. 2 Top Lawyers Lose Argument on War Power. *N.Y. Times*, June 18.

Savage, C., and M. Landler. 2011. War Powers Act Doesn't Apply for Libya, Obama Says. *N.Y. Times*, June 15.

SCAROA. 2022. Study on Cyber-Attack Response Options Act, S.2292-117th Congress (2021–22).

Scheidel, D. 2014. *Profits From the Data Economy: Understanding the Rules of Consumers, Innovators and Regulators in a Data Driven World*. London: Pearson Education.

Scheppele, K. 2019. The Opportunism of Populists and The Defense of Constitutional Liberalism. *German Law Journal* 20 (3): 314–331.

Schlesinger, A. 1973. *The Imperial Presidency*. Boston: Houghton Mifflin.

Schmitt, M. 2014. Below the Threshold Cyber Operations: The Counter Measures Response Option in International Law. *Virginia Journal of International Law* 54 (3): 697–732.

Schutte, S. 2013. *Cached: Decoding the Internet in Global Popular Culture*. New York: New York University Press.

Schwabach, A., and A. Cockfield. 2009. *International Law and Institutions*. EOLSS Publications.

Seidel, J. 2019. Chinese Students Are Stealing Intellectual Property for Beijing, Warns US Intelligence Agency. *News.com.au*, February 3.

Seidman, L. 2014. Constitutional Skepticism: A Recovery and Preliminary Evaluation. ssrn.com/abstract=2387031.

Sen, A. 1970. The Impossibility of a Paretian Liberal. *Journal of Political Economy* 78: 152–157.

Sen, A. 1979. Utilitarianism and Welfarism. *Journal of Philosophy* 76 (9): 463–489.

Sen, A. 1987. *The Standard of Living*. Cambridge: Cambridge University Press.

Sen, A. 1988. The Concept of Development. In *Handbook of Development Economics*, vol. 1, ed. H. Chenery and T.N. Srinivasan. Amsterdam: Elsevier.

Sen, A. 1992. *Inequality Reexamined*. Cambridge: Harvard University Press.

Sen, A. 1999a. *Development and Freedom*. Oxford: Oxford University Press.

Sen, A. 1999b. Equality of What? In *Equal Freedom*, ed. Stephen Darwall. Ann Arbor: University of Michigan Press.

Sen, A. 2002. *Rationality and Freedom*. Cambridge: Harvard University Press.

Sen, A. 2004. Elements of a Theory on Human Rights. *Philosophy and Public Affairs* 32 (4): 315–356.

Sen, A. 2005. Human Rights and Capabilities. *Journal of Human Development* 6 (2): 151–166.

SenateReport. 1983. S. Rep. 98-242 at 7, and Pub. L. No. 98-119, section 2 (b), 57 Stat. 805 (1983).

Setear, J. 1997. Responses to Breach of a Treaty and Rationalist International Relations Theory: The Rules of Release and Remediation in the Law of Treaties and the Law of State Responsibility. *Virginia Law Review* 83 (1): 1–126.

Shackleford, S. 2014. *Managing Cyber Attacks in International Law, Business and Relations: In Search of Cyber Peace*. Cambridge: Cambridge University Press.

Shackleford, Scott J., Charoen Danuvasin, Tristen Waite, and Nancy Zhang. 2019. Rethinking Active Defense: A Comparative Analysis of Proactive Cybersecurity Policymaking. *University of Pennsylvania Journal of International Law* 41 (2): 377–427.

Shantz, J. 2010. *Racism and Borders: Representation, Repression and Resistance*, ed. Jeff Shantz. New York: Algora Publishing.

Sharkey, C. 2013. Trespass Torts and Self-Help for an Electronic Age. *Tulsa Law Review* 44 (4): 677–706.

Shaw, W. 2000. Between Act and Rule: The Consequentialism of G.E. Moore. In *Morality, Rules and Consequences*, ed. B. Hooker, E. Mason, and D. Miller. Lanham: Rowman and Littlefield.

Sherwin, E. 2003. Compensation and Revenge. *San Diego Law Review* 40: 1387–1413.

Shields, R. 1996. *Cultures of the Internet: Virtual Spaces, Real Histories and Living Bodies*. London: Sage.

Shifrinson, Joshua, and Sameer Lalwani. 2014. It's a Common's Misunderstanding: The Limited Threat to American Command of the Commons. In *A Dangerous World?: Threat Perception and Us National Security*, ed. Christopher Preble and John Mueller, 229–244. Washington: Cato Institute.

Shuman, B. 2001. *Issues for Libraries and Information Science in the Internet Age*. Englewood: Libraries Unlimited.
Singh, K. 2005. *World Trade Organizations and the Third World*. New Delhi: Mittel Publications.
Soros, G. 2019. In Defense of Open Society (2019, Hachette Book Group)
Skelrov, M. 2009. Solving the Dilemma of State Responses to Cyberattacks: A Justification for the Use of Active Defenses Against States Who Neglect Their Duty to Prevent. *Military Law Review* 201: 1–85.
Skrabek, Q. 2008. *William Mckinley, Apostle of Professionalism*. New York: Algora Publishing.
Smith, H. 2005. Self-Help and the Nature of Property. *Journal of Law, Economics and Policy* 1 (1): 69–107.
Snyder, J. 2013. *Myths of Empire: Domestic Politics and International Ambitions*. Ithaca, NY: Cornell University Press.
Staake, T., and E. Fleish. 2008. *Counterfeit Trade, Best Practice Strategies and Management Toolbox*. Berlin: Springer.
Standing, A. 2015. Mirage of Pirates: State-Corporate Crime in West Africa's Fisheries. *State Crime Journal* 4 (2): 175–197.
Stapleton, J. 1995. Duty of Care: Peripheral Parties and Alternative Opportunities for Deterrence. *Law Quarterly Review* 111 (2): 301–345.
Stark, H. 2017. A Bipartisan Bill to Strengthen Cyber Security for the Internet of Things. *Forbes.com*, August 20.
Starmer, C., and R. Sugden. 1993. Testing for Juxtaposition and Event-Splitting effects. *Journal of Risk and Uncertainty* 6: 235–254.
Stein, L. 2006. *Speech Rights in America: The First Amendment, Democracy and the Media*. Urbana: University of Illinois Press.
Stein, M. 2009. Nussbaum: A Utilitarian Critique. *Boston College Law Review* 50 (2): 489–531.
Stephen, J. 1874. *Law, Equality, Fraternity*, 2nd ed. London: Elder and Co.
Sterk, S. 2008. Property Rules, Liability Rules, and Uncertainty About Property Rights. *Michigan Law Review* 106 (7): 1285–1335.
Stevens. 2010. US v Stevens, 558 US 460, 480 (2010).
Stiegler, Z., and D. Sprumont. 2013. Framing the Net Neutrality Debate. In *Regulating the Web: Network Neutrality and the Fate of the Open Internet*, ed. Zach Stiegler. Lanham: Rowman and Littlefield.
Stiglitz, J., and A. Charlton. 2000. Aid for Trade. *International Journal of Development Issues* 5 (2): 1–41.
Strauss, P.L. 2011. A Softer Formalism. *Harvard Law Review* 124: 55–61.
Stromseth, J. 1996. Understanding Constitutional War Powers Today: Why Methodology Matters. *Yale Law Journal* 106: 845–915.

Stuxnet. 2010. Iran Confirms Stuxnet Worm Halted Centrifuges. *CBSNEWS.com*, November 29. http://www.cbsnews.com/stories/2010/11/29/world/main7100197.shtml.

Sunstein, C. 1996. Social Norms and Social Roles. *Columbia Law Review* 96 (4): 903–968.

Swan, A. 2012. *Policy Guidelines for the Development and Promotion of Open Access*. UNESCO.

Swygert, M., and K. Yanes. 1998. Focus on Fairness, Efficiency and the Law: The Integration of Fairness into Efficiency. *Washington Law Review* 73 (2): 249–328.

Sykes, G., and D. Matza. 1957. Techniques of Neutralization: A Theory of Delinquency. *American Sociological Review* 22 (6): 664–670.

Tables. 449BC. Twelve Tables, $8.2. Available at http://avalon.law.yale.edu/ancient/hammint.asp.

Tadenuma, K., and Y. Xu. 2012. Extensions of the Fundamental Welfare Theorems in a Non-Welfaristic Framework. *Hitotsubashi Journal of Economics* 53 (1): 107–120.

Tallinn20. 2013. *The Tallinn Manual on the International Law Applicable to Cyber Warfare*, ed. Michael N. Schmitt et al. Cambridge: Cambridge University Press.

Tambini, D., D. Leonardi, and C. Marsden. 2008. *Codifying Cyberspace: Community Self-Governance in the Age of Internet Convergence*. London: Routledge Press.

Tandon, A., and A. Nayyar. 2019. A Comprehensive Survey on Ransomware Attack: A Growing Havoc Cyberthreat. In *Data Management, Analytics and Innovation*, ed. V.E. Balas et al. Singapore: Springer.

Teague. 2011. US v Teague, 646 F.3d 1119 (8th circ 2011).

Temple, J. 2011. Hackers Getting Hacked by Security Firms. *SFChronicle.com*, November 30.

Thomas. 2017. US v Thomas, 877 F.3d. 591 (2017).

Thrifty. 1996. Thrifty-Tel v Benezek, 54 Cal Rptr. 2d 468 (Cal. Ct of Appeals, 1996).

Tinland, O. 2013. The Pragmatist Skepsis as a Social Practice: Skepticism, Irony and Cultural Politics in Rorty's Philosophy, in Pragmatism and the Social Dimension of Doubt: Fresh Perspectives. *European Journal of Pragmatism and American Philosophy* 2: 21–30. Matthias Girel (dir.).

Toft, M. 2017. Why Is America Addicted to Foreign Interventions? *The National Interest*, December 20.

Tonelson, A. 2002. *Race to the Bottom: Why World-Wide Worker Surplus and Uncontrolled a Free Trade Are Sinking American Living Standards*. Boulder: Westview Press.

Toulouse, C., and T. Luke. 2013. *Politics of Cyberspace*. New York: Routledge Press.
Tropina, T., and C. Callahan. 2015. Self- and Co-Regulation. In *Cybercrime, Cyber Security and National Security*. Berlin: Springer.
Tsagourias, N. 2012. Cyber Attacks, Self Defense and the Problem of Attribution. *Journal of Conflict and Security Law* 17: 224–244.
Turner, R. 1991. *Repealing the War Powers Resolution: Restoring the Rule of Law in U.S. Foreign Policy*. Pavilion.
Turner, R. 1994. War and the Forgotten Executive Power Clause of the Constitution: A Review Essay of John Hart Ely's War and Responsibility. *Virginia Journal of International Law* 34: 903–951.
Turner, F. 2010. *From Counterculture to Cyberculture: Steward Brand, the Whole Earth Network and the Rise of Digital Utopianism*. Chicago: University of Chicago Press.
Turner, R. 2013. The War Powers Resolution at 40: Still an Unconstitutional, Unnecessary, and Unwise Fraud That Contributed Directly to the 9/11 Attacks. *Case Western Reserve Journal of International Law* 45 (1): 109–143.
Tussie, D., and N. Woods. 2000. Trade, Regionalism and the Threat to Multilateralism. *Latin American Trade Network*. Working Paper #2.
Uhm, D. 2016. *The Illegal Wildlife Trade: Inside the World of Poachers, Smugglers and Traders*. Cham: Springer.
UNCharter51. 1945. *Charter of the United Nations*. San Francisco.
U.N.T.S. XVI, Art. 51.
Valeriano, B., and B. Jensen. 2019. The Myth of Cyber Offense: The Case for Restraint. Cato Institute Policy Analysis, No. 862, January 15.
Valeriano, B., B. Jensen, and R. Maness. 2018. *Cyber Strategy: The Evolving Character of Power and Coercion*. Oxford: Oxford University Press.
VanBuren. 2021. Van Buren v. United States, 593 U.S ___ (2021), 141 S. Ct. 1648 (2021).
VanBurenBrief. 2020. Brief for petitioner Van Buren, [REPLY BRIEF FOR PETITIONER], Van Buren v US, On Writ of Certiorari to the United States Court of Appeals for the Eleventh Circuit.
VanBurenGov. 2019. Brief for the United States, Van Buren Vs United States, No. 19-783, On Writ of Certiorari to the United States Court of Appeals for the Eleventh Circuit.
Vaseli, A. 2018. The Normative Foundations of the Belt and Road Initiative. In *Normative Reading of the Belt and Road Initiative: Road to New Paradigm*, ed. Wenhua Shay, Kimao Nuotio, and Kangle Zhang. Cham: Springer.
Vasquez. 1999. US v Vasquez, 176 F.3d 1294 (10th Cir. 1999).
Van Der Vossen, B. and J. Brennan. 2018. In Defense of Openness : Why Global Freedom is the Human Solution to Global Poverty (Oxford University Press)

Venugopal, R. 2015. Neoliberalism as a Concept. *Economy and Society* 44 (2): 165–187.

VEP. 2017. *Vulnerability Equities Process and Policy for the United States Government*, November 15.

Verkerke, J. 2002. Is the ADA Efficient? Working Paper No. 02-9, May 13. http://ssrn.com/abstract_Id=311888, at 14.

Verma, A., and A. Verma. 2014. Right to Internet. In *Human Rights and Ethics: Concepts, Methodologies, Tools and Applications*. Hershey: IGI Global.

Vest. 1776. U.S. CONST. art. II, § 1, cl. 1.

Vida, A. 2005. Defining the Parameters of Cyberwar Operations: Looking for Law in All the Wrong Places. *Naval Law Review* 51: 132–174.

Voatz. 2020. Brief of Voatz, Inc., as Amicus Curiae in Support of Respondent in Affirmance of the Decision Below. https://www.supremecourt.gov/DocketPDF/219/19-783/153062/20200903122434600_Voatz%20Amicus%20Brief.pdf.

Von Staveren, I. 2007. The Ethics of Efficiency. *SCEME Working Paper* No. 18. http://www.sceme.org.uk

Von Staveren, I. 2009. Efficiency, in *The Handbook of Economics and Ethics* (ed. by Jan Pell and Irene Van Staveren) (Edward Elgar Publishing).

Waarden, F. 2010. Governing Global Commons: Public-Private of Fish and Forests. *Jerusalem Papers in Regulation and Governance*. Working Paper No. 17.

Waid-Lindberg, C., R. Dobbs, and T. Shelly. 2011. Blame The Media? The Influence of Primary New Source, Frequency of Usage, and Perceived Media Credibility on Punitive Attitudes. *Western Criminology Review* 12 (3): 41–59.

Walker, L. 1986. A Response to Elizabeth M Schneider's Describing and Changing Women's Self-Defense Work and the Problem of Expert Testimony on Battering. *Women's Rights Reporter* 9: 223–236.

Wallace, C. 2001. Xenophobia in Central and Eastern Europe. In *Holding the Course/East-West Institutes*, ed. Peter Rutland. New York: ME Sharpe.

Wallerstein, S. 2005. Justifying the Right to Self-Defense, a Theory of Forced Consequences. *Virginia Law Review* 91: 999–1073.

Walt, S. 2018. *The Hell of Good Intentions: America's Foreign Policy Elite and the Decline of US Primacy*. New York: Farrar, Strauss and Giroux.

WarCrimesA. 1907. Article 75 of the Annex by the Hague Convention IV. *Respecting the Laws and Customs of War on Land*, signed 18 October 1907, Article 23(e).

WarCrimesB. 1977. 18 USC $2441.

Warf, N. 2012. *Global Geography*. Dordrecht: Springer Science and Business Media.

Warford, T. 2019. Trump, May and the Art of Political Brinksmanship. *Financialtimes.Com*, January 11.

Weber, R. 2010. *Shaping Internet Governance: Regulatory Challenges*. Dordrecht: Spring Science and Business Media.

Weber, L. 2015. Leanne Weber, Peace at the Border—A Thought Experiment. In *Rethinking Border Control for a Globalizing World: A Preferred Future*, ed. Leanne Weber. London: Routledge Press.

WEC. 2012. WEC Carolina Energy Sols, LLC v. Miller, 687 F.3d 199 (4th Cir. 2012).

Wendt, A. 1999. *Social Theory of International Politics*. Cambridge: Cambridge University Press.

West, Z. 2012. Note, Young Fella: If You're Looking for Trouble, I'll Accommodate You, Deputing Private Companies for the Use of Hackback. *Syracuse Law Review* 63: 119–146.

Widener, A. 2019. New Restrictions on Foreign Scientists or Fundamental Research Not Needed, Defense Panel Says. *Chemicalandengineering News.Com* 9 (48) (February 22).

Wilentz, S. 2007. *Andrew Jackson: The American Presidents Series—7th President, 1829–1837*. New York: Macmillan Press.

Williams, B. 1987. The Standard of Living: Interests and Capabilities. In *The Standard of Living*, ed. Geoffrey Hawthorn. Cambridge: Cambridge University Press.

Williams, M. 2020. Descartes' Transformation of the Skeptical Tradition. In *The Cambridge Companion to Ancient Skepticism*, ed. R. Bett. Cambridge: Cambridge University Press.

Wohlforth, W. 2008. *World Out of Balance: International Relations and the Challenge of American Primacy*. Princeton: Princeton University Press.

Wong, T. 2017. *The Politics of Immigration: Partisanship, Demographic Change and American National Identity*. New York: Oxford University Press.

Wormuth, F., and E. Firmage. 1986. *To Chain the Dog of War: The War Power of Congress in History and Law*. Dallas: Southern Methodist University Press.

WPR. 1973. The War Powers Resolution, Pub. L. No. 93-248, 87 Stat. 555 (1973) (codified at 50 U.S.C. § § 1541–1548) (1994).

Wrage, S. 1999. Examining the 'Authoritarian Advantage in South East Asian Development in the Wake of Asian Economic Failures. *Studies in Conflict and Terrorism* 22 (1): 21–31.

Wright, T. 2016. Realism's Utopia: What Academic Realism Almost Never Works in Washington. *International Affairs*, January 13.

Wright, R. 2020. Robert Wright, Biden's Foreign Policy Team Is Full of Idealists Who Keep Getting People Killed. *Washingtonpost.com*, December 15.

Wu, T., and C. Yoo. 2007. Keeping the Internet Neutral?: Tim Wu and Christopher Yoo Debate. *Federal Communications Law Journal* 59 (3): 575–604.

Wucker, M. 2016. Move over Black Swan, Here Comes Gray Rhino. *Carnegie Council*, June 14.

Yannakngeous, P., and A. Lowther. 2013. The Prospects for Cyber Deterrence: American Sponsorship of Global Norms. In *Conflict and Cooperation in Cyberspace: The Challenge for National Security*, ed. Pannayotis A. Yannakngeous and Adam B. Lowther. Boca Raton: CRC Press.

Yoo, J. 2002. War and the Constitutional Text. *University of Chicago Law Review* 69 (4): 1639–1674.

Yoo, J. 2003. International Law and the War in Iraq. *The American Journal of International Law* 97 (3): 563–576.

Yoo, C. 2015. Cyber Espionage or Cyber War?: International Law, Domestic Law, and Self-Protective Measures. Faculty Scholarship. At Penn Carey Law 1540.

Young, G. 2003. All the Truth and as Much Justice as Possible. *University of California—Davis Journal of International Law and Policy* 9: 209–227.

Youssef, M., and S. Harris. 2016. Obama Officials Wonder: Why Won't the Boss Stand Up to Putin? *The Daily Beast*, September 29.

Zakaria, F. 1999. *From Wealth to Power: The Unusual Origins of America's World Role*. Princeton: Princeton University Press.

Zargar, S., J. Joshi, and D. Tipper. 2013. A Survey of Defense Mechanism Against Distributed Denial of Service (DDoS) Flooding Attacks. *IEEE Communications and Tutorials* 15 (4): 2046–2089.

Zeiler, T. 1999. *Free Trade, Free World: The Advent of GATT*. EH.Net.

Zipursky, B. 1998. Rights, Wrongs, and Recourse in the Law of Torts. *Vanderbilt Law Review* 51: 1–29.

Zoller, E. 1984. *Peacetime Unilateral Remedies: An Analysis of Countermeasures*. New York: Transnational Publishers.

Index

A. J. Thomas, *The Open World, Hackbacks and Global Justice*,
https://doi.org/10.1007/978-981-19-8132-6

The manufacturer's authorised representative in the EU is Springer Nature Customer Service Centre GmbH, Europaplatz 3, 69115 Heidelberg, Germany. If you have any concerns regarding our products, please contact ProductSafety@springernature.com

Printed and bound by CPI Group (UK) Ltd, Croydon, CR0 4YY
30/09/2026
02225243-0012